Handbook on Liquefaction Remediation of Reclaimed Land

Edited by
Port and Harbour Research Institute, Ministry of Transport, Japan

Translated by
Waterways Experiment Station, US Army Corps of Engineers, USA

Supported by
Coastal Development Institute of Technology, Japan

A. A. BALKEMA / ROTTERDAM / BROOKFIELD / 1997

Published by
A.A. Balkema, P.O. Box 1675, 3000 BR Rotterdam, Netherlands
(Fax: +31.10.4135947)
A.A. Balkema Publishers, Old Post Road, Brookfield, VT 05036-9704, USA
(Fax: 802.276.3837)

ISBN 90 5410 653 0

Printed in the Netherlands

Contents

Preface

Loose sandy deposits sometimes change into a liquid state during earthquakes. This is called liquefaction and poses a serious problem in waterfront areas. When the sandy deposits liquefy, structures built on those deposits are seriously affected; for example, heavy structures settle and light buried structures heave. Reliable remediation with respect to liquefaction is necessary in geotechnical engineering practice.

This book presents methodologies for assessment of liquefaction potential and remedial measures to mitigate liquefaction for reclaimed land. These methodologies are based on continuous research and decade of experience in design and construction of port facilities in Japan, where liquefaction often poses a serious problem in design and construction.

Liquefaction is a complex phenomenon and its complete understanding is difficult. Based on continuous research efforts for three decades, however, many factors related to liquefaction have been identified and understood. The remaining factors will also be understood by accumulating case history data with examples of liquefaction remediation in the field and with further progress in research. In order to include as much information as possible, this book presents not only data which has been established by past experience and research but also current studies.

This book is an English version of a handbook on liquefaction remediation of reclaimed land written in Japanese and published in 1993 by the Coastal Development Institute of Technology, Japan (CDIT), the editor being the Ports and Harbours Bureau, Ministry of Transport (MOT), Japan. Most of the content was produced by the Port and Harbour Research Institute, Ministry of Transport, Japan (PHRI). A specific list of the authors and the committee members for editing are indicated on a consequent page.

Although the handbook was originally written for Japanese engineers and researchers, most of the principles and techniques presented in the book are readily applicable to seismically active regions around the world. The Waterways Experiment Station, US Army Corps of Engineers, USA (WES) and PHRI decided to prepare jointly an English translation of the handbook soon after the original was published. Translation and technical editing of this handbook was a joint efforts of PHRI and WES with additional financial support provided by CDIT. Efforts of Mr R.H. Ledbetter (WES) for translation; Drs K. Zen, T. Uwabe, T. Sugano, Messrs H. Yamazaki, K. Ichii, E. Iizuka and M. Miyata (PHRI) for technical editing; and Messrs K. Oikawa and F. Kitamura (CDIT) for publication, are gratefully acknowledged. In addition, we also gratefully acknowledge the technical editing accomplished by Mr C.R. Heidengren, Past President ASCE Japan Section and Dr Stephen Altobelli, the Lovelace Institutes, New Mexico, USA.

We hope that this book will be a useful reference for engineers and researchers involved in liquefaction remediation in seismically active regions around the world.

March 1996

Susumu Iai
A.G. Franklin

Committee for liquefaction remediation of reclaimed land (as of March 1993)

Chairman
Y. Umehara. Executive Director, CDIT

Members
S. Noda. Deputy Director General, PHRI.
K. Zen.* Chief, Soil Dynamics Laboratory, Geotechnical Engineering Division, PHRI.
H. Yamazaki.* Senior Research Engineer, Geotechnical Engineering Division, PHRI.
T. Inatomi. Chief, Structural Dynamics Laboratory, Structural Engineering Division, PHRI.
T. Uwabe.* Chief, Earthquake Disaster Prevention Laboratory, Structural Engineering Division, PHRI.
S. Iai.* Chief, Geotechnical Earthquake Engineering Laboratory, Structural Engineering Division, PHRI.
M. Kazama.* Senior Research Engineer, Structural Engineering Division, PHRI.
M. Shiomi. Chief, Design Standard Laboratory, Planning and Design Standard Division, PHRI.
H. Ouchi. Supervisor for Port Engineering, Engineering Division, Ports and Harbours Bureau, MOT.
S. Higashiyama. Deputy Director, Engineering Division, Ports and Harbours Bureau, MOT.
S. Ishiyama. Former Director, Disaster Countermeasure Office, Coast Administration and Disaster Prevention Division, Ports and Harbours Bureau, MOT.
T. Ito. Director, Disaster Countermeasure Office, Coast Administration and Disaster Prevention Division, Ports and Harbours Bureau, MOT.
S. Miyazaki. Former Chief, First Disaster Prevention Planning Section, Coast Administration and Disaster Prevention Division, Ports and Harbours Bureau, MOT.

T. Negi. Chief, First Disaster Prevention Planning Section, Coast Administration and Disaster Prevention Division, Ports and Harbours Bureau, MOT.

K. Toyama. Member, First Disaster Prevention Planning Section, Coast Administration and Disaster Prevention Division, Ports and Harbours Bureau, MOT.

Y. Koyano. Former Deputy Director of Administrative Planning Section, Second District Port Construction Bureau, MOT.

K. Kawasaki. Deputy Director of Administrative Planning Section, Second District Port Construction Bureau, MOT.

H. Kato. Special Assistant to Director, Yokohama Investigation and Design Office, Second District Port Construction Bureau, MOT.

Secretaries

M. Shikamori. Director, Second Investigation and Research Division, CDIT.

K. Shiota. Senior Research Engineer, Investigation and Research Division, CDIT.

*Authors

CHAPTER 1

Introduction

1.1 BACKGROUND

Many large reclaimed land projects have been planned in recent years, including waterfront developments, offshore artificial islands, and offshore airport construction. For conventional reclamation work in open water, the density of the reclaimed land has not been controlled. This conventional practice resulted in many cases in the sand in the reclaimed land being a loose sediment, having an N-value, from a standard penetration test (SPT), less than 10. The loosely sedimented saturated sandy soil tends to liquefy under earthquake loading, as often seen in past earthquake disasters as noted in Chapter 2. In order to maintain the safety of structures and facilities in reclaimed land areas, due consideration should be given to the problem of liquefaction.

Since the Niigata earthquake of 1964, many geotechnical/earthquake engineers have made efforts to understand the mechanism of liquefaction and to establish standards and guidelines on liquefaction based on case history data, in-situ, and laboratory testing. In the first half of the seventies, procedures for evaluating liquefaction potential were specified in 'Seismic design specifications for highway bridges and commentaries' (Japan Road Association), 'Technical standards for port and harbour facilities and commentaries' (Japan Port and Harbour Association), 'Design standards for national railway structures (foundations and retaining structures)' (Japanese National Railway), and 'Design standards of building foundations (revised 1974)' (Architectural Institute of Japan). These design standards and codes have been revised several times in order to incorporate research results and technology developments as shown in Appendix 1. These standards and codes, however, incorporate only the well established facts and technologies.

Liquefaction is a complex and not fully understood phenomenon. In engineering practice, the most rational and appropriate liquefaction remediation measures for reclaimed land are needed. Thus, there is a pressing need for a state-of-the-art handbook to serve as a reference for practicing engineers involved in liquefaction remediation of reclaimed land.

1.2 OBJECTIVE AND USE

This handbook discusses methods for evaluating liquefaction potential and remediation measures for liquefaction in reclaimed land. Facilities constructed on reclaimed land are various and diverse, and in many cases, the tasks and procedures for investigation regarding liquefaction differ depending on the type of facility. In this handbook, the major part of the discussion was originally developed through research on liquefaction for designing port and harbour facilities such as quay walls, bulkheads, and the reclaimed land behind these structures. The discussion, therefore, emphasizes liquefaction remediation measures for port and harbour facilities. It was intended in this handbook, however, to contribute to understanding liquefaction remediation measures appropriate for foundation soils in general, including facilities and structures constructed within reclaimed areas.

This handbook is meant to serve as a supplement to the so-called standards and codes. It presents the-state-of-the-art technologies and useful practical tips as a reference for understanding and fully utilizing the standards and codes. In addition, it includes results of on-going research and development on new liquefaction remediation measures and related research trends.

The style of the handbook is descriptive; theoretical details are omitted unless they are absolutely necessary for practicing engineers.

1.3 TECHNICAL CONTENTS

This handbook is composed of Chapters 1-5, Technical notes, and Appendices. In Chapter 2, the liquefaction phenomenon is briefly explained, and then causes of damage and damage patterns for various structures depending on the type of structure are presented, referring to earthquake case histories. In Chapter 3, the basic strategy of liquefaction remediation of reclaimed land is discussed. In Chapter 4, major attention is given to prediction/determination methods for liquefaction of reclaimed land, including soil survey/testing and methods for evaluating dynamic deformation/strength characteristics of soil, and analysis of seismic shear stress in the subsoil. In Chapter 5, various liquefaction remediation measures are discussed, including useful details often needed for field implementation. In the Technical notes, on-going research trends are discussed with respect to liquefaction prediction/damage degree evaluation and liquefaction remediation measures. In the Appendices, useful supplemental information is provided, including current Japanese standards on liquefaction.

CHAPTER 2

Liquefaction phenomenon and examples of damage

2.1 LIQUEFACTION PHENOMENON

When subjected to earthquake shaking, saturated sandy soil may suddenly change into a liquid like muddy water. This phenomenon is called liquefaction. In a broader sense, the phenomenon in which a granular material changes to a liquid state as a result of strong vibration, whether the soil is saturated with water or not, is called generically 'liquefaction'. In this handbook, liquefaction of saturated soil will be the main subject of discussion.

In sandy soil, sand particles maintain mutual contact before an earthquake and forces can be transmitted through these contacts. This allows the shear resistance of soil to support a structure resting on the ground surface. This is shown schematically in Figure 2.1a. When sandy soil deforms due to shear stress caused by vibration during an earthquake, contact between the particles is lost as shown in Figure 2.1b. Then, the force originally supported in a vertical direction through the contact points is instead transmitted through the pore water. This condition corresponds to a state of 'liquefaction'. In this state, contact between the sand particles is lost so that the shear resistance of the sandy soil is lost, and the sandy soil manifests a behavior similar to a liquid having the unit weight of the saturated soil. After soil liquefaction, contact between the sand particles is eventually re-established as the pore water flows out, resulting in the state shown in Figure 2.1c; namely, the soil stabilizes again, but settling has occurred. The volume decrease in the settled soil is the same as the volume of the pore water which has flowed out.

Actual soils are much more complex, composed of irregular assemblages of sand particles with various grain sizes. Therefore, even when the soil undergoes shear deformation, all of the contact points are not lost at once; some sand particles may lose contact, but the rest maintain contact to preserve part of the soil skeleton. In this case, excess pore water pressure (water pressure defined by deducting hydrostatic pressure from current pore water pressure) is generated in proportion to the relative number of lost contacts.

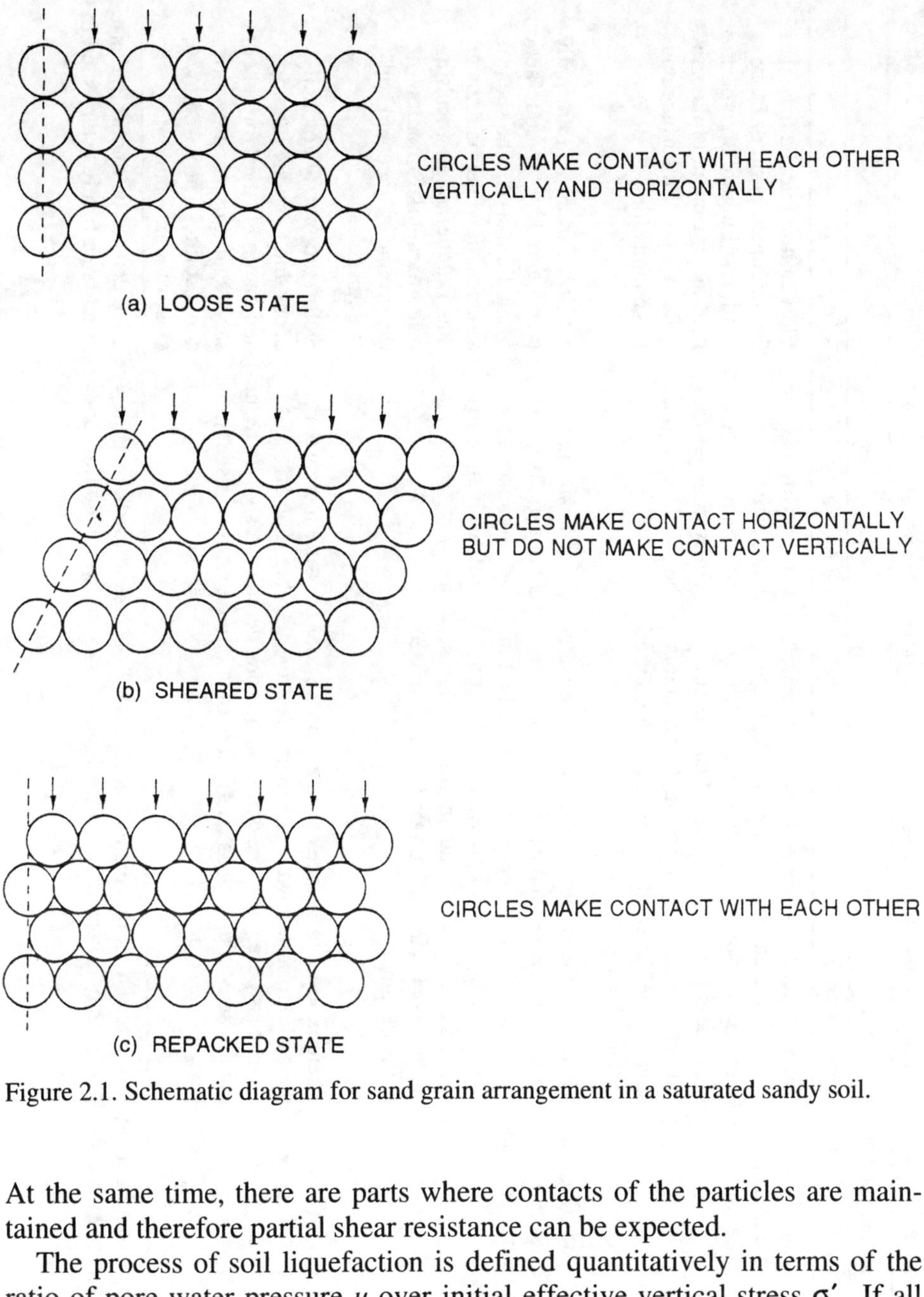

Figure 2.1. Schematic diagram for sand grain arrangement in a saturated sandy soil.

At the same time, there are parts where contacts of the particles are maintained and therefore partial shear resistance can be expected.

The process of soil liquefaction is defined quantitatively in terms of the ratio of pore water pressure u over initial effective vertical stress σ'_v. If all the effective stress σ'_v originally transmitted through the contact points between the particles before shearing is received by the pore water when all the contacts have been lost, then $u = \sigma'_v$, so that the excess pore water pressure ratio becomes $u/\sigma'_v = 1.0$. Conversely, the state in which there is no excess pore water pressure at all corresponds to $u/\sigma'_v = 0$. A state in which a

Table 2.1. Deformation characteristics of sand based on density and excess pore water pressure ratio.

Density of sand	Excess pore water pressure ratio u / σ'_v			
	$u / \sigma'_v = 0$	$0 < u / \sigma'_v < 0.5$	$0.5 \le u / \sigma'_v < 1.0$	$u / \sigma'_v = 1.0$
Loose sand	No influence of excess pore water pressure	There is some influence of excess pore water pressure increase on the soil; the soil, however, remains relatively stable. The effect of the excess pore water pressure can be considered in design by reducing the constants such as coefficient of subgrade reaction used in the design	This transitional state is not usually stable and, in most cases, swiftly changes to the $u / \sigma'_v = 1.0$ state	Flow failure may be triggered by the liquefaction of loose sand, resulting in large permanent displacement of the subsoils and ground surface and large landslides
Medium sand	No influence of excess pore water pressure	There is some influence of excess pore water pressure increase on the soil; the soil, however, remains relatively stable. The effect of the excess pore water pressure can be considered in design by reducing the constants such as coefficient of subgrade reaction used in the design	This transitional state is not usually stable and, in most cases, swiftly changes to the $u / \sigma'_v = 1.0$ state	No flow failure will normally be triggered by the liquefaction of medium sand. The shear resistance, however, is reduced due to liquefaction, resulting in a significant deformation of the ground surface and soil structures
Dense sand	No influence of excess pore water pressure	There is some influence of excess pore water pressure increase on the soil; the soil, however, remains relatively stable. The effect of the excess pore water pressure can be considered in design by reducing the constants such as coefficient of subgrade reaction used in the design		Deformation may gradually occur due to cyclic loading, and caution is therefore necessary regarding this aspect of soil stability. The soil, however, remains stable

A rough criterion for sand density as referred to in this table is as follows: Loose sand/medium sand = relative density below 80%; dense sand = relative density over 80%. The density criterion which distinguishes loose from medium sands has not been firmly established in practice.

portion of the particle contacts is lost corresponds to $0 < u/\sigma'_v < 1.0$. In this handbook, a state in which $u/\sigma'_v = 1.0$ or a state close to it will be called the state of 'liquefaction'.

The shear resistance and deformation characteristics of saturated soils affected by earthquakes are basically characterized by the excess pore water pressure ratio. It is important to note, however, that the shear resistance and deformation characteristics also depend on the density of the soil. These facts can be summarized as shown in Table 2.1.

As shown in Table 2.1, even for the state of $0 < u/\sigma'_v < 1.0$ where liquefaction does not occur, the deformation characteristics of the soil vary according to soil density. When the soil is dense, stability can be maintained even when excess pore water pressure increases; however, when the soil is loose, an unstable state often occurs when the excess pore water pressure ratio exceeds about 0.5; i.e. the excess pore water pressure tends to increase suddenly, resulting in liquefaction.

It is also noted that, even in the $u/\sigma'_v = 1.0$ state, namely, for a state of 'liquefaction', the deformation characteristics of the soil differ noticeably according to the density of soil as indicated in the table. This fact is explained in more detail as follows:

1. In contrast to loose sand, dense sand temporarily restores the particle contacts when sheared even after the excess pore water pressure ratio reaches 1.0, so that it may support a static shear stress; i.e. dense sand does not change to a completely liquid state. In designs involving liquefaction, it is important to take into account the favorable characteristics of dense sand.

2. When the density of the sand becomes particularly low, however, permanent displacement (flow failure) of subsoil similar to a large landslide or mud flow is generated even if the grade of the ground surface is very flat. In this case, one cannot rely on the residual shear resistance of the subsoil which generally contributes to the stability of structural foundations. In addition, a new external force due to permanent displacement of the ground surface may result in damage to buried structures and pile foundations. In design involving liquefaction, it is necessary to consider carefully permanent displacement of loosely deposited soils.

The deformation characteristics of the soil explained in this section should be the basis for design considerations of liquefaction remediation.

2.2 EXAMPLES OF DAMAGE

There are many examples of serious damage caused by liquefaction in earthquakes. Some typical examples are discussed below.

Figure 2.2. Damage to a quay wall at Ohama Wharf, Akita Port caused by the Nihonkai-Chubu earthquake (1983).

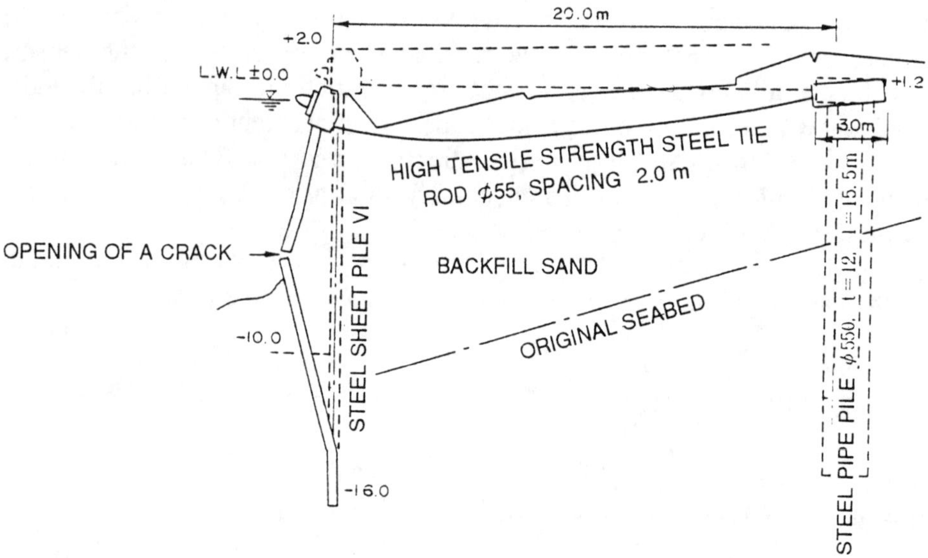

Figure 2.3. Cross section of a quay wall at Ohama Wharf, Akita Port (unit *m*).

Figure 2.4. Damage to a quay wall at Shimohama Wharf, Akita Port caused by the Nihonkai-Chubu earthquake (1983).

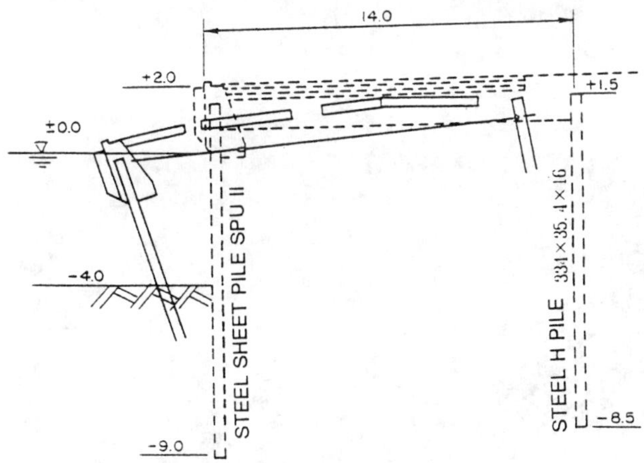

Figure 2.5. Cross section of a quay wall at Shimohama Wharf, Akita Port (unit *m*).

Quay walls

Damage to Ohama Wharf in Akita Port caused by the 1983 Nihonkai-Chubu earthquake is shown in Figure 2.2. The quay wall was a steel sheet pile type as shown in Figure 2.3. The damage shown here was caused by a large stress applied to the sheet pile wall due to liquefaction of the sand backfill. For this quay wall, the foundation soils supporting its anchor did not liquefy and the top part of the sheet pile wall supported by the anchor was barely displaced. The bending moment on the sheet pile wall, however, became excessively large, resulting in damage to the sheet pile wall.

Damage to Shimohama Wharf in Akita Port during the same earthquake is shown in Figure 2.4. This wharf was a steel sheet pile type similar to that of the Ohama Wharf as shown in Figure 2.5. The foundation soil supporting the anchor also liquefied, however, resulting in the entire wharf deforming noticeably seaward.

These two examples indicate that the mode of damage to sheet pile type quay walls differs greatly according to the fixation condition of the anchors.

Buildings

An example of damage to buildings in Niigata City during the Niigata earthquake in 1964 is shown in Figure 2.6. This building was a four-story reinforced concrete structure with spread foundation on sandy subsoil. As shown in the figure, noticeable settlement occurred due to the reduced bearing capacity of the foundation soil. The eccentric load from a penthouse on the roof resulted in tilting.

Figure 2.6. A building that sank/tilted due to soil liquefaction during the Niigata earthquake (1964).

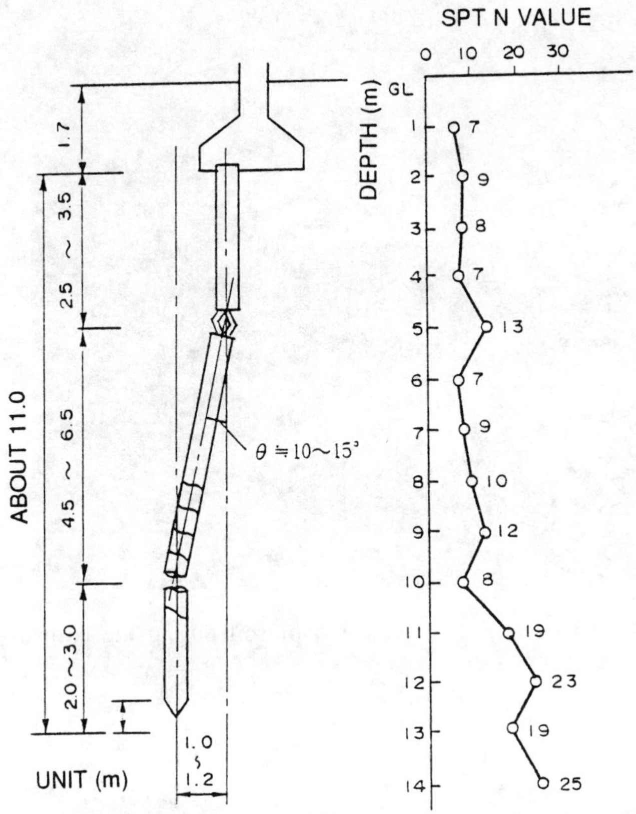

Figure 2.7. Deformation and damage to foundation piles caused by soil liquefaction during the Niigata earthquake (1964) (Kawamura et al. [1]).

Damage to buildings supported by pile foundations was minor during this earthquake. About 20 years after the earthquake, however, noticeable damage to the pile foundation was discovered as shown in Figure 2.7 [1]. The cause of this damage was permanent ground displacement due to liquefaction of loosely deposited sand [2].

Bridges

Damage to Showa Bridge during the Niigata earthquake is shown in Figure 2.8. The cause of the damage was considered to be liquefaction of the foundation soil. Recent study reveals that soil liquefaction caused excessively large permanent ground displacement from the bulkhead toward the center of the river [2]. In addition to reduction or loss of the subgrade reaction for the bridge pile foundation, the permanent soil displacement began to act as a new external force on the pile foundation, resulting in yielding of the foundation piles.

Damage to an approach to the Showa Bridge during the same earthquake is shown in Figure 2.9. Even if noticeable damage is not induced in the main

Figure 2.8. Damage to Showa Bridge due to liquefaction of soil during the Niigata earthquake (1964).

Figure 2.9. Damage to the Showa Bridge approach caused by the Niigata earthquake (1964).

body of a bridge, damage to the approach often occurs, resulting in impairment of the total function of the bridge. Most of the damage mechanisms to the approach are common to those which occur in quay walls and embankments.

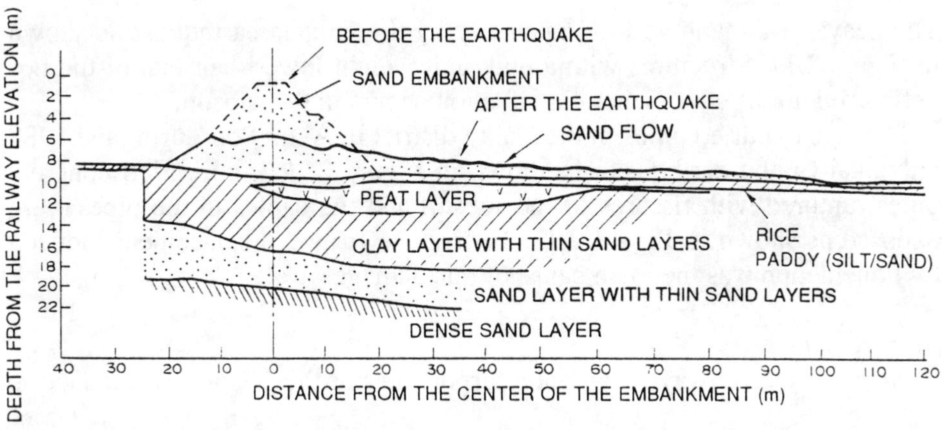

Figure 2.10. Damage to a railway embankment caused by the Niigata earthquake (1964).

Figure 2.11. Manhole lifted up due to liquefaction of soil during the Niigata earthquake (1964).

Embankments
Damage to a railway embankment caused by the Niigata earthquake [3] is shown in Figure 2.10. In this example, soil liquefaction triggered the flow failure, resulting in serious damage.

Buried structures
The heave of a manhole in Niigata city by the Niigata earthquake is shown in Figure 2.11. Structures with a bulk unit weight lower than that of the liquefied soil are lifted up when liquefaction occurs in the subsoil.

Damage to buried pipes in the Gaiko district in Akita Port during the 1983 Nihonkai-Chubu earthquake is shown in Figure 2.12. The buried concrete pipes ruptured with fissures of the subsoil and the cables in the pipes were exposed as shown in Figure 2.12. Permanent ground displacement induced by liquefaction was the main cause of this damage.

Airport runways
Damage to the runway of Niigata Airport during the Niigata earthquake is

Figure 2.12. Damage to buried pipes caused by the Nihonkai-Chubu earthquake (1983).

Figure 2.13. Damage to the Niigata Airport caused by the Niigata earthquake (1964).

shown in Figure 2.13. Excessively large displacement and strain induced by the earthquake motion in the liquefied soil and permanent soil displacement associated with it were the main causes of the damage. Differential settlement caused by liquefaction and sand boiling were also the causes of the damage.

2.3 CAUSE AND MODE OF DAMAGE TO STRUCTURES IN RECLAIMED LAND

2.3.1 *Classification of damage induced by liquefaction considering the cause of damage*

Damage to various structures caused by liquefaction shown in Section 2.2 can be roughly classified into two types based on the cause of damage.

In the first case, reduced shear resistance of soil due to liquefaction is not sufficient to maintain the stability of structures which should be able to resist external forces such as gravity and inertia force. The failure modes of this type of damage are closely related to the equilibrium conditions for the forces acting on the structure. Typical examples of damage in this category are settlement/tilting of structures due to reduction in bearing capacity of the foundation, deformation of earth structures due to reduction in slope stability, deformation of port and harbour structures caused by an increase in active earth pressures and reduction in passive resistance of subsoil, and heave of buried structures due to a reduction in the constraining force of the soil.

In the second case, the soil originally contributing to the stability of structures before an earthquake begins to act as a new external force on the structure due to excessively large displacement of the liquefied subsoil dur-

Table 2.2. Damage induced by liquefaction classified according to the cause of damage.

Cause of damage			Mode of damage	Type of structures	Remarks
Reduction in soil shear resistance due to liquefaction (reduction in resistance against external force)	Generation of excess pore water pressure	Generation of an uplift force	Uplift of buried structures	Water pipe; sewage pipe; gas pipe; underground electric power and telecommunications lines; oil pipeline; submerged tunnel; gas station; buried tank; manhole; underground sewage tank; underground water tank	Excess pore water pressure increase may be due to the liquefaction phenomenon itself, and changes induced in the soil properties are a result of excess pore water pressure increase; the excess pore water pressure increase can be considered as the cause of all damage. The pore water pressure increase here, however, indicates the form of damage which is directly caused by an increase in excess pore water pressure
	Reduction in the subsoil stability	Reduction in the bearing capacity of foundation subsoil	Settlement, tilting, and overturning of spread and pile foundation structures	Bridge; tank; port and harbour cargo crane; buildings	In contrast to the heave of relatively light underground structures, structures on ground result in settlement
		Reduction in the soil strength	Destruction and embankment instability	Soil structures	Reduction in the soil resistance is the main cause
	Change in earth pressure	Increase in earth pressure	Seaward displacement of wharf and bulkhead	Wharf/bulkhead	—
		Reduction in the horizontal resistance of the foundation subsoil	Damage to the pile foundation	Trestle type pier; bridge; tank; foundation piles for buildings	Damage to foundation may lead to the destruction of the superstructure

Table 2.2. Continued.

Cause of damage			Mode of damage	Type of structures	Remarks
Excessive large displacement due to liquefaction (generation of new external force)	Generation of horizontal permanent displacement	Generation of horizontal external force	Rupture and breakage of buried pipes	Water pipe; sewage pipe; gas pipe; underground electric power and telecommunication lines	Extensive research is being conducted on these types of damage
			Destruction of superstructure due to movement or destruction of foundation	Damage to foundation; damage to superstructure; collapse of bridge	
		Excessive subsoil strain (due to generation of localized liquefaction)	Destruction of superstructure due to movement or destruction of foundation	Water pipe; sewage pipe; gas pipe; underground electric power and telecommunication lines; oil pipeline; submerged tunnel	Strain in nonuniform subsoils due to localized liquefaction
	Generation of excessively large displacement amplitude	—	Damage to the superstructure due to excessively large displacement in the foundation	Bridge; buildings	This mode of damage may be difficult to differentiate from damage caused by strong shaking without liquefaction
		A long period motion	Sloshing of liquid in a tank	Oil tank; tanks for liquefied natural gas (LNG) and liquefied propane gas (LPG)	This mode of damage is cited as a possibility although an actual case history has not yet been obtained.
	Water spouting and sand boil	—	Sand boil and water spouting	Man-made beach, parks and athletic fields	Sand boiling, water spouting and cracking of the ground surface may damage an area used as an evacuation field during emergencies.
	Settlement and cracking of the ground surface	—	Sand boil and water spouting	Soil structures in general such as embankments	—

Table 2.3. Damage classified according to structure type.

Type of structure	Damage mode	Cause of damage	Structures
Port and harbour structures	Seaward displacement of the wharf/bulkhead	Increase in the earth pressure in the backfill soil	Gravity type wharf/bulkhead
	Settlement/tilting of the wharf/bulkhead	Reduction in the bearing capacity of the foundation soil	Gravity type wharf/bulkhead
	Deformation/yielding of the wharf/bulkhead and yielding of the anchor	Reduction of soil resistance in front of the anchor	Anchored sheet pile type wharf/bulkhead
	Yielding/breaking of the sheet pile wall at the wharf/bulkhead and yielding/breaking of the tie rod	Increase in the earth pressure in backfill soil behind the sheet pile wall	Anchored sheet pile type wharf/bulkhead
	Deformation of sheet pile wall at the wharf/bulkhead	Reduction in soil resistance for the embedded portion of the sheet pile wall	Anchored sheet pile type wharf/bulkhead
Spread foundation	Settlement/tilting/overturning of spread foundation structures (including those with pile foundations that do not reach the nonliquefied layer)	Reduction of the bearing capacity of the foundation soil	Trestle type pier; piers of bridge (abutment); tank; foundations for cargo cranes at ports and harbours
Pile foundation	Breaking/destruction of the foundation pile	Reduction in soil resistance in the liquefaction layer and permanent ground displacement	Piers of bridge (abutment); tank; foundations for cargo cranes at ports and harbours; building foundations (railway terminals, airport terminals, and warehouses)

Table 2.3. Continued.

Type of structure	Damage mode	Cause of damage	Structures
Structures on the ground surface	Damage/destruction of the superstructure	Generation of excessive displacement amplitude	Superstructure of bridge and expansion joints;
		Permanent ground displacement in the horizontal or vertical direction	Railway tracks; protection wall against oil spilling; superstructure of bridge and expansion joints; buildings
Soil structures	Settlement of the embankment Lateral spread of the embankment	Reduction in shear resistance	Embankments (highway, railway)
	Deformation of the ground surface such as movement, cracking, depression, settlement, sand boiling, water spouting, heave caused by liquefaction	Permanent ground displacement and settlement/sand boil	Airport runway, highway, railway and park
Buried structures	Rupture/bending of the buried pipe	Excessively large ground strain at the boundary between liquefied and non-liquefied layers Excessive external force due to permanent ground displacement	Buried pipes (water supply, sewage, gas, underground electrical power and telecommunication lines)
	Heave of the buried structure	Increase in uplift force due to excess pore water pressure	Underground storage tank; Buried pipe; manhole; gas station; buried tank

Table 2.4. Schematic diagram on modes of damage.

Type of structure — Port and harbour structure

Schematic diagram of the damage mode

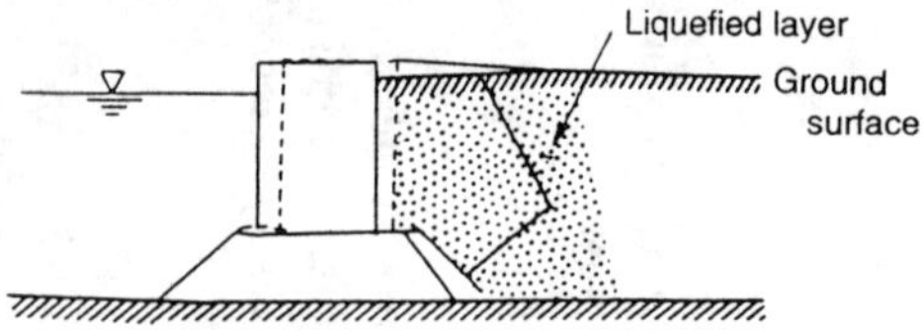

Mode of damage	Sliding toward the sea
Cause of damage	Increase in earth pressure in the backfill soil
Remark	Liquefaction of backfill soil behind a gravity type structure

Type of structure — Port and harbour structure

Schematic diagram of the damage mode

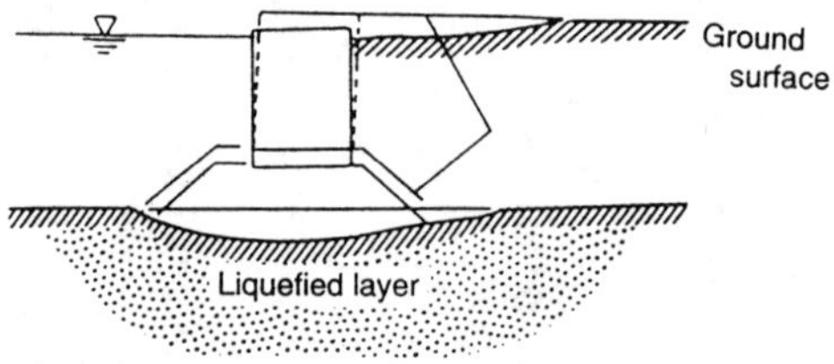

Mode of damage	Sliding toward the sea/tilting/settlement of the structure
Cause of damage	Reduction in the bearing capacity of the foundation soil
Remark	Liquefaction of foundation soil below a gravity type structure

Type of structure — Port and harbour structure

Schematic diagram of the damage mode

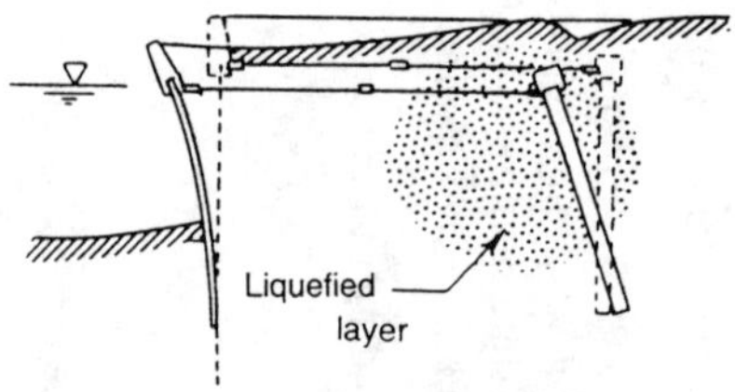

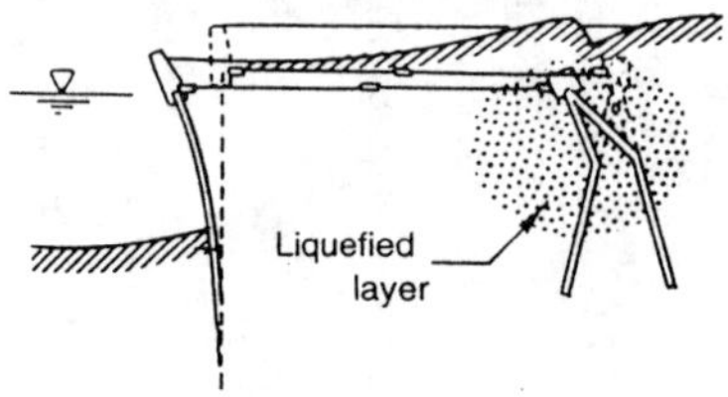

Mode of damage	Deformation/failure of the structure and yielding of the anchor
Cause of damage	Reduction in soil resistance in front of the anchor
Remark	Liquefaction in front of the anchor of a sheet pile type structure

Table 2.4. Continued.

Type of structure	Port and harbour structure
Schematic diagram of the damage mode	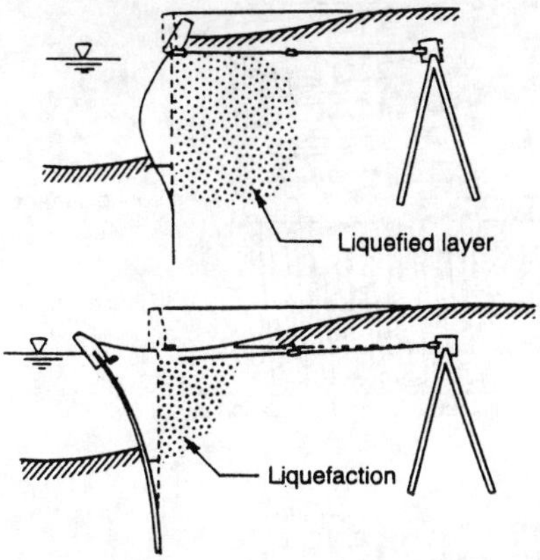
Mode of damage	Yielding of the sheet pile wall and tie rod
Cause of damage	Earth pressure increase in the sand backfill
Remark	Liquefaction of soil backfill for a sheet pile type structure

Type of structure	Port and harbour structure
Schematic diagram of the damage mode	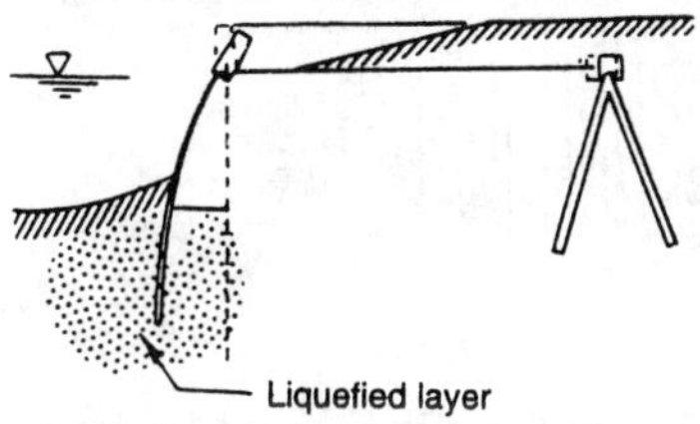
Mode of damage	Deformation in the sheet pile wall
Cause of damage	Reduction in the soil resistance for the embedded portion of the sheet pile structure
Remark	Liquefaction of the subsoil in the embedment portion of the sheet pile structure

Type of structure	Spread foundation
Schematic diagram of the damage mode	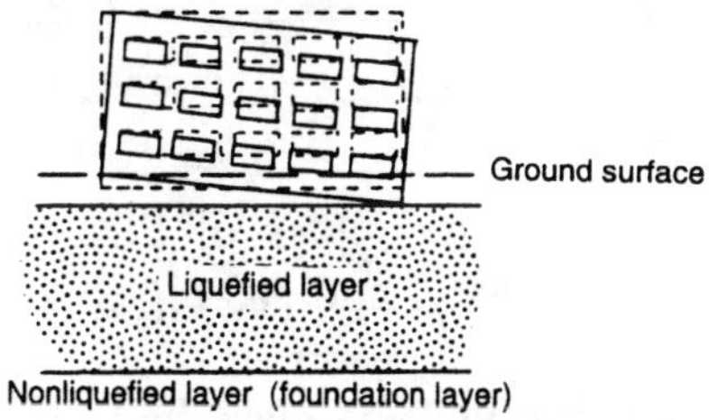
Mode of damage	Settlement/tilting/overturning of the spread foundation
Cause of damage	Reduction in the bearing capacity of the foundation soil
Remark	

Table 2.4. Continued.

Type of structure	Pile foundation
Schematic diagram of the damage mode	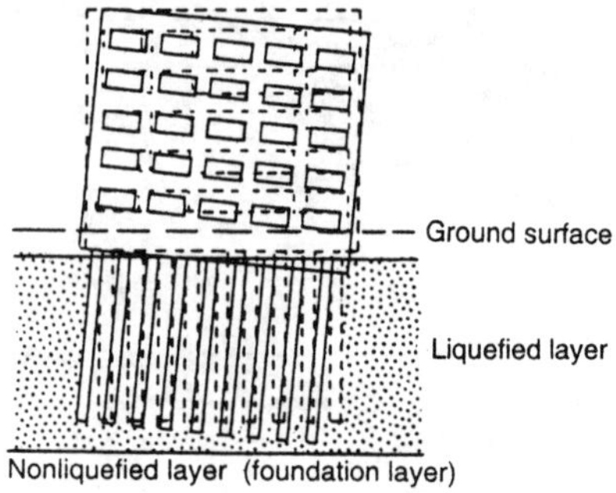 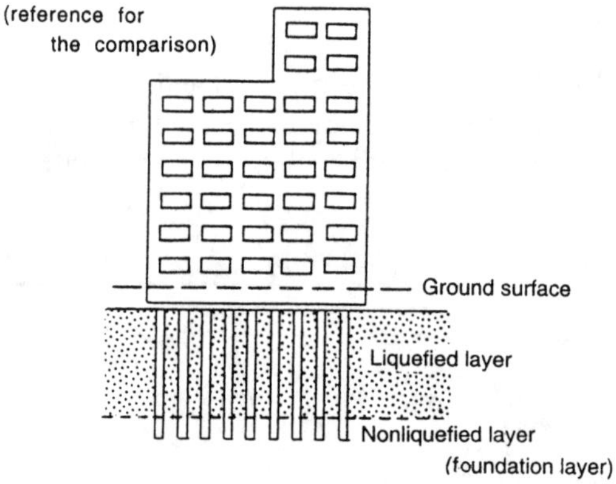
Mode of damage	Settlement/tilting of the pile foundation due to insufficient length of pile
Cause of damage	Reduction in the bearing capacity of the foundation soil
Remark	The length of piles is not sufficient (top drawing) (Reference for the comparison) Pile foundation is sufficiently deep and reaches the nonliquefied layer (the form of damage mentioned above does not occur but takes the form of damage to the pile foundation) (Refer to the next item)

ing or after an earthquake. Typical examples of this type of damage are damage to airport runways, buried water and gas pipes, and pile foundations for bridges and buildings.

Damage induced by liquefaction may be classified by the cause of damage as shown in Table 2.2.

Table 2.4. Continued.

Type of structure	Pile foundation
Schematic diagram of the damage mode	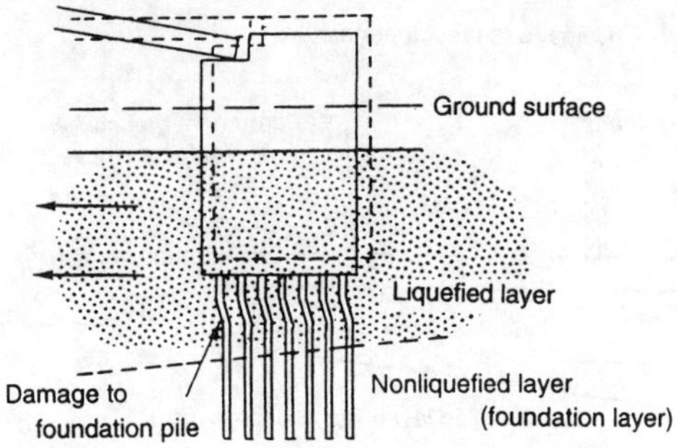
Mode of damage	Bending of piles in the pile foundation
Cause of damage	Permanent displacement of the ground and reduction in horizontal soil resistance in the liquefied layer
Remark	Permanent ground displacement and reduction in the horizontal soil resistance of the liquefied layer occurs

Type of structure	Structure on the ground surface
Schematic diagram of the damage mode	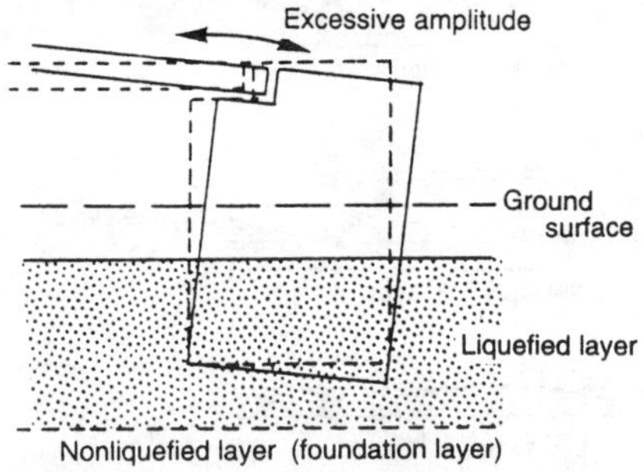
Mode of damage	Damage/failure of the superstructure
Cause of damage	Generation of excessive displacement amplitude
Remark	Displacement of excessive amplitude is generated at the ground surface

2.3.2 *Classification of modes of damage according to structural types*

The causes of damage mentioned earlier are a good means of classification. This is especially true when one of the causes of damage becomes a dominant factor for one type of structure. There are, however, many cases when damage is induced by a combination of two or more causes. In these cases, classification of damage becomes more practical if it is done according to

Table 2.4. Continued.

Type of structure	Structure on the ground surface
Schematic diagram of the damage mode	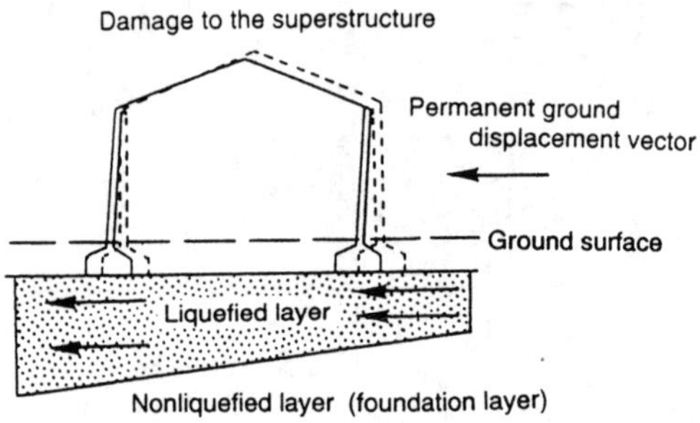
Mode of damage	Damage/failure of the superstructure
Cause of damage	Permanent soil displacement in the horizontal or vertical direction
Remark	Vertical and horizontal nonuniform distribution of permanent displacements is induced by liquefaction

Type of structure	Soil structure
Schematic diagram of the damage mode	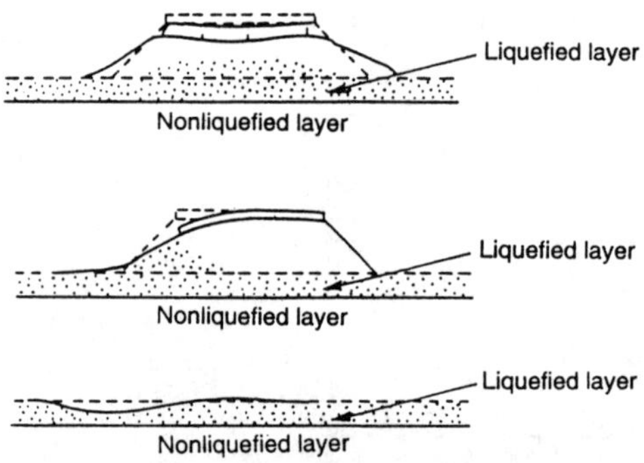
Mode of damage	Settlement of the embankment and lateral displacement
Cause of damage	Deformation of the ground surface induced by sand boil, water spouting, depression, cracking, heave and movement caused by liquefaction
Remark	Permanent displacement, settlement and sand boil

the type of structures constructed on reclaimed land. This type of classification of damage is presented in Tables 2.3 and 2.4.

To account for liquefaction in design, it is important to grasp the mode and cause of damage for each structural type as a basis for choosing suitable remediation measures.

Table 2.4. Continued.

Type of structure	Buried structure
Schematic diagram of the damage mode	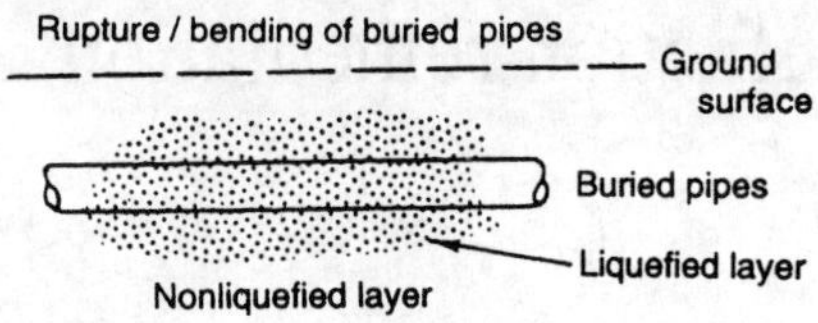
Mode of damage	Rupturing/bending of buried pipes
Cause of damage	Excessive strain in the subsoil at the boundary between liquefied and nonliquefied layers
Remark	Localized liquefaction of the ground is generated or thickness of the liquefied layer varies

Type of structure	Buried structure
Schematic diagram of the damage mode	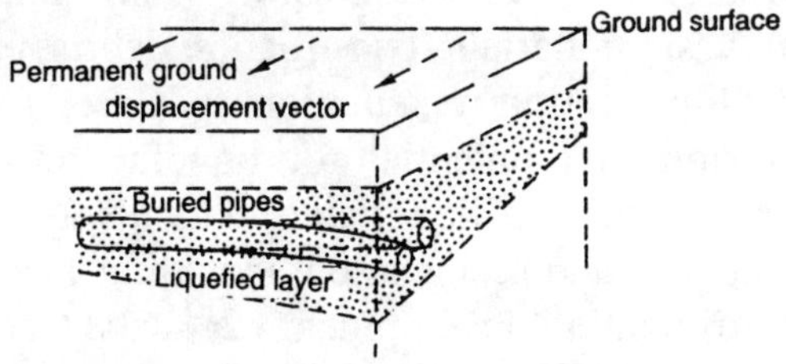
Mode of damage	Rupture/bending of buried pipeline
Cause of damage	Excessive external force due to permanent ground displacement
Remark	Permanent displacement of the ground results in nonuniform distribution

Type of structure	Buried structure
Schematic scheme of the damage mode	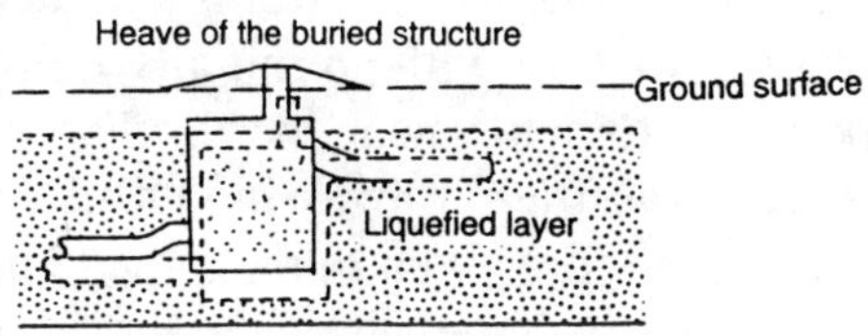
Mode of damage	Heave of the buried structure
Cause of damage	Increase in uplift pressure due to the generation of excess pore water pressure
Remark	The bulk unit weight is larger than that of the liquefied ground around buried structures

CHAPTER 3

Strategy for liquefaction remediation of reclaimed land

3.1 RECLAIMED LAND AND LIQUEFACTION

Reclaimed land refers to a man-made land area constructed by depositing fill soils in open water areas such as oceans, lakes and harbours. The commonly used landfill materials are 1) soils originating from land, 2) soils dredged from seabed and 3) waste materials (construction debris, excavated soils and rubbish). Regarding liquefaction remediation, it is preferable to use landfill material which has minimal potential for liquefaction. Although landfill material is preferable to be non-liquefiable, it is always necessary to consider the overall plans for land reclamation, including 1) availability of landfill material from residential development, 2) seabed soils dredged from channels, and 3) the possibilities for combining waste material treatment with land reclamation.

It should be noted, however, that all reclaimed land will not always undergo liquefaction during an earthquake. According to a study based on 22 earthquakes, from the period between the Edo earthquake of 1855 and the Chibaken-East-Off earthquake of 1987, 85% of the 203 liquefaction sites were found in land reclaimed within 40 years ago, and other 15% of liquefaction of reclaimed land occurred in land older than 40 years. This fact suggests that the occurrence of liquefaction is strongly related to the age of the landfill [1]. Even in ground filled in with an initially uniform soil, there is a tendency for coarse particles to settle near the discharge end of the placement pipe and fine particles to settle far away in landfills constructed by hydraulic fill techniques [2]. Therefore, even in reclaimed land areas filled with a single material, various types of liquefaction can be expected based on the construction methods and age. An appropriate investigation which considers the various characteristics of reclaimed land is necessary. The conditions under which liquefaction will occur are explained in detail in Chapter 4.

3.2 SCALE AND TIMING OF LIQUEFACTION REMEDIATION

Liquefaction remediation of reclaimed land is part of overall land reclamation planning. In planning liquefaction remediation, it is necessary to determine the geometrical scale and timing of remediation by taking into account the land use plan, anticipated consequences of liquefaction, and the financial resources required for liquefaction remediation. The strategy for liquefaction remediation can be classified according to a geometrical scale as follows:

– Large scale remediation: Remediation of an entire reclaimed land area or of each large subdivision;

– Localized remediation: Remediation for each structure or facility as an integrated structural body.

Combinations of these two can also be used in geotechnical engineering practice.

The strategy for liquefaction remediation can be further divided into two categories considering the timing of the remediation. The first strategy can be divided into the following:

a) Remediation during land reclamation;

b) Remediation after completion of land reclamation;

The second strategy can similarly be divided as follows:

c) Remediation before or during construction of structures

d) Remediation after construction of structures (remediation for existing structure).

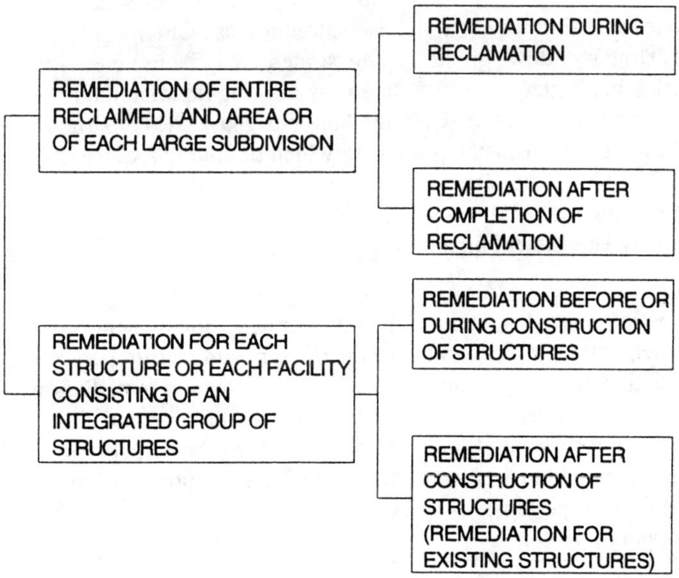

Figure 3.1. Scale and timing of liquefaction remediation.

These divisions can also be applied to structures such as quay walls and bulkheads which constitute the perimeter facilities of reclaimed land. These strategies can be summarized as shown in Figure 3.1.

Specific design procedures and construction methods for liquefaction remediation are explained in detail in Chapter 5.

3.2.1 *Geometrical scale of liquefaction remediation*

A strategy for determining the geometrical scale of liquefaction remediation is chosen after a careful comparison of the advantages and disadvantages of structure-by-structure remediation versus wide area remediation. The principal factors are summarized in Table 3.1 based on technical and execution considerations.

As shown in this table, consistent liquefaction remediation measures must be used for structures and facilities which have a common connector. Some examples of possible problems when this factor is not considered are shown in Table 3.2.

Table 3.1. Characteristics of strategies for liquefaction remediation of reclaimed land based on the geometrical scale.

	Remediation of an entire reclaimed land area or of each large subdivision	Remediation for each structure or facility consisting of an integrated group of structures
Technical aspects	This strategy assures uniform safety with respect to liquefaction over a wide area Compared to remediation for each structure, this strategy allows greater freedom in selecting liquefaction remediation measures It is not necessary to reconsider liquefaction when the land-use plan changes in the future This strategy may be efficient on a large scale	This strategy allows site specific and structure specific remediation tailored to the characteristics of each structure or facility This strategy may be more efficient by limiting liquefaction remediation to those sites or structures which definitely require it
Execution aspect	When a connection exists between structures, which belong to different owners, negotiation and cooperation may be required. In contrast, this strategy, which allows remediation of the entire reclaimed land area, does not require this type of negotiation and cooperation	For this strategy, the owner of each structure and facility performs the liquefaction remediation and thereby negotiation and adjustment among the owners for the cost of remediation is minimal

Table 3.2. Examples in which the connection between the facilities becomes a weak point in liquefaction remediation.

Facilities	Examples of problems in which the connection between the facilities becomes a weak point in terms of liquefaction remediation
Connection between buildings and buried pipes for water, sewage, gas, electricity and telecommunication	When a pile foundation is used as a liquefaction remediation measure for a building, settlement due to liquefaction will not occur at the building, but when remediation measures which allow settlement of buried pipes are used, there is a possibility of excessive relative displacement in the connection to the building
Roadways and buried structures	When a remediation measure which permits settlement to occur is used for a roadway but a remediation measure which does not allow settlement is used for a structure buried under the road (or vice versa), there is a possibility of noticeable differential settlement during an earthquake
Bridge and its access	When the bridge is supported on piles and the access is an embankment, there is a possibility of differential settlement at the access to the bridge due to settlement of the embankment

Table 3.3. Characteristics of liquefaction remediation for reclaimed land according to execution timing.

Remediation during land reclamation	Remediation after completion of land reclamation
Landfill materials can be chosen with regard to soil liquefaction When ground improvement other than liquefaction remediation measures, such as remediation measures for consolidation of a clay layer, is not necessary, then the reclamation process can be done more quickly than liquefaction remediation performed after the completion of a landfill Efficient construction may be possible with simultaneous liquefaction remediation and landfill placement	Rational remediation based on the geotechnical data from the actual subsurface soils is possible after completion of land reclamation. In particular, data on the actual subsoil conditions may be obtained based on in-situ testing results, leading to reliable and efficient remediation of the reclamation area; whereas, in the case of remediation during reclamation, it is necessary to assume the characteristics of the subsoils before reclaimed land is constructed

3.2.2 *Execution of liquefaction remediation*

Referring to strategies a-d described in Section 3.2 and Figure 3.1, when a decision is made to execute large-scale remediation, either strategy a or b may be selected. The characteristics of these strategies are summarized in Table 3.3. If remediation for each structure or each facility is to be performed, strategy c is automatically selected. Strategy d is usually reserved for cases when the need for liquefaction remediation of an existing structure is later discovered.

CHAPTER 4

In-situ and laboratory testing and assessment of liquefaction potential

4.1 DESIGN CONSIDERATIONS RELATED TO LIQUEFACTION AND OUTLINE OF METHODS FOR ASSESSING LIQUEFACTION POTENTIAL

4.1.1 *Design considerations related to liquefaction*

General design considerations related to liquefaction are shown in a flow chart in Figure 4.1. Initially, liquefaction potential is assessed; then, the extent of damage to (or amount of deformation of) structures due to liquefaction is evaluated; and finally, liquefaction remediation measures are determined.

In this handbook, two phases in assessing liquefaction potential during design consideration are distinguished. The first phase is *prediction of liquefaction*, referring to the assessment of liquefaction potential obtained directly from the simplified procedures presented in this book. Prediction of liquefaction can, therefore, be accomplished as a routine practice. The second phase is *determination of liquefaction*, referring to an engineering decision during design as to whether or not the subsoil in question will liquefy. In determining liquefaction, it is necessary to take into account various factors such as how much safety margin should be allowed for the structure concerned.

For the steps shown in Figure 4.1, engineering technology for estimating the extent of damage to (or amount of deformation of) structures due to liquefaction has not yet been well established. A detailed explanation on the current level of development of this technology is given in Technical note 1. In current practice, the decision to do liquefaction remediation is, in many cases, based on overall engineering considerations including the importance of the structure in concern.

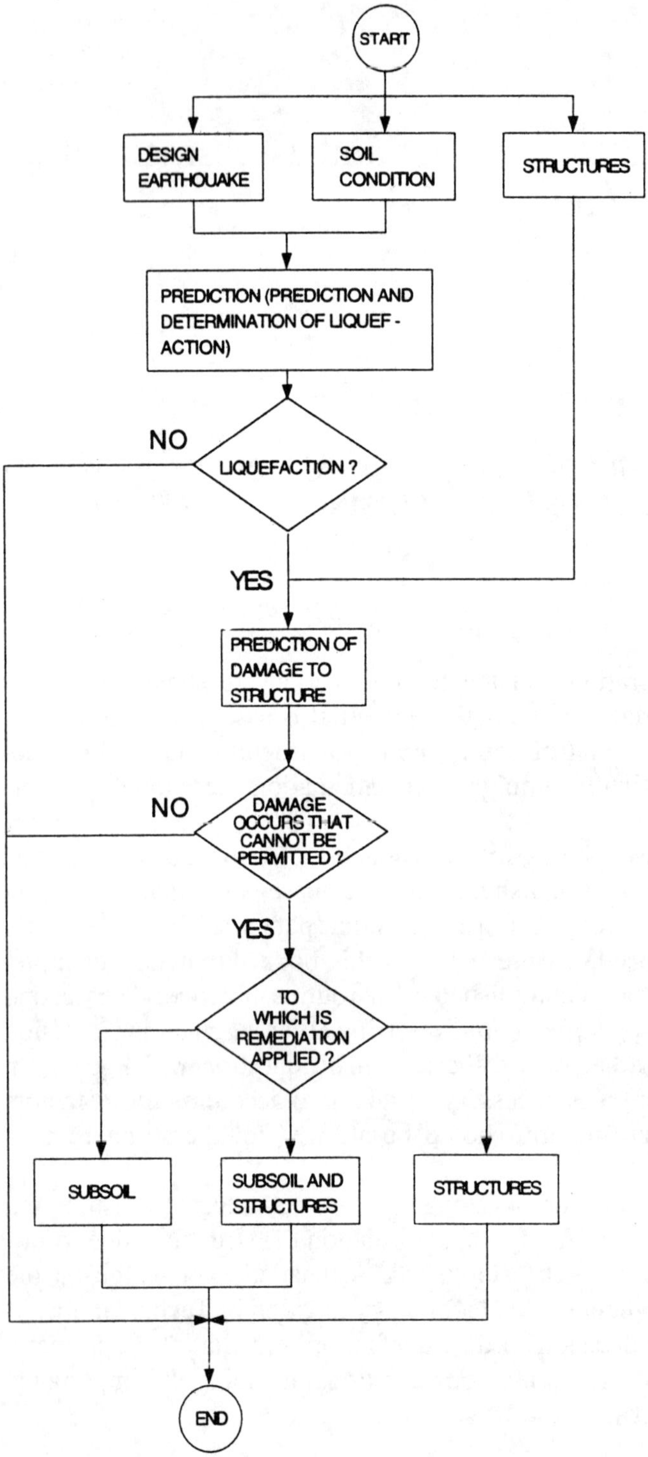

Figure 4.1. Flow chart of design consideration on liquefaction.

Table 4.1. Classification of various methods for assessing liquefaction potential.

Number	Evaluation of liquefaction strength of soil	Evaluation of earthquake effects	Examples of prediction methods
1	—	Richter magnitude and epicentral distance	Kuribayashi & Tatsuoka [1], Wakamatsu [2]
2	Geomorphology	—	Kotoda et al. [3]
3	Gradation/SPT *N*-value	—	Standards for regulation of hazardous materials [4]
4	Gradation/SPT *N*-value	Maximum acceleration on the ground surface	Specifications for highway bridges [5]
5	Gradation/SPT *N*-value	Maximum acceleration on the ground surface and Richter magnitude	Recommendations for the design of building foundations [6]
6	Gradation/SPT *N*-value	Total stress model	Technical standards for port and harbour facilities and commentaries [7]
7	Cyclic triaxial test	Total stress model	Seed & Idriss [8], Ishihara [9]
8	Cyclic triaxial test	Effective stress model	Finn et al. [10], Iai et al. [11]

4.1.2 *Methods for assessing liquefaction potential*

An assessment of liquefaction potential is made in principle by comparing earthquake induced shear stress to the liquefaction strength of the soil. Various methods for assessing liquefaction potential [1-11] thus far proposed may be classified according to the methods for evaluating shear stress and liquefaction strength as shown in Table 4.1.

Methods based on Richter magnitude of earthquake and epicentral distance [1, 2] shown in 1 of Table 4.1 are used for stipulating regions not subject to liquefaction, provided that the Richter magnitude and location of the earthquake considered in design are known in advance. A method based on geomorphology [3] shown in 2 of Table 4.1 is also used for designating regions not subject to liquefaction. This method requires classification of the site in terms of geological and geomorphological data such as former river channel, natural levee and swamp. This method is particularly suitable for developing countries, or for areas where in-situ testing such as the standard penetration test and laboratory testing are difficult.

Methods based on gradation and SPT *N*-value [4-7] shown in 3-6 of Table

4.1 have been adopted in the design standards and codes in Japan; they are collated under Appendix 1 in this handbook. The principle of these methods is to assess the liquefaction potential by comparing the liquefaction strength evaluated from gradation and SPT *N*-values of subsoils with the shear stress ratio defined by the ratio of the earthquake induced shear stress over the effective vertical stress. If the shear stress ratio exceeds the liquefaction strength of the subsoils, liquefaction is predicted; if not, nonliquefaction is predicted.

Among the methods based on cyclic triaxial tests, the methods based on the total stress model [8, 9] shown in 7 of Table 4.1 evaluate the earthquake induced shear stress from earthquake response analysis, ignoring the effects of excess pore water pressure during shaking. The methods based on the effective stress model shown in 8 of Table 4.1 [10, 11] incorporate the effects of excess pore water pressure increase during shaking, computing various outputs relevant to liquefaction phenomena. The total stress model is used when the occurrence of liquefaction is predicted as either affirmative or negative; the effective stress model is used to predict the extent of liquefaction as specified by the excess pore water pressure ratio. The method based on the total stress model is explained in detail in Section 4.5 and the method based on the effective stress model in Technical note 1.

4.1.3 *Liquefaction prediction method in this handbook*

In this handbook, the liquefaction prediction/determination method [7] used for port and harbour structures is explained. This method was developed as a result of observations of strong earthquakes in port and harbour areas in Japan for a quarter of a century, as well as case histories of liquefaction in reclaimed lands in Japan. Applicability of this method to the liquefaction of reclaimed land has been verified by many liquefaction case histories [12, 13]. The process, which is described in Section 4.5, consists of two methods: one based on gradation and SPT *N*-value, to be called the *gradation/SPT N-value method*; and the other based on cyclic triaxial test results, to be called the *cyclic triaxial method*. The gradation/SPT *N*-value method corresponds to the method indicated in 6 of Table 4.1, based on total stress analysis and gradation/SPT *N*-value. The cyclic triaxial method corresponds to the method indicated in 7 of the table, based on the total stress analysis and the cyclic triaxial test.

4.2 SOIL SURVEY AND TEST

4.2.1 *Types of investigation and the procedure*

A soil survey and test on reclaimed land is classified into four categories

Table 4.2. Guideline for geotechnical survey and test for port and harbour facilities.

Classification	Investigation method	Investigation objective	Investigation contents	Sampling	Major test items
Preliminary investigations	Data collection	Arrangement plan for structures, general investigation plan	Boring and sounding data, recording of ground water level, topographic map, geologic investigation map, record of settlement/failure of structures, construction work report, aerial photograph, *liquefaction case history*	—	—
	Site reconnaissance	Same as above	Topography, geology, record of failure and settlement of structures	Only the surface layer	Physical test
General investigation	Boring, sampling, soil testing, sounding, geophysical probing	Arrangement plan for structures, design of structures, detailed investigation plan	Foundation depth, subsoil stratification, strength, consolidation characteristics, soil type, compaction characteristics, ground water level, permeability, *liquefaction strength, dynamic deformation characteristics*	Obtain one undisturbed sample per depth of 1-1.5 m in clay soil *Obtain sample at a depth of 1.5-2.0 m as a standard in sandy subsoils*	Obtain SPT N-value for each depth of 1-2 m in sandy soil, q_u, W, γ_t in clay soil *Cyclic triaxial test, dynamic deformation test, elastic wave exploration*
Detailed investigation	Boring, sampling, soil testing, sounding, test pit	Design report for structures, construction work control record	Same as above	Same as above	Same as above
Supplementary investigation	Same as above	Same as above	Same as above	—	—
Special testing	Slip failure test, field test, model test, in-situ permeability test	Same as above	Bearing capacity, slope stability, consolidation characteristics, permeability, *items related to liquefaction remediation measures*	—	—

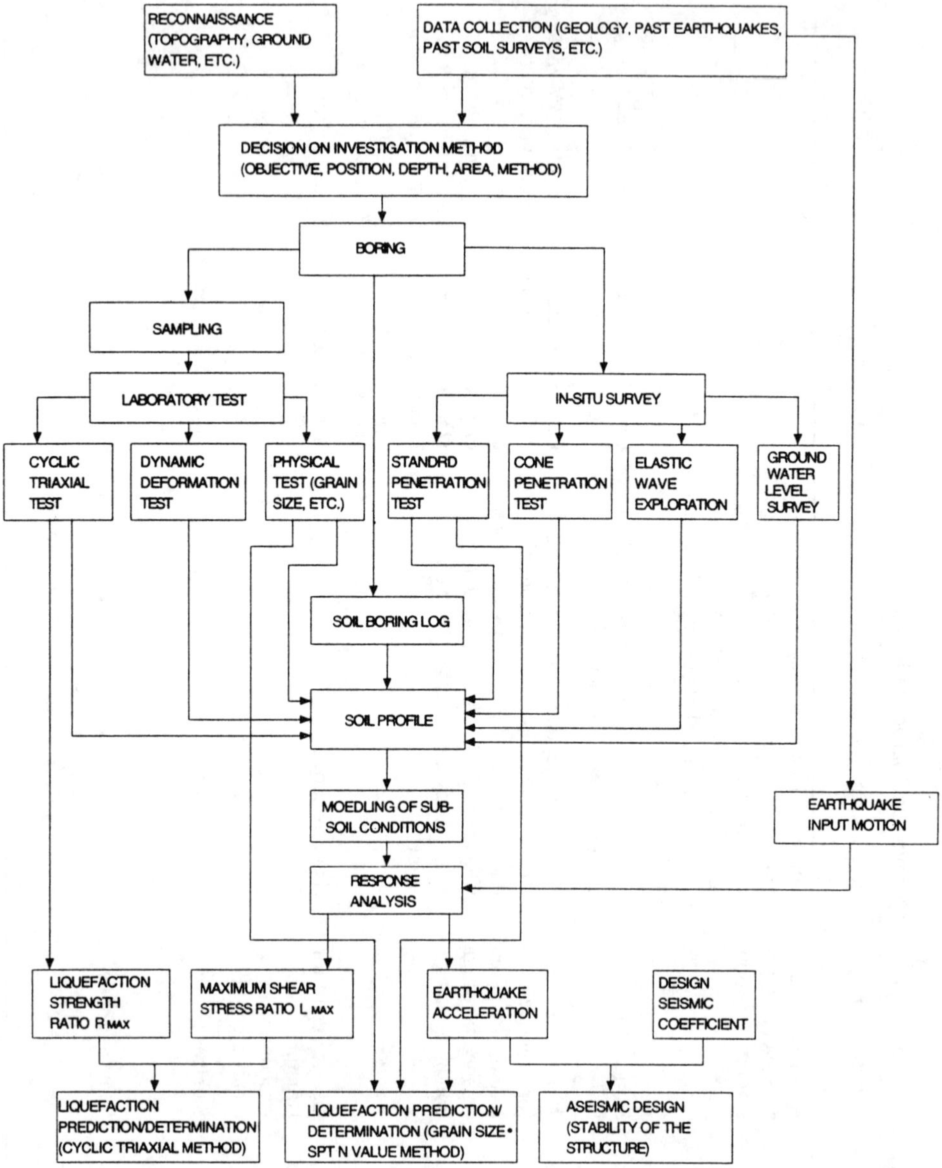

Figure 4.2. Soil survey flow chart.

based on the order and timing of the investigation:

1. Preliminary investigation;
2. General investigation;
3. Detailed investigation;
4. Supplementary investigation.

The preliminary investigation is performed prior to land reclamation mainly for determining the plan for the general and detailed investigations which will be performed after the land reclamation. The results of the preliminary investigation are used to determine the investigation objectives, position, depth, area, method, etc. for the general and detailed investigations. The general and detailed investigations are the main means of evaluating the geotechnical conditions of the reclaimed land. The supplementary investigation is an additional investigation performed when unexpected results are obtained in the detailed investigation or when data obtained from the investigation are found to be insufficient.

Items of these investigations necessary for aseismic evaluation of reclaimed land are shown in Table 4.2. This table is based on the standard items of geotechnical investigations specified in the Port and Harbour Investigation Manual [14], supplemented by the specific items necessary for aseismic evaluation as designated by the italics in Table 4.2. The specific items required for aseismic evaluation are as follows:

1. *Elastic wave exploration.* The elastic wave (*P*- and *S*-waves) velocity in the small strain level is determined by a field survey such as a down hole or cross hole method;

2. *Liquefaction test.* The liquefaction strength of the sandy soil is determined by a laboratory test such as an undrained cyclic triaxial test;

3. *Dynamic deformation test.* The strain dependent shear modulus (also called shear stiffness) and damping factor (also called damping ratio) of the soil are obtained from a laboratory test such as a cyclic triaxial test.

The special tests designated in the lower most row in Table 4.2 are those which become necessary in designing remediation measures against liquefaction.

Figure 4.2 shows the investigation flow chart for aseismic evaluation of reclaimed land including prediction/evaluation of liquefaction.

4.2.2 *Layout of survey points and depth requirements*

4.2.2.1 *Layout of investigation points*

A practical layout is made, locating points of investigation by considering the required accuracy, time schedule, and cost. In a soil survey of port and harbour areas, a geotechnical investigation plan is recommended with the intervals specified in Table 4.3 [15]. The arrangement shown in Table 4.3 is a standard that can be modified to suit individual conditions of the subsoil, the type/form of the structure, and the objective of the investigation. In a subsoil profile with complex stratification, the data investigation points become closer to each other vertically and horizontally. Complex stratification is often encountered in Japan in areas close to a river flowing into a harbour or in those areas of reclaimed land close to a discharge gate of overflow

Table 4.3. Layout of survey points (Unit: m).

1. *When the stratification is relatively uniform both horizontally and vertically*

		Along the face of quay wall		Perpendicular to the face of quay wall			
		Designated intervals		Designated intervals		Distance from the face of quay wall (maximum)	
		Boring	Sounding	Boring	Sounding	Boring	Sounding
General investigation	Wide area	300~500	100~300	50	25	50~100	50~100
	Small area	50~100	20~50	50	25	50~100	50~100
Detailed investigation		50~100	20~50	20~30	10~15	50~100	50~100

2. *When the stratification is complex*

	Along the face of quay wall		Perpendicular to the face of quay wall			
	Designated intervals		Designated intervals		Distance from the face of quay wall (maximum)	
	Boring	Sounding	Boring	Sounding	Boring	Sounding
General investigation	Below 30	15~20	20~30	10~15	50~100	50~100
Detailed investigation	10~30	5~10	10~20	5~10	50~100	50~100

Note: Some soundings require a borehole and some do not. Only those which do not require a borehole are shown in the columns of 'sounding'. Sounding that requires a borehole is done at the same interval as boring.

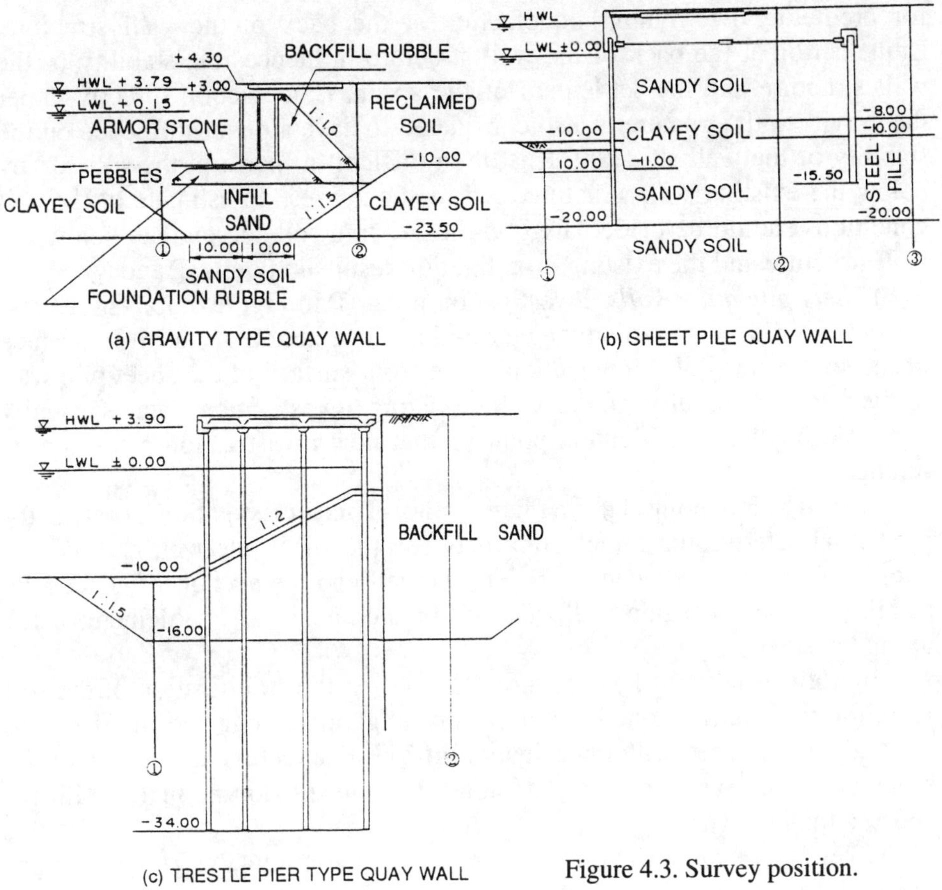

Figure 4.3. Survey position.

water temporarily used during the land filling process.

In determining the points of investigation for an existing quay wall, a traverse at right angles to the wall is fixed at a suitable position by taking the stratification of the subsoil profile into consideration. When the stratification is presumed to be uniform in the horizontal direction, the interval should be about one traverse per 100 m along the face line of the quay wall.

The investigation position along one traverse should be determined, with reference to Figure 4.3, according to the structure of the quay wall. The investigation position shown in Figure 4.3 is determined based on the following considerations.

a) *Gravity quay walls*. Investigation point ① in Figure 4.3a: This investigation point is used for predicting/determining liquefaction at the front end of the wall structure. Liquefaction of the front end of the gravity wall affects the stability of the wall structure.

– Investigation point ② in Figure 4.3a: This investigation point is used for

for predicting/determining liquefaction at the back of the wall structure. Liquefaction at the back of the wall structure influences the stability of the wall structure and the settlement of the apron. Liquefaction directly under the gravity wall structure is also considered to have a major influence on the stability of the wall structure. It is often difficult to perform the soil survey below the existing wall structure, and in such cases, an estimate of the soil conditions should be made based on the existing soil survey data below the wall structure and the existing investigation results in points ① and ②.

b) *Sheet pile quay walls*. Investigation point ① in Figure 4.3b: This investigation point is for predicting/determining liquefaction at the front surface of the sheet pile wall. Liquefaction at the front surface of the sheet pile wall influences the stability of the wall. Existing investigation data is usually available for this investigation point so that new investigation can often be omitted.

– Investigation point ② in Figure 4.3b: This investigation point is for predicting/determining liquefaction between the sheet pile wall and the anchor. Liquefaction between the sheet pile wall and the anchor influences the stability of the sheet pile wall, stability of the anchor and settlement of the apron.

– Investigation point ③ in Figure 4.3b: When the stratification of the soil is uniform in the horizontal direction, investigation designated in ③ can be omitted and replaced with investigation in ②. Liquefaction at the back of the bracing has the same influence as noted in investigation point ② with respect to the facility.

c) *Trestle type piers*. Investigation point ① in Figure 4.3c: This investigation point is for predicting/determining liquefaction at the front surface of the trestle type pier. Liquefaction at the front surface of the trestle type pier influences the stability of the trestle type pier structure and stability of the soil retaining wall.

– Investigation point ② in Figure 4.3c: This investigation point is for predicting/determining liquefaction at the back of the soil retaining wall. Liquefaction at the back of the soil retaining wall influences the stability of the soil retaining wall and settlement of the apron.

4.2.2.2 *Survey depth*

The depth for subsurface boring and sampling can generally be established to be the depth where an adequate base layer is encountered. In the conventional design of structures with respect to static loads, a base layer is defined as a soil layer which has a sufficient strength to support surcharge loads from landfill or a structural load. In soil surveys for port and harbour areas for example, the base layer with respect to static loads is evaluated from the SPT *N*-value as follows. For a relatively small structure or a structure without a pile foundation, an SPT *N*-value of 30 or more is required to define a base layer. For a large structure with a pile foundation, an SPT *N*-value of 50 or more is required to define a base layer. When a layer with an SPT *N*-value of 50 or more is not encountered, the investigation depth must be determined by referring to the relationship between SPT *N*-value and the base layer thickness listed below [15].

– When SPT *N*-value is 10 to 20: over 5 m;

– When SPT *N*-value is 20 to 30: over 3 to 5 m;

– When *N* is 30 to 50: over 2 to 3 m.

Standard penetration tests (SPT) are normally performed at 1 m intervals to obtain *N*-values. When bedrock is encountered, diamond rock core drilling is usually conducted with cores retained for at least 1 m and the investigation is not considered complete until bedrock properties are determined by visual and/or geotechnical testing.

The investigation depth related to liquefaction generally differs from that for static bearing design mentioned earlier. For evaluation of liquefaction potential, the depth of geotechnical investigation is determined as follows.

a) *Depth of base layer for seismic response analysis*. For evaluation of liquefaction, seismic response analysis of the subsoil profile is necessary to evaluate shear stress in the subsoils. The base layer for the seismic response analysis is defined as the shallowest soil layer having shear wave velocity (V_s) over 300 m/s. When the shear wave velocity is not obtained, the bedrock or soil layer with an SPT *N*-value of 50 or more in sand and/or unconfined compressive strength (q_u) of over 65 tf/m^2 in clay should be the base layer. When determining the base layer from soil survey results, use existing data to verify that a soft soil layer is not encountered at a depth below the base layer. When a layer with an SPT *N*-value of 50 or more does not appear, the depth of the base layer for seismic response analysis is normally greater than the depth of the base layer for static bearing design.

These conditions can be summarized as follows:

– When shear wave velocity is measured, soil survey is performed down to a layer with a shear wave velocity of 300 m/s or higher. A layer with a lower velocity should not appear at a depth below this layer.

– When shear wave velocity is not measured, soil survey is performed to the bedrock, or a sandy soil layer with an SPT *N*-value of 50 or higher, or a clayey soil layer with an unconfined compressive strength of 65 tf/m^2 or higher.

b) *Depth of liquefaction investigation*. It has been reported [16] that there are no case histories of liquefaction at a depth of 20 m below ground surface. The theoretical and quantitative bases for this fact has yet to be established but a certain deduction may be possible to explain this fact. At a depth of more than 20 m, the effective vertical pressure becomes relatively large, and hence the shear stress ratio induced by an earthquake is likely to become small. It is also possible that loosely sedimented sand layers may not have maintained their original loose state if they underwent the large effective vertical pressure for a long time, and were subjected to several earthquakes. Thus, it has become a standard practice in Japan that the geotechnical investigation for predicting/determining liquefaction is limited to a depth of 20 m below the ground or seabed surface. When there is specific evidence, however, that indicates liquefaction susceptibility of a soil layer at a depth greater than 20 m, the investigation depth should be greater than 20 m.

4.2.2.3 *Investigation procedure*

There may be many cases in which the depth of the base layer would exceed 20 m and it may not be possible to investigate down to the base layer in all bore holes. In such cases, a limited number of representative points are initially selected for detailed investigation/test for seismic response analysis. At these representative points, standard penetration tests, elastic wave exploration, and undisturbed sampling should be performed to evaluate the shear modulus/damping factor of each layer.

When the subsurface stratification is relatively uniform, it is possible to establish an investigation plan pointed toward a reduction in the number of borings which reach the base layer.

Depending on the conditions at the site, such as the required investigation accuracy, cost, and time period, it would be possible to adopt a simplified investigation procedure. This could be done by estimating the shear modulus/damping factor with an empirical equation based on SPT *N*-value and physical properties of the soil to be shown in Section 4.3.

A soil survey for evaluation of liquefaction potential is conducted primarily by the standard penetration test (SPT). This survey aims at predicting liquefaction using the grain size and SPT *N*-values. When it is difficult to evaluate the liquefaction potential using the grain size and SPT *N*-values, the liquefaction strength is evaluated by implementing undisturbed sampling and cyclic triaxial tests in the supplementary investigation. The results of the cyclic triaxial tests can also be used for calibrating the evaluation results based on the grain size and SPT *N*-values.

4.2.3 *Parameters of interest and sampling density*

4.2.3.1 *Parameters of interest*

In order to evaluate the liquefaction potential, the subsoil shear stress profile and the liquefaction strength of the soil must be known. The object of the soil survey/test is to evaluate the soil constants necessary for evaluating these items. The measured parameters and methods of measurements are shown in Table 4.4. Among these items, the elastic wave exploration, liquefaction test, and dynamic deformation test determine the dynamic characteristics of the soil.

Table 4.4. Survey methods and items classified by survey objective.

Survey method	Survey item	Investigation for estimating the shear stress during earthquake	Investigation for liquefaction prediction: Gradation/SPT *N*-value method	Investigation for liquefaction prediction: Cyclic triaxial method
Boring	Boring log	+	+	+
	Ground water level	+	+	+
Sounding	Standard penetration test (SPT *N*-value)	+	+	
PS logging	*P*-wave velocity	+		
	S-wave velocity	+		
Undisturbed sampling	Sampling of clayey soil	–		
	Sampling of sandy soil			+
Dynamic deformation test	Liquefaction strength			+
	Dynamic deformation characteristics	+		
Physical test	Unit weight	–		+
	Specific gravity	+	+	+
	Water content	+	+	
	Grain size	+	+	+
Unconfined compression test	Unconfined compressive strength	–		

Note: + = used as a standard; – = use depends on the actual case.

a) *Survey for seismic response analysis.* The shear stress in subsoil during earthquakes is computed by performing a response analysis of the subsoil profile with respect to input earthquake motion as indicated in Section 4.4. The response analysis of the subsoil profile requires information on subsoil layer formation, depth of the base layer, density of each layer, shear modulus/damping factor, ground water level, and the stress state in the soil. In order to estimate the shear stress in the soil, it is necessary to obtain not only the soil constants of the sand layer which is studied for liquefaction potential but also the soil constants related to the deeper layers affecting seismic response of subsoil profile, including the clay layers below the liquefiable sand layers.

b) *Survey for obtaining the liquefaction strength.* As indicated in Section 4.5, the liquefaction potential is evaluated based on the gradation and SPT *N*-value and the cyclic triaxial test. For the gradation/SPT *N*-value method,

Table 4.5. Survey/test items and intervals.

Survey method	Survey/test item		Soil type	Survey interval	Survey objective
Sounding	Standard penetration test	SPT *N*-value	Sand/sandy soil	1 m	Liquefaction prediction based on gradation/SPT *N*-value
		Grain size distribution (gradation)	Sand/sandy soil	1 m	
In-situ survey	Elastic wave exploration*	*P*-wave velocity *S*-wave velocity	Sand-clay	1 m	Seismic response analysis Shear modulus in very small strain range
Undisturbed sampling	Liquefaction test	Liquefaction strength	Sand/sandy soil	1.5~2 m	Liquefaction strength in cyclic triaxial method
	Dynamic deformation test*	Shear modulus/ damping factor	Sand-clay	For each layer	Seismic response analysis Strain dependence curve in medium-large strain range
	Density test*	Unit weight	Sand-clay	For each layer	Calculation of effective overburden pressure

*Empirical equations can also be used for simplified evaluation of soil parameters.

the SPT N-value and gradation for each sand layer are necessary as shown in Table 4.4. For the cyclic triaxial method, the liquefaction strength is determined by the cyclic triaxial test using undisturbed samples obtained from the site. Since the liquefaction strength is influenced by disturbance in the sample, it is important to use a sampling method which minimizes sample disturbance as discussed in Section 4.2.4.

4.2.3.2 *Sampling density*

Studies must be conducted at depth intervals chosen so that soil constants can be evaluated accurately. Of the items listed in Table 4.4, the standard penetration test, grain size test, unit weight test, elastic wave exploration, undisturbed sampling, liquefaction test, and the dynamic deformation test should be conducted at the intervals shown in Table 4.5. Conducting the standard penetration tests at 1 m intervals is the standard practice in Japan and is found suitable for liquefaction potential evaluation. Grain size tests should be conducted on each of the samples obtained during the standard penetration test.

Elastic wave velocity is normally measured at 1 to 2 m intervals in Japan, using the bore holes in which the standard penetration tests were performed. Measurement at 1 m intervals is recommended in order to take into account the influence of the overburden pressure change within a continuous soil layer.

With regard to sampling intervals in sandy soil for performing liquefaction tests, 1.5 to 2.0 m intervals is suitable. When the fine particle portion, less than 74 μm in diameter, is minimal, however, the sampling should be conducted in two steps. When the sandy soil layer is thick and homogenous, it is possible to perform sampling in the upper part, center, and bottom part within the continuous homogenous layer being investigated. Specimens used for the liquefaction test should include 3 to 4 specimens at each depth.

For the samples used for dynamic deformation tests to evaluate the strain dependent modulus and damping factor, more than one sample is necessary for each sand and clay layer. The standard number of specimens used in the dynamic deformation test is 1 to 3 specimens from each layer.

4.2.4 *Investigation and test methods*

4.2.4.1 *Standard penetration test*

The standard penetration test in Japan is performed according to JIS A 1219 *Standard penetration test method for soil*. The standard penetration test can be used for a relatively wide range of soil properties. It cannot be used, however, for soils including rocks, cobblestones, and pebbles. The standard penetration test can be performed on soft clay but the reliability of the result for evaluating dynamic properties of the soil should be regarded as low. It is better not to use the standard penetration test in soft clay because, when the standard penetration test is made, the soil at the base of the borehole is considerably disturbed. For such a case, the strength of the soil should be evaluated by obtaining undisturbed samples and conducting laboratory tests.

4.2.4.2 *Elastic wave exploration*

The soil constants necessary for seismic response analysis are the shear modulus, damping factor and unit weight of soil in each layer. The shear modulus and damping factor of soil are significantly influenced by shear strain level and they should be evaluated over a strain range from a very small shear strain level (about 10^{-6}) to a failure strain level (of the order of 10^{-2}).

The shear modulus/damping factor at a very small shear strain level can be measured either by elastic wave exploration with PS logging or a laboratory test using an undisturbed sample. From the elastic wave exploration, *S*-wave velocity V_s and *P*-wave velocity V_p are obtained, but a method for obtaining the damping factor has not yet been adequately established. To evaluate the damping factor, conducting separate dynamic deformation tests is preferable.

For the simplified estimation, the method using the existing empirical equations given in Section 4.3 may be used. For example, shear modulus (G_o) at a very small shear strain level is calculated with Equation (4.4) in Section 4.3.

Elastic wave exploration (PS logging) measures the propagation speed of a *P*-wave (or compression wave) and a *S*-wave (shear wave) in the soil. The methods of measurement are indicated in Figure 4.4. Although not shown in

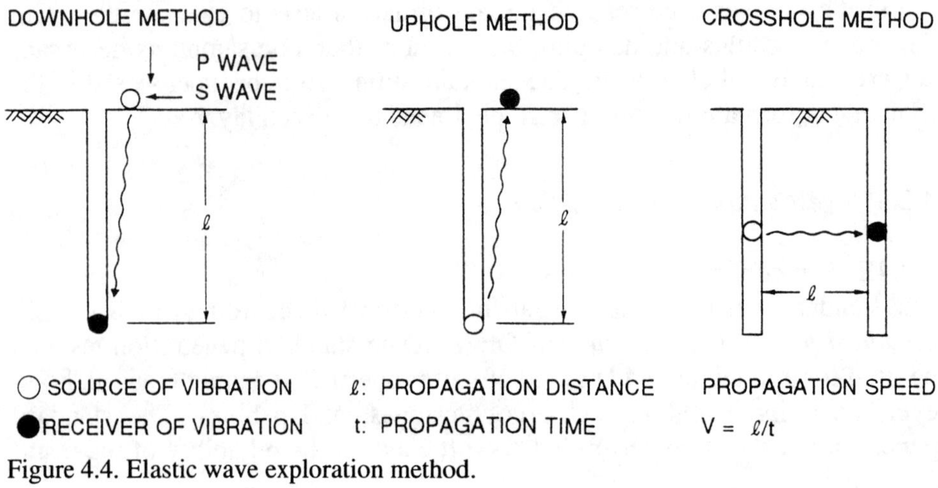

Figure 4.4. Elastic wave exploration method.

Figure 4.4, a relatively new method, the so called suspension type logging method, measures the elastic wave velocity by suspending equipment with a vibration generator and vibration receiver attached on both ends within a borehole.

Of the methods shown, the downhole method is often used in soil surveys. This method measures the arrival time of the seismic wave with a vibration receiver set within the hole from a vibration source at the ground surface. Normally, a *P*-wave is generated by striking the soil surface vertically with a wooden mallet and the *S*-wave is generated by striking a plate set horizontally on the ground surface (plate striking method). From the propagation time and the distance between the ground surface and the vibration receiver at each depth, the velocity of the elastic wave propagation is obtained by depicting a time distance curve. The propagation velocity obtained in this case is an average value for each layer. When it is necessary to consider the influence of the overburden pressure change in subsoil, analysis should be performed by dividing the layer into finer elements.

The uphole and crosshole methods are used when it is difficult to generate vibrations at the ground surface such as in the subsoils at seabed level in open water. The crosshole method has the advantage of obtaining the wave propagation velocity for each depth of investigation and of checking the influence of the overburden pressure. The cross hole method uses two boreholes separated by a few meters.

Elastic wave exploration is a non-destructive investigation method and has the advantage that it does not involve soil disturbances. When the effective vertical pressure is changed due to structure or embankment construction after elastic wave exploration, correction should be made in the *S*-wave with Equation (4.18) in Section 4.4. Examples of elastic wave exploration conducted in the past at port and harbour regions in Japan are shown in Table 4.6 [17].

4.2.4.3 *Undisturbed sampling*

In order to evaluate the in-situ liquefaction strength of sandy soil, undisturbed samples must be obtained. Undisturbed sampling of sandy soil has more widely attempted since the liquefaction phenomenon during the Niigata earthquake of 1964. Many types of samplers were devised thereafter. Compared to undisturbed sampling of clayey soil, undisturbed sampling of sandy soil is difficult. As an engineering practice, however, it has become possible to obtain sandy soil with little disturbance [18].

Applicability of the sand samplers presently used differs slightly depending on the properties of soils, and it is hence necessary to have a basic understanding of the properties of the soil being investigated before the selection of the sampler. The results of the standard penetration test conducted before the sampling should be very useful in selecting an appropriate sand sampler.

Table 4.6. Case histories of elastic wave exploration.

Survey point	Land/ seabed	Elastic wave exploration method	Wave propagation mode	Vibration source	Position of source vibration	Position of receiver	Casing	Borehole diameter	Survey depth	Measurement interval
Yokosuka city, Kurihama testing field (1)	Land TBM +0.44 m	PS logging	Down-hole	*P*-wave: wooden maul *S*-wave: plate striking	Ground surface, 5 m of distance from a borehole	Received at 22 points within a borehole	Polyvinyl chloride (PVC) pipe Down to 42.5 m	86 mm	2.5~42.5 m	1 m
	Land TBM +0.44 m	Cross-hole	Cross-hole	Strike a rod appended with a circular plate (ϕ 50 to 70 mm) with a maul	Horizontal distance 5 m received at 15 points within a borehole		Polyvinyl chloride (PVC) pipe Down to 42.5 m	86 mm	3~42 m	3 m
Yokosuka city, Kurihama testing field (2)	Land TBM +0.63 m	PS logging	Down-hole	*P*-wave: wooden maul *S*-wave: plate striking	Vicinity of a borehole	Received at 26 points within a borehole	Casing Down to 4 m	86 mm	4~30 m	1 m
	Land TBM +0.63 m	Cross-hole	Cross-hole	Connect a rod to a packer tube appended with a steel backing plate and strike it vertically	Vibration generated at borehole No. 1, received at boreholes No. 2 and 3 Horizontal distance between each hole is 3 m		Casing Down to 4 m	86 mm	4~30 m	1 m
Kobe port, Port Island	Land CDL +4.2 m	PS logging	Down-hole	*P*-wave: air gun and wooden maul *S*-wave: plate striking	Vicinity of a borehole	Received at 40 points within a borehole	—	66~150 mm	2~80 m	2 m
Kobe port, Rokko Island (1)	Land CDL +3.7 m	PS logging	Down-hole	*P*-wave: air gun and wooden maul *S*-wave: plate striking	Vicinity of a borehole	Received at 51 points within a borehole	—	66~150 mm	2~102 m	2 m

Table 4.6. Continued.

Survey point	Land/ seabed	Elastic wave exploration method	Wave propagation mode	Vibration source	Position of source vibration	Position of receiver	Casing	Borehole diameter	Survey depth	Measurement interval
Kobe port, Rokko Island (2)	Land CDL +4.5 m	PS logging	Downhole	*P*-wave: wooden maul *S*-wave: plate striking	2 m to borehole No. 58.1, 7 m to borehole No. 58.2, 9 m to borehole No. 58.3, received vibration at 3 boreholes		About 19 m in fill layer	65~85 mm	20.4~32.4 m	1 m
	Land CDL +4.5 m	Crosshole	Crosshole	Shear hammer (fix with hydraulic jack)	Vibration generated within casing of borehole No. 58.1, 5 m and 7 m from borehole No. 58.1 to borehole No. 58.2 and 58.3, respectively received at 13 points		Vibration source within casing	65~85 mm	20.4~32.4 m	1 m
Yokohama port, Daikokufuto (1)	Seabed CDL −12.4 m	PS logging	Downhole	Connect a rod to a gas pipe appended with a steel plate (50 × 50 × 0.45 cm) and strike it vertically and horizontally	Seabed surface 3 m away from a borehole	Received at 20 points for every point within a borehole	About 1 m from seabed surface	127 mm	1~20 m	1 m
Yokohama port, Daikokufuto (2)	Seabed CDL −13.1 m	Combination of crosshole and PS logging	Uphole	Bundle of 1 to 3 electric detonators	Vibration generated at every point at prescribed length of borehole No. 4, respectively received at 2 points at depths 3 m and 10 m within supplementary borehole No. 9 separated by 5 m (measured separately with borehole No.4 without casing for subsoils shallower than 15 m)		15 m from seabed surface (No. 4)	127 mm	1~30.7 m	1 m

Table 4.6. Continued.

Survey point	Land/ seabed	Elastic wave exploration method	Wave propagation mode	Vibration source	Position of source vibration	Position of receiver	Casing	Borehole diameter	Survey depth	Measurement interval
Yokohama port, Daikokufuto (2)	Seabed CDL –13.1 m	Cross-hole	Cross-hole	Connect a rod to a packer tube appended with a steel backing plate and strike it in the vertical direction with a wooden mallet	Vibration generated at borehole No. 9, received at borehole No. 2 separated by 7 m and borehole No. 4 separated by 3 m from the vibration source, received at 12 points		About 1 m from seabed surface	127 mm	1.6~10 m	1 m
Hiroshima port, Hatsukaichi area	Seabed CDL –8.3 m	PS logging	Down-hole	Connect a rod to a gas tube appended with a steel plate (50 × 50 × 0.45 cm) and strike it vertically and horizontally	Seabed surface, separated by 3 m from a borehole	Every point in a borehole, received at 10 points	About 15 m from seabed surface	86 mm	2~20 m	1 m
Kamaishi harbour	Seabed CDL –61.2 m	PS logging	Uphole	Insert electric detonator mounted with iron rod into the hole	Set 9 vibration sources at the same time within a borehole	Received at 2 points on the seabed surface separated by 3 m and 5 m from a borehole	—	135 mm	2~10 m	1 m

Table 4.7. Standard sampling specifications.

Sampling tube diameter	70~75 mm	
Sample length per sampling	Fines content less than 5% Fines content between 5 and 20%	40 cm 75 cm
Sampling interval	1.5~2.0 m	
Sample preservation	Fines content less than 5% Fines content over 5%	Freezing method Confined method

a) *Sampling plan*. Methods for undisturbed sampling of sandy soil are not as well established as those for undisturbed sampling of clay as mentioned earlier. The principles of sampling, however, are common to clay and sand and therefore undisturbed sampling of sandy soil may be basically conducted by referring to sampling methods for clay (refer to Port and Harbour Investigation Manual [14]).

The difference in the sampling of sandy soil from the sampling of clayey soil is in the type and diameter of the sampler used, length of the sample that can be obtained, sampling interval, and transport and preservation methods of the obtained sample. The standard specifications are shown in Table 4.7.

b) *Diameter of the sampling tube*. The diameter of the sampling tube obviously must be larger than that of the specimen used for the cyclic triaxial tests. The sample disturbance becomes smaller with larger diameter sampling. Sampling becomes easier, however, with smaller diameter sampling because the frictional force and arching effect to retrieve soil samples are larger for smaller samples in relative terms. The diameter of the sampling tube should be determined by considering these factors.

In sampling of clayey soil, ϕ 75 mm is the normal dimension [18]. In sampling sandy soil, there is no qualitative data related to diameter and disturbance of the sample. Samplers currently used for sampling sandy soils with 51 to 200 mm diameter are shown in Table 4.8.

The standard specimen diameter for the cyclic triaxial test is 50 mm. The diameter of the undisturbed sample obtained for the cyclic triaxial test is, therefore, about 70 to 75 mm when the triaxial test conditions are considered.

c) *Length of the sampling*. For the cyclic triaxial test, a sample length of at least 40 to 60 cm is necessary because a minimum of three specimens (with a height of 10 to 12.5 cm) are necessary for the cyclic triaxial tests for each soil layer.

The length of undisturbed sand samples is also restricted to some degree by the sample diameter. Clean sand with a fines content of less than 5% can be sampled up to about 5 times the sample diameter. Sandy soil with a fines content of about 20% can be sampled up to about 10 times the sample di-

Table 4.8. Type of sand sampler and its application.

Penetration method	Type and name	Sample holding mechanism				Sample diameter (mm)	Borehole diameter (mm)	Soil conditions*	Remarks
		Structure	Piston	Catcher	Bottle check				
Static penetration method	Fixed piston sampler (piston rod type)	S	+			75	86	N < 5	Suitable when the fines content is high
	Fixed piston sampler (water pressure type)	S	+			75	100	N < 15	This sampler is also called Osterberg sampler. Penetration process of sampler can be monitored and recovered sample length can be measured
	Improved Bishop type sampler	W	+	+		53	100	N < 15	Suitable for homogeneous soil with a minimum amount of shells and gravel (specimen formation is unnecessary)
	Twist sampler	W	+	+		70	97	N < 25	A large penetration force is required (applicable to soils with wide range of SPT *N*-value)
	Special sampler with core catcher	T	+	+		75	143	N < 10	With spring basket type core catcher. This sampler needs large device for penetration
	Piston sampler with core catcher	S	+	+		68, 81	85, 116	N < 10	Modified from fixed piston sampler for clayey soil with core catcher appended (easy handling)
	Large diameter sampler	S		+		200	355	N < 10	9-12 specimens can be obtained at one time. Large facility is required

Table 4.8. Continued.

Penetration method	Type and name	Sample holding mechanism: Structure	Piston	Catcher	Bottle check	Sample diameter (mm)	Borehole diameter (mm)	Soil conditions*	Remarks
Rotary method	Rotary foil sampler	T	+			93	210	30 < N < 60	Applicable to gravelly sand, cemented soil and soft rock. Large facility is required
	Dennison type piston sampler	W	+			68	116	20 < N < 30	Applied to special soils such as Shirasu (volcanic ashes) and Masado (decomposed granite)
	Triple tube sampler	T			+	82	127	25 < N < 60	Applicable to dense and cemented sand
	Rotary type sampler	T			+	51	90	10 < N < 20	Suitable for homogeneous soil with a minimum amount of shells and gavels (specimen formation is unnecessary)

Note: Structure for holding samples, S = single tube; W = double tube; T = triple tube. *The range of SPT *N*-values indicated for the applicable soil is an approximate value. Undisturbed sampling of very loose sandy soil (e.g. having SPT *N*-value of about one) and gravelly soil is difficult and for these soils, sampling by in-situ freezing is recommended.

ameter [18]. If the 75 mm diameter sampler is used, these conditions lead to sampling lengths of about 40 cm for clean sand and 75 cm for sandy soil with a fines content of about 20%.

It is usually possible to obtain samples for the required length with one continuous sampling. When the fines content is small, it is preferable to make sampling in two steps with a shorter length.

d) *Sampling interval.* It is possible to determine sampling interval from the sampling length for one sampling, number of samples, and sampling interval. Standard interval for sampling is 1.5 to 2.0 m.

e) *Sample preservation.* The obtained samples must be preserved with minimum disturbance during transportation to the laboratory from the sampling site and in the laboratory until the start of laboratory testing. In order to preserve the sample during transportation to the laboratory, one can use either the freezing method or application of constraining pressure (called the constraint method).

The freezing method is used when the fines content is small. When the fines content exceeds 5%, influence of the freezing/melting appears [19] so that it is better not to perform freezing for such samples. When the freezing method is used, it is necessary to reduce the water expansion caused by freezing by placing a cap with a hole on the bottom end of the sampling tube, reducing the amount of pore water in the sample by gravity drainage, and performing freezing with fixation between the particles with a minimum amount of water. When the degree of saturation is less than 85 to 90%, it is possible to ignore the disturbance in the sample caused by expansion during freezing. This is achieved by draining about 150 to 180 ml of water from a 80 cm sample within the sampling tube with a diameter of 75 mm.

The time required for draining the water can be short when the fines content is low, but when the fines content is high, it increases. For example, in one case, water had to be drained for about 3 days. After completing the water drainage from the sampling tube, the sample is frozen. The freezing method involves inserting the sampling tube into a box made of styrofoam and filling it with pulverized dry ice or directly spraying it with liquid nitrogen.

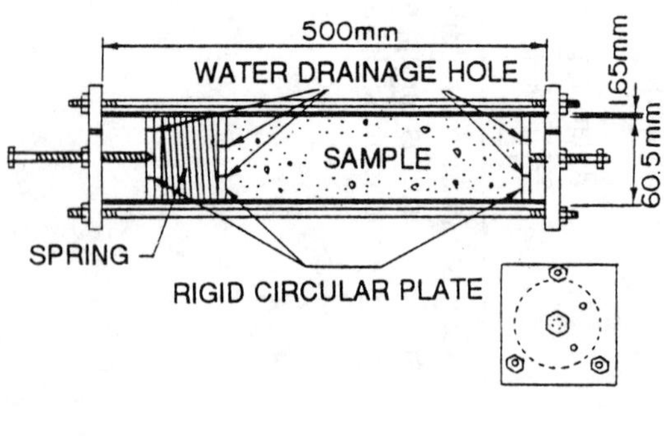

Figure 4.5. Example of a sample confining device

The constraint method for sample preservation is used when the fines content of the soil sample is high. This method uses a spring or beloframe cylinder [20], applies constraining pressure corresponding to the effective vertical pressure of the sampling depth, and maintains and preserves the state in-situ. One example [18] of a sample pressurization device is shown in Figure 4.5.

If the freezing method is used during transportation, the sample should be preserved in the laboratory in a freezer until the laboratory test begins. If the constraint method is used in transportation, the sample should be preserved on a shelf with minimum disturbance in the laboratory. A place with constant temperature and high humidity is preferable for storage. If the sample must be preserved for a long period of time, sample dehydration should be prevented.

f) *Transportation of sample.* During transportation from the sampling site to the laboratory, sample disturbance must be minimized. The sample should first be carefully carried by hand, carefully placed in a box, and then transported by a vehicle such as a passenger car (e.g. wagon, van) which has soft springs.

g) *Type of sand sampler and its application.* The sampling technique can be classified into static penetration methods and rotary coring methods. The former methods are further classified into a piston type and an open type. The piston type includes those with and without a core catcher device. The types and specifications of sand samplers are shown in Table 4.8, which was prepared by taking documented data [18] into consideration.

Though not mentioned in Table 4.8, undisturbed sampling using in-situ freezing techniques is another method [21]. This method performs sampling by freezing the soil in-situ before sampling. It is an expensive method. Since a good quality sample is obtained by this method, however, it is used for investigation of foundation soil for important structures. For details on the characteristics of each sampling method, refer to 'Soil survey method' (see Chapter 7) issued by the Japanese Society of Soil Mechanics and Foundation Engineering [18].

h) *Quality evaluation of the sample.* A method for evaluating the disturbance in sand sampling has not yet been established. Several attempts have been made, however, for evaluating the quality of the sample. Comparing the sample length obtained and the length of insertion in the sampling tube is one method. Another method evaluates the quality of sample using an X-ray photography technique [22, 23]. An attempt to evaluate disturbance based on the shear modulus at small strain level can also be made [24]. When there is a large variation in the test data, it is necessary to evaluate the test results based on the quality of the sample.

4.2.4.4 *Liquefaction test*

In order to evaluate the liquefaction strength of saturated sandy soil, a cyclic shear test is performed. This test simulates the liquefaction phenomenon during earthquakes by applying cyclic shear to the saturated sandy soil under undrained conditions. There are several cyclic shear tests used in liquefaction testing of saturated sandy soils, including the cyclic triaxial test, cyclic simple shear test, and cyclic torsion shear test. Each of these tests has advantages and disadvantages. The one which is most commonly used and reproducible is the cyclic triaxial test. The cyclic triaxial test is performed according to the Japanese Society of Soil Mechanics and Foundation Engineering 'Cyclic triaxial test method for soil' (JSF T 541-199) [25].

In the cyclic triaxial test, an effective confining pressure is isotropically applied to an undisturbed sample. If the sample is obtained from a normally consolidated deposit, it is preferable to apply the effective vertical pressure as an isotropic cell pressure instead of an average effective principal stress at the depth where the sample was obtained. This is necessary to avoid the effects of artificial overconsolidation in the laboratory.

4.2.4.5 *Dynamic deformation test*

There are three methods in laboratory dynamic deformation testing to evaluate strain dependent moduli and damping factor of soils. The ultrasonic pulse method uses the same principle as elastic wave exploration. The resonance method utilizes the resonance of the specimen. The cyclic triaxial method is based on the stress-strain curve obtained from a cyclic loading test.

The cyclic triaxial method which can accommodate a wide range of strains has been used in recent years. Axial loading is applied in steps to the specimen and the shear modulus/damping factor is obtained from the relationship between shear strain and shear stress at various levels of loading. Five to ten levels of shear strain amplitudes are chosen from the range of 10^{-5} to 10^{-2} for testing. Loading of 10 waves is applied continuously for each step in an undrained condition. After each step, the specimen is left in a drained condition for the next step. For summarizing the test results, the stress-strain curve of the 10th wave is normally used. The method for obtaining the specific dynamic deformation characteristics from the stress-strain curve is noted in Section 4.3.1.

Dynamic deformation characteristics are influenced by the effective confining pressure during the test. When an undisturbed sample of normally consolidated soil is obtained, the effective vertical pressure at the depth of sampling is isotropically applied by cell pressure to avoid the influence of overconsolidation as mentioned earlier. In order to obtain the in-situ shear modulus of the soil from the laboratory test results, correction of these results is necessary so that a shear modulus corresponding to the average effective principal stress at the sample depth is obtained. When the effective vertical pressure of the soil changes due to construction of a new structure, the value after the change should be used for seismic response analysis.

4.2.4.6 *Density test*

Density of soil can be either determined by directly measuring the unit weight of the undisturbed sample or referring to typical numbers of unit weight given according to the soil types (for specific numbers, refer to Section 4.4). When an undisturbed sample is obtained, the unit weight is determined from this sample. If block sampling is performed, the density test is made based on the Japanese Society of Soil Mechanics and Foundation Engineering 'Wet density test' (JSF T 191-1990) [26].

4.2.4.7 *Other tests*

If liquefaction is predicted at a site and remedial measures are necessary, a soil survey/test for design and construction of the remedial measures is necessary. In this case, the items to be evaluated are:

- Selection of the liquefaction remedial measure;
- Installation range of the remedial measures;
- Installation method of the remedial measures;
- Installation control of the remedial measures;
- Verification of the results of the remediation.

The investigation/test methods for liquefaction remediation are similar to those for evaluation of liquefaction potential. Depending on the type of liquefaction remedial measures, a special investigation/test may be necessary. Chapter 5 should be referred to for particular soil survey/tests for each remedial measure.

4.3 DYNAMIC DEFORMATION/STRENGTH CHARACTERISTICS OF SOIL

4.3.1 *Shear modulus and damping factor of soil*

4.3.1.1 *Model of stress/strain curve*

The dynamic shear stress-shear strain relationship for a soil is evaluated by analyzing a skeleton and hysteresis curves as shown in Figure 4.6. The skeleton curve exhibits noticeable nonlinearity when the shear strain amplitude becomes large. There are many models for simulating this type of behavior of soils in seismic response analysis, including the equivalent linear, bilinear, hyperbolic (Hardin-Drnevich model), and Ramberg-Osgood models [9]. The bilinear, Ramberg-Osgood, and hyperbolic models are shown in Figures 4.7a-4.7c. These models simulate the stress strain relationship with

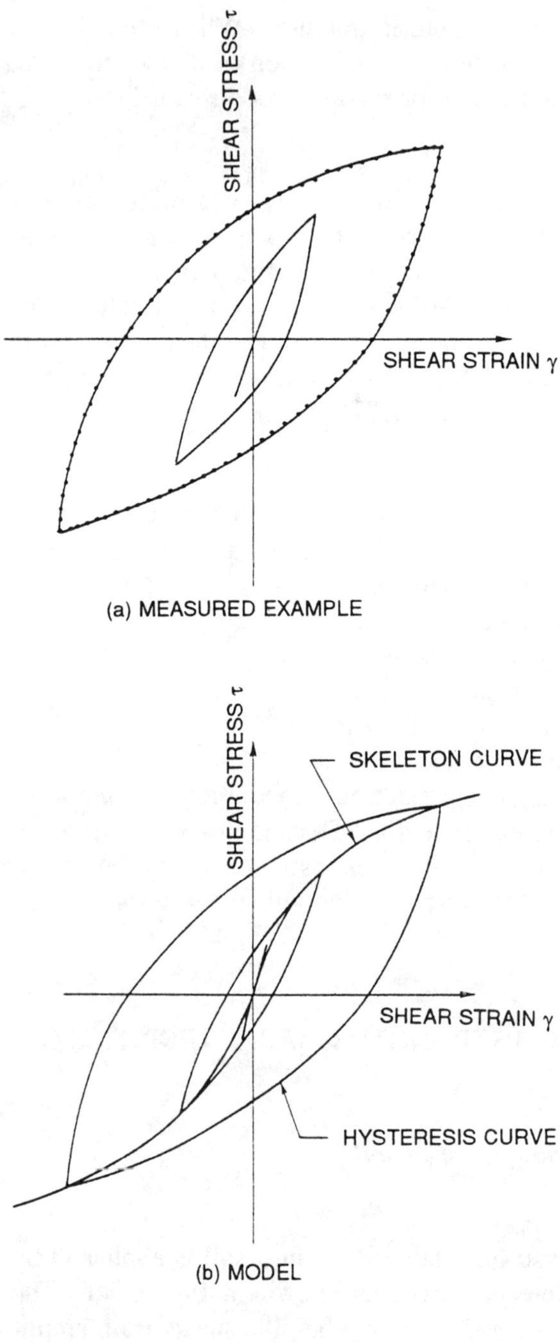

Figure 4.6. Stress-strain curve.

functions shown in Table 4.9. It is possible to introduce these models in the equivalent linear analysis or directly introduce them in the step-by-step non-linear analysis in the seismic response analysis of a subsoil profile. In Table 4.9, the $G/G_o - \gamma$, $h - \gamma$ relationship are computed from the non-linear curves of each model. The three to four parameters in these models are determined

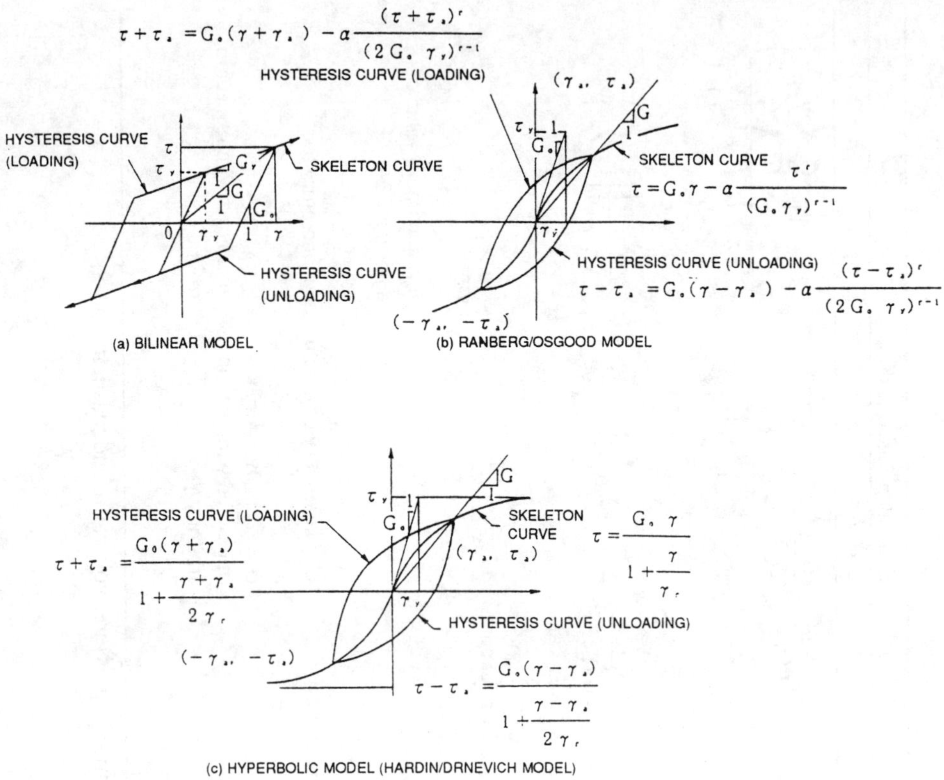

Figure 4.7. Model of stress-strain curve.

from the soil tests. In actual seismic response analysis, an equivalent linear analysis is often used because of the computation time required and stability in the analysis results.

4.3.1.2 *Representation of deformation characteristics in equivalent linear model*

In order to reasonably evaluate the behavior of subsoil during earthquakes, it is necessary to characterize and idealize the nonlinearity of the dynamic stress-strain relationship of the soil with respect to the shear strain level over a wide range. In the equivalent linear model, the dynamic stress-strain relationship of soil is expressed with two parameters; shear modulus and damping factor. Shear modulus G and damping factor h are defined in Equations (4.1) and (4.2) with respect to the hysteresis loop shown in Figure 4.8.

$$G = \frac{\tau}{\gamma} \tag{4.1}$$

Table 4.9. Functional forms of stress-strain models.

Model name	Parameter*	Stress-strain model: Skeleton curve	Stress-strain model: Hysteresis curve	Equivalent linearization: $G/G_o \sim \gamma$	Equivalent linearization: $h \sim \gamma$
Bilinear model	G_o, G_y, γ_y	$\tau = G_o\gamma$ ($\gamma \le \gamma_y$: yield strain) $\tau = G_y(\gamma - \gamma_y) + G_o\gamma_y$ ($\gamma \ge \gamma_y$)	Loop of parallelogram composed of line segments having slopes of G_o, G_y, drawn from points of $(\pm\gamma_a, \pm\tau_a)$	$\frac{G}{G_o} = \frac{G_y}{G_o}\left(1 - \frac{\gamma_y}{\gamma}\right) + \frac{\gamma_y}{\gamma}$	$h = \left(\frac{2}{\pi}\right) \frac{\left(1 - \frac{G_y}{G_o}\right)\left(\frac{\gamma}{\gamma_y} - 1\right)}{\frac{G_y}{G_o}\left(\frac{\gamma}{\gamma_y} - 1\right) + 1} \frac{\gamma_y}{\gamma}$
Ramberg-Osgood model	α, r, G_o, γ_y	$\tau = G_o\gamma - \alpha \frac{\tau^r}{(G_o\gamma_y)^{r-1}}$ γ_y : yield strain	$\tau \pm \tau_a = G_o(\gamma \pm \gamma_a) - \alpha \frac{(\tau \pm \tau_a)^r}{(2G_o\gamma_y)^{r-1}}$ r, α : empirical constants	$\frac{G}{G_o} = 1 - \alpha\left(\frac{G}{G_o}\right)^r \left(\frac{\gamma}{\gamma_y}\right)^{r-1}$	$h = \frac{2}{\pi}\frac{\gamma - 1}{\gamma + 1}\left(1 - \frac{G}{G_o}\right)\alpha$
Hyperbolic model (Hardin-Drnevich model)	G_o, γ_r, h_o	$\tau = \frac{G_o\gamma}{1 + \gamma/\gamma_r}$ γ_r : reference strain	Not explicitly specified by a function	$\frac{G}{G_o} = \frac{1}{1 + \gamma/\gamma_r}$	$h = h_o\left(1 - \frac{G}{G_o}\right)$ h_o : maximum value of damping factor
Equivalent linear model (direct modeling from measured stress-strain relationship)	–	Measured stress-strain curve	Measured stress-strain curve	G_o : laboratory or in-situ test $G = \tau/\gamma$: laboratory test (measured with respect to arbitrary γ)	h : laboratory test $h = \frac{1}{\pi}\frac{\Delta W}{W}$ (measured with respect to arbitrary γ)

*These parameters differ depending on many factors such as a confining pressure, void ratio, and plasticity index.

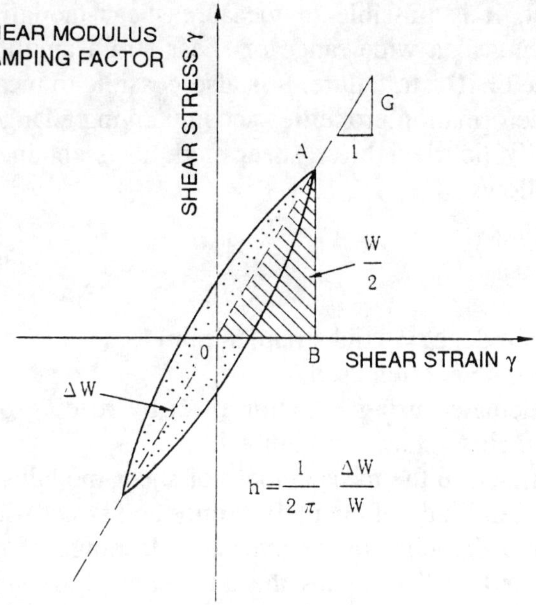

Figure 4.8. Equivalent linear modeling of stress-strain relationship.

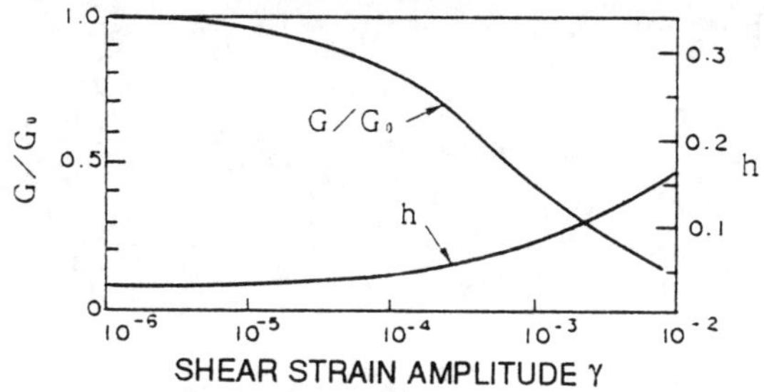

Figure 4.9. Shear strain dependence of shear modulus/damping factor.

$$h = \frac{1}{2\pi}\frac{\Delta W}{W} \tag{4.2}$$

where: τ = shear stress amplitude, γ = shear strain amplitude, W = strain energy, ΔW = hysteretic damping energy.

Shear modulus G and damping factor h are obtained with respect to the shear strain amplitude γ and indicated as the $G/G_o - \gamma$ curve and $h - \gamma$ curve as shown in Figure 4.9, where G_o is the shear modulus at $\gamma = 10^{-6}$.

4.3.1.3 *Measurement of shear modulus and damping factor*

The shear modulus and damping factor are obtained from an in-situ investigation using an elastic wave such as the PS logging method, or laboratory tests such as the resonant and cyclic triaxial tests as mentioned earlier. For the laboratory test, it is necessary to obtain undisturbed samples from the field. In the laboratory test, it is possible to measure shear moduli and damping factors corresponding to a wide range of shear strain amplitudes from a shear strain amplitude of 10^{-6} to failure. It is also possible to measure the change in the dynamic deformation properties accompanying changes in overburden pressure. In the cyclic triaxial test, shear modulus is obtained by using Poisson's ratio ν as follows:

$$G = \frac{\sigma_a}{2\varepsilon_a(1+\nu)} \tag{4.3}$$

where: σ_a = axial stress amplitude, ε_a = axial strain amplitude.

The Poisson's ratio of $\nu = 0.45$ is often used.

The damping factor is calculated using Equation (4.2) by reading off W and ΔW from the stress-strain curve shown in Figure 4.8.

The in-situ survey is restricted to the measurement of shear modulus corresponding to a shear strain amplitude of about 10^{-6} only and is not used in measuring shear modulus and damping factor over a wide range of shear strain amplitudes as mentioned earlier. It has the advantage, however, of measuring the in-situ values directly. The elastic constant of the soil is calculated using Equations (4.4)-(4.6) from the elastic wave velocity measured by elastic wave exploration utilizing boreholes.

$$G_o = \rho V_s^2 = \frac{\gamma_t}{g} V_s^2 \tag{4.4}$$

$$E_o = 2(1+\nu)\, G_o \tag{4.5}$$

$$\nu = \frac{(V_p / V_s)^2 - 2}{2\{(V_p / V_s)^2 - 1\}} \tag{4.6}$$

where: V_p = P-wave velocity (m/s), V_s = S-wave velocity (m/s), G_o = shear modulus (tf/m^2), E_o = Young's modulus (tf/m^2), ν = Poisson's ratio, ρ = density (t/m^3), γ_t = total unit weight (tf/m^3), g = acceleration of gravity (m/s^2).

In elastic wave exploration for a soft seabed soil, caution is required in the measurement with respect to the protection method of the hole wall, accurate measurement of waveform, and vibration generating/receiving methods of elastic wave (P-wave, S-wave). A comparison of the shear moduli G_o measured by the in-situ investigation and by laboratory tests is shown in Figure 4.10 [17]. Shear modulus G_o (in-situ) ranges from about one half to twice the shear modulus measured in the laboratory G_o (lab).

4.3.1.4 *Simplified estimation of shear modulus and damping factor*

When a laboratory test or an in-situ survey cannot be performed, it is possi-

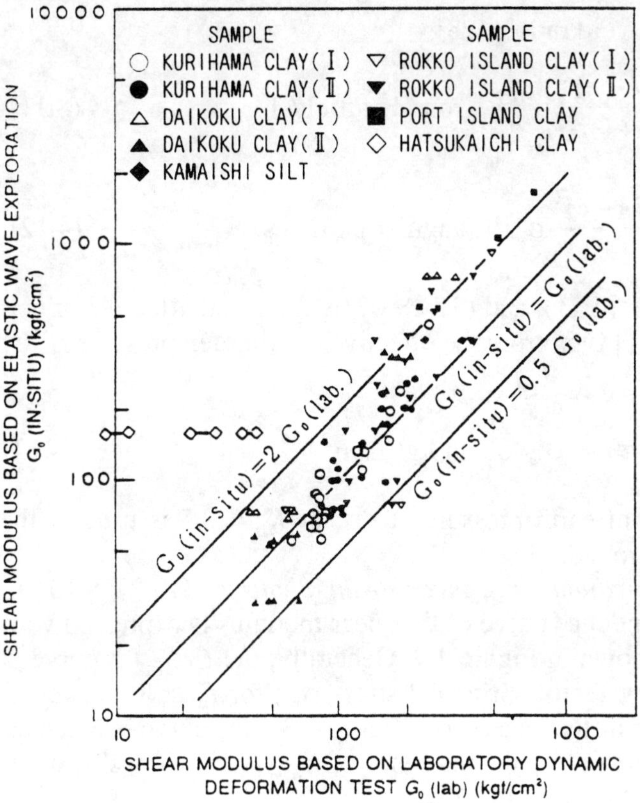

Figure 4.10. Comparison of shear moduli.

ble to estimate the approximate value of shear modulus and damping factor of the soil based on the following empirical equation.

a) *Shear modulus at very small strain amplitude* ($\gamma = 10^{-6}$) G_o (kgf/cm^2).

– Clayey soil:

1. Relationship with the unconfined compressive strength q_u [17]

$$G_o = 170\, q_u \tag{4.7}$$

2. Relationship with the plastic index I_p and void ratio e

$$G_o = (285 - 2I_p)\sigma'_m \quad (I_p \geq 30) \tag{4.8}$$

$$G_o = (1.6 I_p + 185)\frac{(2.973 - e)^2}{1 + e}(\sigma'_m)^{0.5} \tag{4.9}$$

$$(I_p < 30,\ \text{if}\ \ I_p = \text{NP then}\ \ I_p = 0)$$

– Sandy soil:

1. Relationship with SPT N-value [27]

$$G_o = 144\, N^{0.68} \tag{4.10}$$

2. Relationship with void ratio e [28]

$$G_o = 700 \frac{(2.17 - e)^2}{1+e} \sigma_m'^{0.5} \quad \text{(round particles)} \tag{4.11}$$

$$G_o = 330 \frac{(2.973 - e)^2}{1+e} \sigma_m'^{0.5} \quad \text{(angular particles)} \tag{4.12}$$

In Equations (4.8), (4.9), (4.11), and (4.12), σ'_m is the mean effective principal stress in the soil. It is given from the effective overburden pressure σ'_v by the following equation.

$$\sigma'_m = \frac{1+2K_o}{3} \sigma'_v \quad (\sigma'_m, \sigma'_v; \text{ in kgf / cm}^2) \tag{4.13}$$

Normally, a coefficient of earth pressure at rest of $K_o = 0.5$ is used in the horizontally deposited ground.

b) *Shear modulus at medium/large shear strain amplitude* ($10^{-6} < \gamma\ 10^{-2}$).

The shear strain dependence curve of the shear modulus is expressed with the $G/G_o - \gamma$ curve as shown in Figure 4.9. Generally, the $G/G_o - \gamma$ curve is influenced by the mean effective principal stress σ'_m; for $I_p \geq 30$, however, the $G/G_o - \gamma$ curve is not influenced by σ'_m. For $I_p < 30$, the curve is obtained with Equation (4.14) which takes the influence of σ'_m [17] into consideration as

$$\frac{G}{G_o} = \bar{A}(I_p, \gamma)\, \sigma_m'^{\bar{n}(I_p, \gamma)} \tag{4.14}$$

Table 4.10. $\bar{A}$ (I_p, γ) and $\bar{n}$ (I_p, γ) at representative strain levels.

Shear strain amplitude γ	Plasticity index I_p: N.P. to 9.4		from 9.4 to 30		30 or more	
	$\bar{A}$ (I_p, γ)	$\bar{n}$ (I_p, γ)	$\bar{A}$ (I_p, γ)	$\bar{n}$ (I_p, γ)	$\bar{A}$ (I_p, γ)	$\bar{n}$ (I_p, γ)
10^{-6}	1	0	1	0	1	0
10^{-5}	0.93	0.01	0.96	0	0.97	0
5×10^{-5}	0.83	0.03	0.91	0.01	0.93	0
10^{-4}	0.75	0.05	0.84	0.02	0.89	0
2.5×10^{-4}	0.56	0.10	0.74	0.05	0.82	0
5×10^{-4}	0.43	0.16	0.59	0.09	0.70	0
10^{-3}	0.30	0.22	0.45	0.16	0.58	0
2.5×10^{-3}	0.15	0.30	0.26	0.22	0.42	0
5×10^{-3}			0.12	0.26	0.28	0
10^{-2}					0.18	0

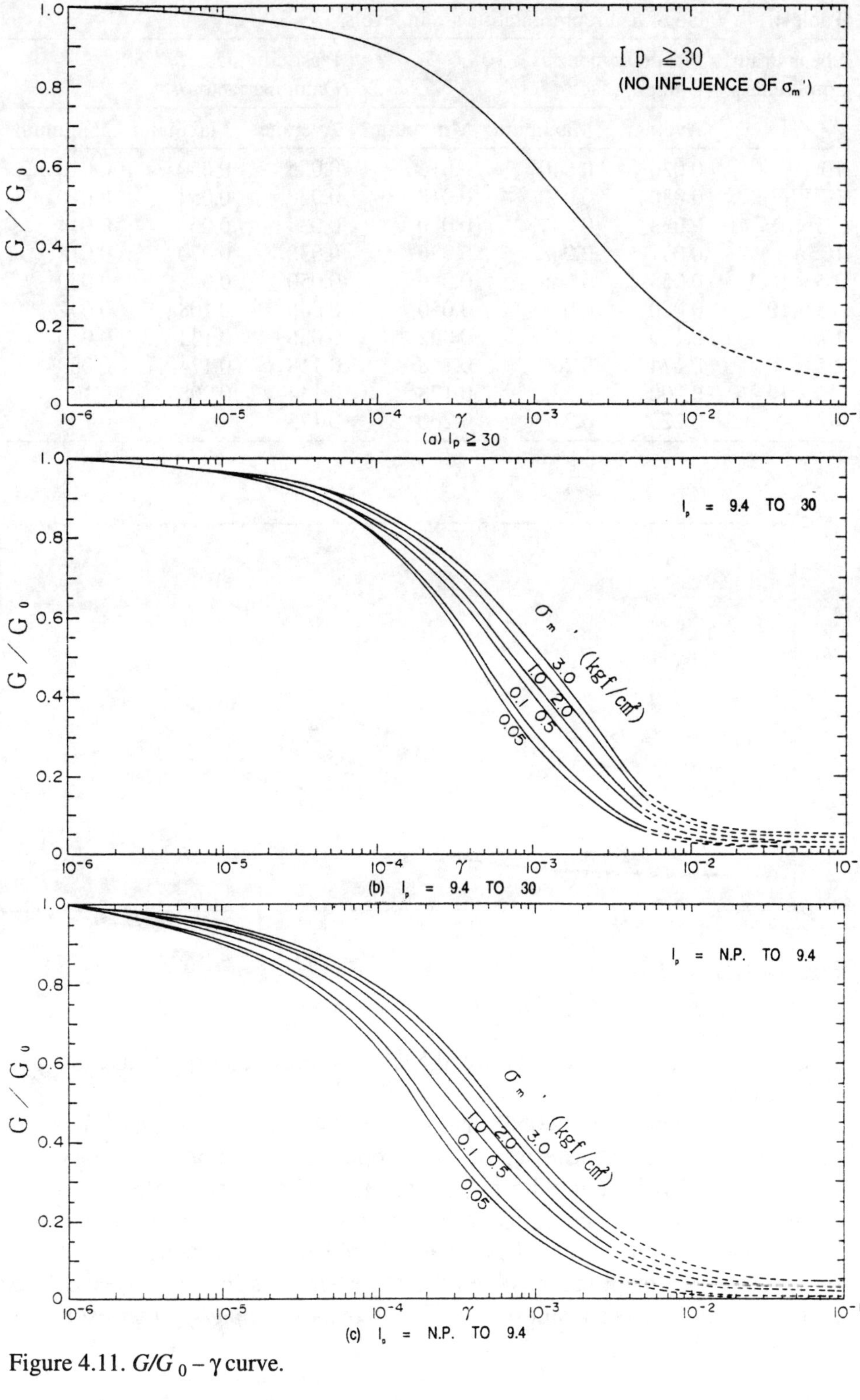

Figure 4.11. $G/G_0 - \gamma$ curve.

Table 4.11. Value of h at representative strain levels.

Shear strain amplitude γ	Plasticity index $I_p < 30$ Damping factor h			Plasticity index $I_p \geq 30$ Damping factor h		
	Average	Maximum	Minimum	Average	Maximum	Minimum
10^{-6}	0.026	0.040	0.016	0.025	0.050	0.010
10^{-5}	0.030	0.040	0.018	0.030	0.054	0.010
5×10^{-5}	0.033	0.042	0.020	0.034	0.062	0.014
10^{-4}	0.037	0.048	0.026	0.038	0.070	0.018
2.5×10^{-4}	0.055	0.068	0.040	0.050	0.088	0.030
5×10^{-4}	0.080	0.098	0.060	0.066	0.108	0.042
10^{-3}	0.120	0.145	0.092	0.086	0.133	0.056
2.5×10^{-3}	0.174	0.200	0.148	0.118	0.174	0.080
5×10^{-3}	0.200	0.222	0.178	0.144	0.208	0.100
10^{-2}	0.220	0.240	0.200	0.175		0.125

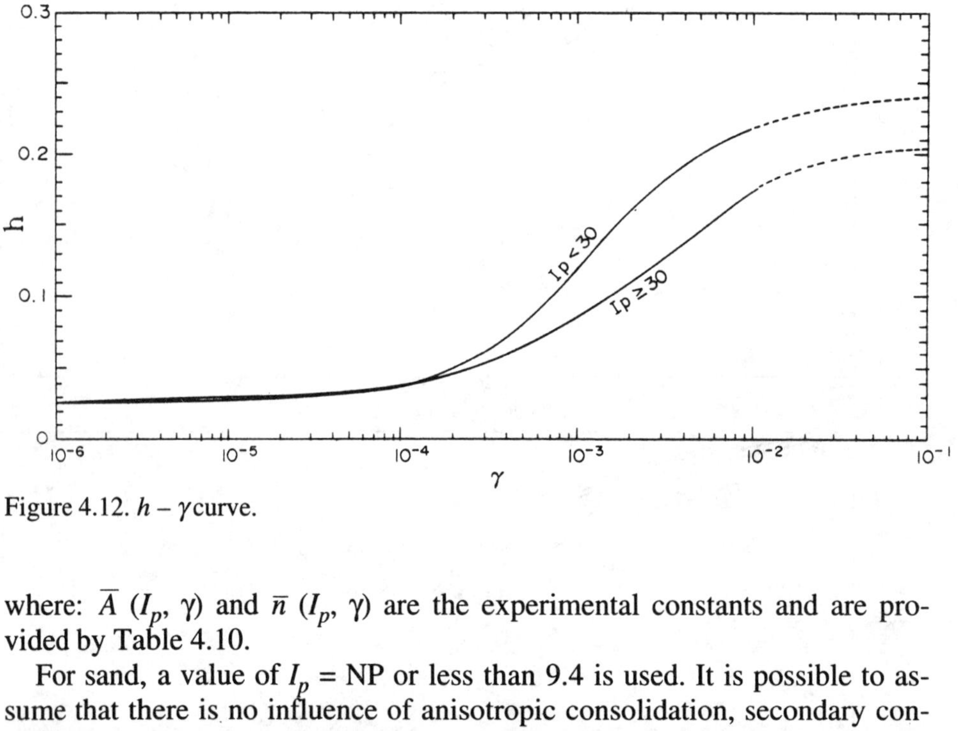

Figure 4.12. $h - \gamma$ curve.

where: $\bar{A}$ (I_p, γ) and $\bar{n}$ (I_p, γ) are the experimental constants and are provided by Table 4.10.

For sand, a value of I_p = NP or less than 9.4 is used. It is possible to assume that there is no influence of anisotropic consolidation, secondary consolidation, and void ratio with respect to the $G/G_o - \gamma$ relationship in practical use. The $G/G_o - \gamma$ curve using Equation (4.14) is shown in Figure 4.11. The broken line in Figure 4.11 is not based on experimental results but rather is an estimated value. When the shear strain is in the range indicated by the broken line as a result of the seismic response analysis, the result is an approximate value.

c) *Damping factor.* With regard to damping factor, the influences of the secondary consolidation, void ratio, and average effective principal stress are not well established. The values with respect to the representative strain level for soil of $I_p \geq 30$ and $I_p < 30$ are shown in Table 4.11. Figure 4.12 shows the $h - \gamma$ curve obtained based on Table 4.11.

4.3.2 *Liquefaction characteristics*

4.3.2.1 *Representation of liquefaction strength*

Liquefaction strength of sand in the cyclic triaxial test is expressed as the relationship between the magnitude of applied cyclic shear stress τ_l and the number of load cycles N_l needed to cause liquefaction. The occurrence of liquefaction is defined by the occurrence of *initial liquefaction* when the excess pore water pressure generated by undrained cyclic shearing equals the initial effective confining pressure. For a clean loose sand, it is known that liquefaction defined with a double amplitude axial strain of 5% and initial liquefaction are almost equivalent. An example of the relationship between the number of load cycles N_l and cyclic shear stress τ_l thus defined is shown in Figure 4.13. The liquefaction strength at an arbitrary number of load cycles can be read off from Figure 4.13.

When the initial liquefaction cannot be clearly defined in the test result due to the cyclic mobility of dense sand or sandy soil with a large fines content, liquefaction is defined by the state when the double amplitude of axial strain reaches some designated value. In these cases, the double amplitude axial strain of 10% is often used to define liquefaction.

4.3.2.2 *Factors affecting the liquefaction strength*

The factors affecting liquefaction strength are classified into either load, soil, stress, or other test conditions. These conditions are further classified as

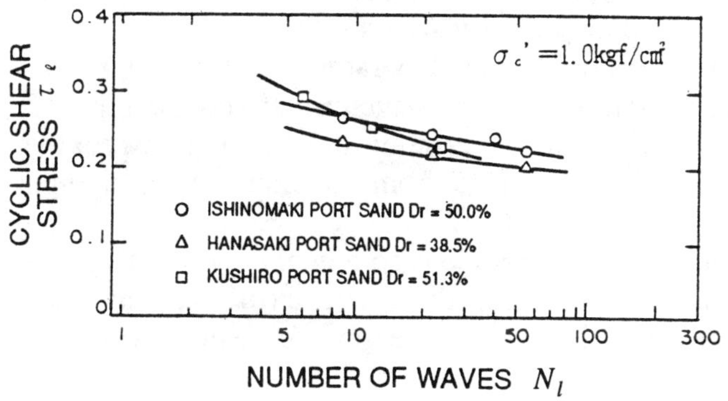

Figure 4.13. An Example of cyclic shear stress and number of load cycles.

Table 4.12. Influential factors on liquefaction strength.

1. Load conditions	Cyclic shear stress amplitude Waveform Vibrational frequency Irregularity in the waveform Multidirectional shear Reversal of shear stress
2. Soil conditions	Density (relative density) Grain size distribution Degree of saturation Structure of soil skeleton Cyclic shear history Cementation effect
3. Stress conditions	Average effective principal stress Overconsolidation ratio Initial shear stress
4. Test conditions	Specimen dimension Specimen preparation method Consolidation time Membrane penetration *B*-value Type/structure of test equipment

shown in Table 4.12. The degree of influence of these factors on the liquefaction strength is shown below.

Load conditions

a) *Cyclic shear stress amplitude.* The liquefaction strength is a function of the magnitude of cyclic shear stress amplitude. In the cyclic triaxial test, several different levels of cyclic shear stress amplitude are applied and the number of load cycles necessary to induce liquefaction is measured. An example of these test results is shown in Figure 4.13. In most cases, cyclic shear stress ratio τ_l/σ'_c is plotted on the vertical axis.

b) *Waveform.* The waveforms used in the laboratory test are sine, rectangular, triangular, and trapezoidal waves. The waveform affects the liquefaction strength. There is a report that the liquefaction strength with respect to a rectangular wave decreases by about 15% from the liquefaction strength with respect to a sine wave [29].

A cyclic shear stress of uniform amplitude is commonly applied as a standard practice. The difficulty of predicting the cyclic shear stress in actual subsoils during future earthquake is one of the reasons for using a waveform of uniform amplitude.

c) *Frequency*. In the frequency range corresponding to an earthquake motion, the influence of frequency on the liquefaction strength can be ignored. If other conditions regarding clean sand are the same, a similar relationship between the cyclic shear stress ratio and the number of load cycles is obtained regardless of the frequency used, in the range (0.05 Hz to 12 Hz).

d) *Irregularity of waveform*. The waveform of cyclic shear stress generated within soil by an actual earthquake is irregular. The liquefaction strength of sand subjected to irregular shear stress cannot be defined in the same manner as the liquefaction strength with respect to uniform cyclic shear stress discussed earlier. In the laboratory liquefaction test, therefore, liquefaction strength is evaluated by loading with regular cyclic shear stress and conversion is made to obtain equivalent liquefaction strength with respect to irregular cyclic shear stress. One method is to obtain an equivalent cyclic stress τ_{eq} and an equivalent number of load cycles N_{eq} through the concept of cumulative damage commonly used in the evaluation of fatigue failure of metals [30, 31]. Another method is to obtain an equivalent cyclic stress τ_e which causes liquefaction at 20 load cycles based on the maximum shear stress τ_{max} of the irregular cyclic shear stress-waveform. The factor used for the conversion depends on the waveform; the waveform is classified into either a shock or vibratory type [9].

e) *Multidirectional shearing*. The cyclic shear stress generated within sand during earthquakes is not always uni-directional. This results in a reduction of about 10 to 15% off from the liquefaction strength obtained with uni-directional cyclic shearing [32].

f) *Reversal of shear stress*. For an embankment slope, a static initial shear stress τ_o exists before the loading of cyclic shear stress. When the cyclic shear stress amplitude is smaller than the value of τ_o, the total shear stress in the soil does not reach 0 and the stress is unidirectional. When the cyclic shear stress amplitude exceeds τ_o, however, the reversal of stress occurs. Because a number of factors related to the effect of shear stress reversal in liquefaction strength are not clearly understood, a laboratory test is usually conducted by assuming the state of $\tau_o = 0$ in horizontal layered soil.

Soil conditions

a) *Density (relative density)*. The liquefaction strength and relative density of soil are proportional for loose to medium sands. In the case of clean sand with a small coefficient of uniformity, there is an average relationship of $R_{l20} = 0.0042D_r$ between liquefaction strength R_{l20} at 20 load cycles (in decimal) and relative density D_r (%) [33]. Figure 4.14 shows an example of this relationship between D_r and R_{l20} for undisturbed samples obtained in port and harbour regions in Japan.

b) *Grain size distribution*. Liquefaction strength differs according to the grain size distribution of the sand. For clean sand, a simplified procedure for

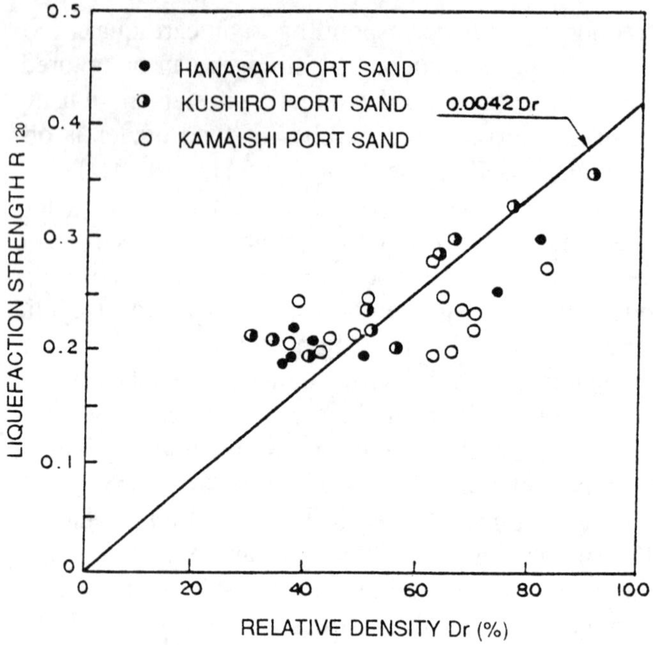

Figure 4.14. Relative density and liquefaction strength.

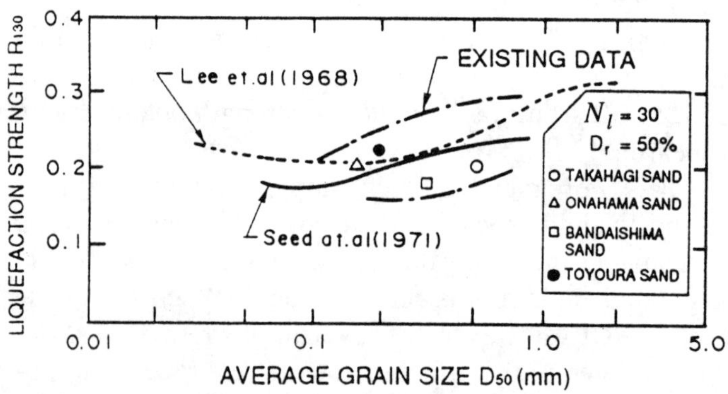

Figure 4.15. Average grain size and liquefaction strength.

liquefaction evaluation is well established [34]. When the average grain size D_{50} is used as a representative index of the grain size, liquefaction strength of sand with an average grain size of about 0.1 mm is noted to be the smallest as shown in Figure 4.15 [35]. With regard to sandy soil containing fines (grain size of less than 74 μm), there is a tendency for liquefaction strength to increase with higher fines content but a quantitative relationship to fines content has not yet been fully established.

c) *Degree of saturation.* When the degree of saturation decreases, the liquefaction strength increases. When conducting liquefaction tests, it is necessary to make the degree of saturation in the specimen close to 100%. In order to confirm the degree of saturation, it is only necessary to check the pore water pressure coefficient designated as the B-value. The B-value corresponds to 1.0 when the saturation degree is 100%. It is difficult to achieve the B-value of exactly 1.0 but it is preferable to confirm that the B-value is at least greater than 0.95.

d) *Others.* Research related to cementation effects on cyclic hysteresis and the soil particle skeleton are also being conducted. These factors can be effectively incorporated in the liquefaction strength by using undisturbed samples. It is necessary, therefore, to obtain samples of as high quality as possible.

Stress conditions

a) *Average effective principal stress.* The liquefaction strength τ_l of clean sand is proportional to the mean principal stress σ'_m irrespective of the anisotropic conditions of stress (to be specified by an earth pressure coefficient at rest K_o). Therefore, the curve specified by τ_l/σ'_m in Figure 4.13 does not depend on K_o values. In cyclic triaxial tests, an effective confining pressure σ'_c is applied isotropically (mean effective principal stress in this case is $\sigma'_m = \sigma'_c$) ignoring the anisotropy of stress. In this case, the effective overburden pressure σ'_v is often used instead of σ'_m in-situ. This is to prevent the specimen from becoming overconsolidated as mentioned earlier. When applying the result of the cyclic triaxial test to horizontally layered soil, it is necessary to consider the anisotropic stress state by assuming a reasonable K_o value.

b) *Overconsolidation ratio.* The liquefaction strength of sand in an overconsolidated state is known to increase significantly in comparison to the liquefaction strength of sand in a normally consolidated state. Sand in the overconsolidated state is, therefore, harder to liquefy than sand in the normally consolidated state. When a liquefaction test is performed on an undisturbed sample under an isotropic consolidated condition with effective confining pressure corresponding to effective overburden pressure, the strength may be underestimated. The in-situ overconsolidation ratio, however, is usually not considered due to the difficulty of evaluating it accurately. The liquefaction strength, when the influence of overconsolidation is not considered, is a conservative estimate.

c) *Initial shear stress.* The effect of the initial shear stress is closely related to the effect of stress reversal mentioned earlier. When the liquefaction strength of semi-infinite horizontal subsoil conditions is obtained, however, the initial shear stress is 0 and its influence need not be considered.

Other test conditions

In addition to the factors affecting liquefaction strength, the influence of specimen dimensions, method of specimen preparation, time of consolidation, membrane penetration, and type of testing apparatus should also be considered for accurate liquefaction evaluation.

4.4 SHEAR STRESS DURING EARTHQUAKES

4.4.1 *Outline of shear stress evaluation in subsoil*

A wave generated at a seismic fault propagates through the earth's crust, passes into the subsoil profile from a base layer, and then reaches the earth's surface. In this process, the seismic wave noticeably changes within the sub-

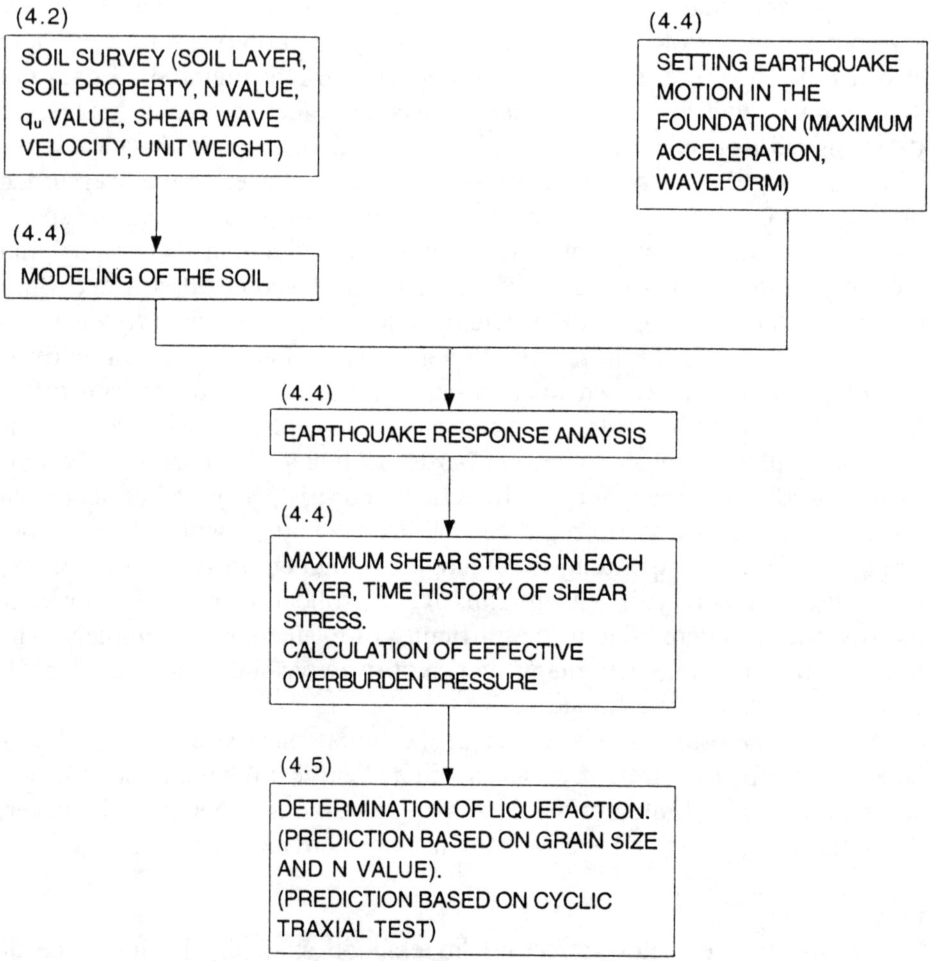

Figure 4.16. Evaluation procedure of shear stress during earthquakes.

soil profile. The object of the earthquake response analysis for liquefaction potential evaluation is to evaluate this change within the subsoil profile. This response analysis is performed in the sequence shown in Figure 4.16.

As mentioned earlier, the term 'base layer' refers to a bedrock or a sandy soil layer with an SPT *N*-value of 50 or more, a clay layer with an unconfined compressive strength (q_u-value) of 65 tf/m^2 or more, or a soil layer with a shear wave velocity of 300 m/sec or more. The subsoil profile refers to soil deposits above the base layer.

Earthquake motion estimated in the base layer for each site investigated is determined by considering the regional seismic activity and waveform characteristics of the earthquake motion as shown in Section 4.4.2.

In the earthquake response analysis, the dynamic soil properties must be evaluated based on the soil data obtained according to Sections 4.2 and 4.3. The details are shown in Section 4.4.4.

4.4.2 *Earthquake motion in the bedrock*

4.4.2.1 *Outline*

In this handbook, the input earthquake motion in the base layer is specified by the maximum acceleration given in Figure 4.17 and Table 4.13 and the acceleration waveforms shown in Figure 4.18. Figure 4.17 and Table 4.13 were obtained based on the degree of seismic hazard expected to occur one

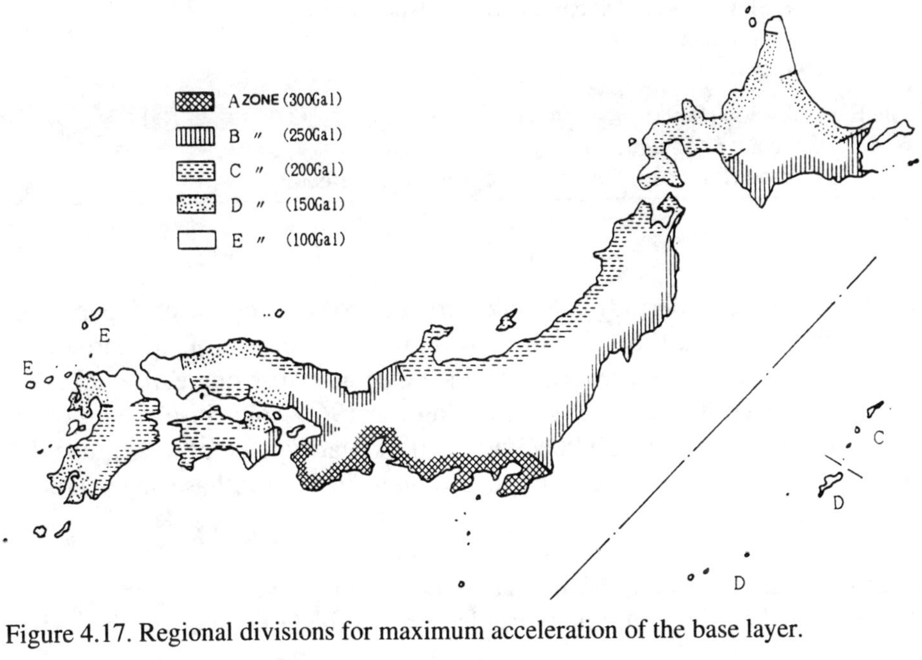

Figure 4.17. Regional divisions for maximum acceleration of the base layer.

Table 4.13. Regional maximum accelerations in the base layer.

Zone	Geographical region	Maximum acceleration (Gal)
Zone A	Kanto (Chiba, Tokyo, Kanagawa) Chubu (Shizuoka, Aichi) Kinki (Mie, Wakayama)	300
Zone B	Hokkaido (Nemuro, Kushiro, Tokachi, Hidaka) Tohoku (Pacific coast in Aomori prefecture, Iwate, Miyagi, Fukushima) Kanto (Ibaragi) Chubu (Fukui) Kinki (Osaka, Hyogo, Kyoto) Shikoku (Tokushima)	250
Zone C	Hokkaido (Iburi, Oshima) Tohoku (Region except for Pacific coast in Amori prefecture, Akita,Yamagata) Chubu (Niigata, Toyama, Ishikawa) Chugoku (Tottori, Hiroshima) Shikoku (Kochi, Ehime) Kyushu (Oita, Miyazaki, Amami Islands of Kagoshima prefecture, Kumamoto)	200
Zone D	Hokkaido (Abashiri, Hiyama, Shiribeshi, Ishikari, Rumoi) Chugoku (Okayama, Shimane) Shikoku (Kagawa) Kyushu (Region except for the islands of Nagasaki prefecture, Shiga, region except for Amami Islands of Kagoshima prefecture) Okinawa	150
Zone E	Hokkaido (Soya) Chugoku (Yamaguchi) Kyushu (Fukuoka, the islands of Nagasaki prefecture)	100

in 75 years (return period) calculated for the coastal region of Japan using past earthquake data [36]. For the acceleration waveform of the earthquake motion in the base layer, records obtained during two strong earthquakes are used. These records were obtained during the 1968 Tokachi-oki earthquake of magnitude 7.9 and the 1978 Miyagi-ken-oki earthquake of magnitude 7.4. Details of maximum acceleration and waveforms in the base layer are explained below.

4.4.2.2 *Maximum acceleration of earthquake motion in base layer*

As mentioned earlier, maximum acceleration of the base layer for the coastal

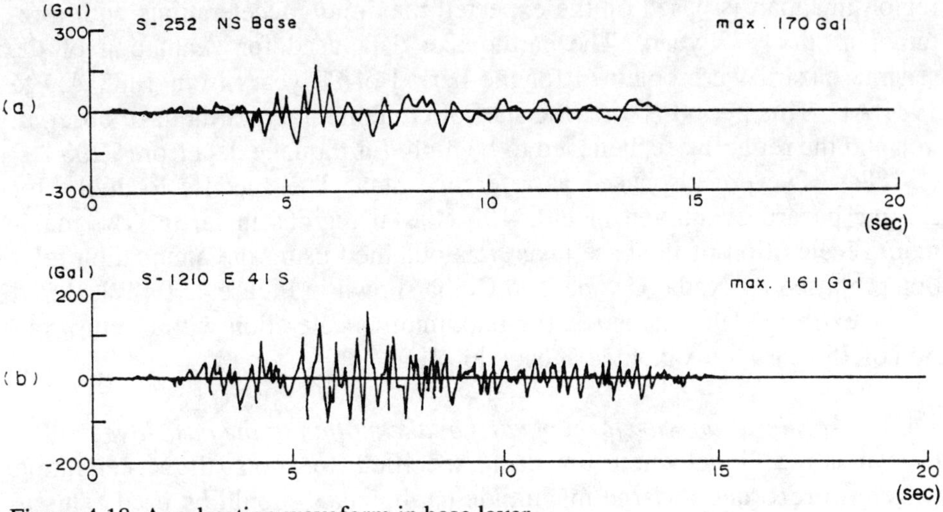

Figure 4.18. Acceleration waveform in base layer.

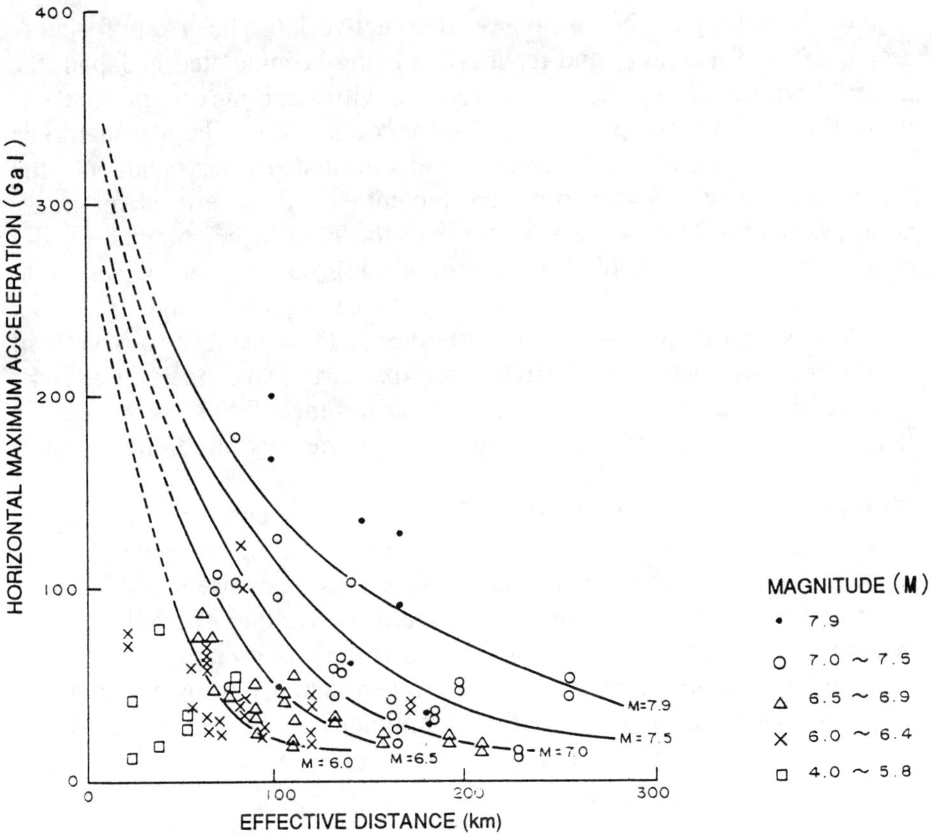

Figure 4.19. Relationship between the effective distance and maximum acceleration in base layer.

region in Japan is based on the expected maximum accelerations with a return period of 75 years. The earthquake data used for evaluation of the seismic hazard were obtained for the period of 97 years from 1885 AD to 1981 AD. This period is after the start of contemporary earthquake observation and the reliability is believed to be high. Earthquake data from 1200 AD to 1884 AD were also used as reference data. The specific locations for seismic hazard evaluation include 190 coastal regions in Japan. The maximum acceleration of the base layer was obtained using the attenuation relation proposed by Noda, Uwabe, and Chiba shown in Figure 4.19 [37]. Based on the extreme value statistics, the maximum acceleration with a return period of 75 years was obtained at each location.

4.4.2.3 *Acceleration waveform of earthquake motion in the base layer*

In evaluating liquefaction for each specified location, the acceleration waveform recorded in large magnitude earthquakes should be used. This is because the longer the earthquake motion continues in the earthquake with larger magnitude, the larger the influence on liquefaction becomes, provided the maximum acceleration is the same.

Although a strong earthquake observation network has been kept for more than a quarter of a century and records are being accumulated in Japan, the number of strong earthquake motion records with large maximum acceleration is limited. In fact, for earthquake records with a large magnitude of about 8.0, the number of records is very limited. In this handbook, the following two types of waveforms are recommended to be used for liquefaction evaluation. One is the waveform in the base layer obtained by deconvolution from the recorded waveform at the ground surface observed at Hachinohe Port during the 1968 Tokachi-oki earthquake (magnitude 7.9); see 'S-252 NS Base' in Figure 4.18. Another is the waveform observed in Ofunato Port during the 1978 Miyagi-ken-oki earthquake (magnitude 7.4), see 'S-1210 E41S' in Figure 4.18. The magnitude of this earthquake is smaller but this waveform was selected since it was measured directly on a rock outcrop.

The waveforms differ in the predominant frequency contents, as is apparent from the spectra shown in Figure 4.20. Since the subsoil conditions at each location have different resonance frequencies, response analysis using at least two types of waveforms is considered reasonable. The value of the maximum acceleration shown in Figure 4.18 is the observed or deconvolved maximum acceleration at the base layer. When actually using these acceleration waveforms in Figure 4.18, their amplitudes are scaled in accordance with the expected maximum acceleration with a return period of 75 years at the base layer for each location.

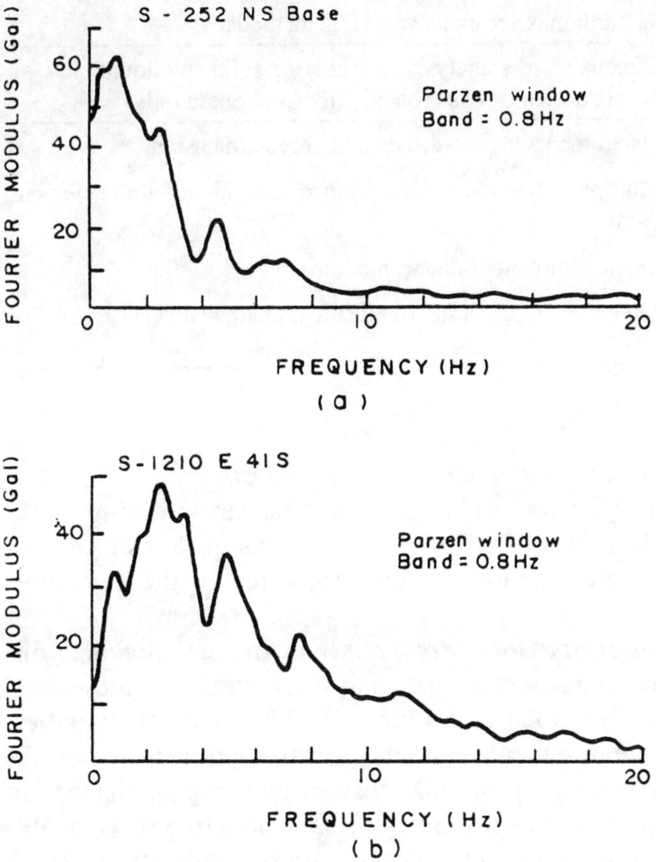

Figure 4.20. Fourier Spectra of earthquake motions for response analysis.

4.4.2.4 *Precautions for determining the earthquake motion separately from this handbook*

It is possible to establish the earthquake motion in the base layer by a site specific study instead of using the earthquake motions mentioned earlier. In this case, sufficient caution is necessary regarding reasonable results in the assessment of liquefaction potential. For example, if a waveform to be used for the site specific study contains only one large cycle of maximum amplitude as in a record of an earthquake with small magnitude, it is not considered reasonable for the waveform to be scaled with the maximum accelerations presented in Table 4.13.

4.4.3 *Methods for earthquake response analysis of the subsoil profile*

As mentioned earlier, the amplitude and waveform of the earthquake motion change through propagation from the earthquake source to the earth's surface. It is necessary to use a reasonable response analysis method which can represent the change with sufficient accuracy.

Table 4.14. Classification of earthquake response analysis methods.

Analysis method (modeling of saturated soil)	Effective stress analysis (considering solid and liquid phases) Total stress analysis (considering solid phase only)
Dimension	One dimension, two dimension, three dimension
Analytical model	Multiple reflection model, lumped mass model, finite element model
Material characteristics	Linear, equivalent linear, nonlinear
Time or frequency domain	Time domain analysis, frequency domain analysis

4.4.3.1 *Types of methods for earthquake response analysis*

Earthquake response analysis for evaluating the shear stress during earthquakes can be classified as shown in Table 4.14. Various methods for earthquake response analysis are explained below according to the categories shown in Table 4.14.

a) *Effective stress analysis and total stress analysis*. In prediction/determination of liquefaction, earthquake response analysis is classified into the effective stress method and the total stress method. When the soil liquefies, excess pore water pressure is generated and the effective stress decreases. As a result, the stiffness and damping of soils change, resulting in changes in the response characteristics of the subsoil profile. The effective stress analysis can express this condition by computing the excess pore water pressure increase in the soil. In the total stress analysis, excess pore water pressure is not computed so that a change in the earthquake response caused by effective stress change cannot be considered. Therefore, when the excess pore water pressure ratio is increased over 0.5, there is a large possibility that the computed result using the total stress method differs considerably from the actual earthquake response. A comparison of the results of the effective stress and total stress analyses will be discussed later in d.

b) *One, two, and three dimensional analyses*. For earthquake response analysis, one to three dimensional analyses may be performed. For a horizontally layered subsoil profile, the one-dimensional analysis is performed. When the soil-structure interaction is considered for such structures as quay walls, the two-dimensional analysis is performed. For other complex structures of three dimensional geometry, three-dimensional analysis may be performed. Presently the use of three-dimensional analysis, however, is limited due to restrictions in the time required for computation.

c) *Types of models*.

– *Multiple reflection model*. The multiple reflection model, shown in Figure 4.21, idealizes the soil as a stack of horizontal soil layers. The shear

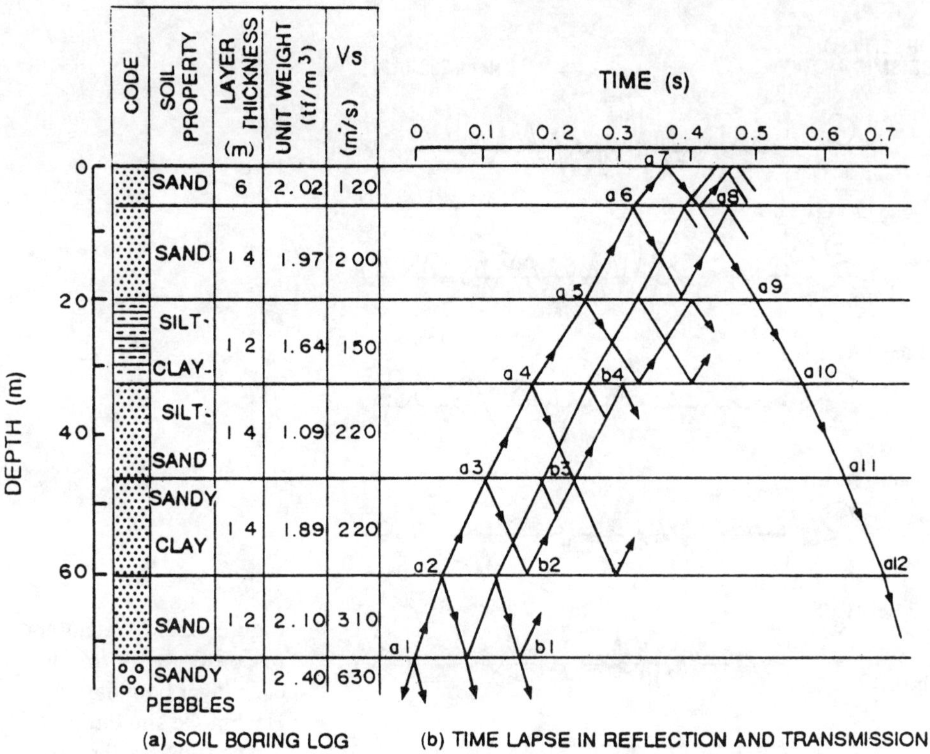

Figure 4.21. Propagation of seismic wave based on multiple reflection model.

wave propagates from the base layer upward through the stack of soil layers, repeating transmission and reflection at the boundary of each layer. Figure 4.22 shows one example of the time history in the earthquake motion at the boundary of each layer. In this example, the stress-strain relationship of the soil is idealized as being linear. Strain dependent stiffness and damping are used to idealize the nonlinearity of the soils in the equivalent linear analysis. The computer program SHAKE [38] is often used.

– *Lumped mass model.* The lumped mass model simulates the subsoil profile with a combination of lumped masses, springs and damping mechanism as shown in Figure 4.23. In this method, the computer program is relatively simple and it is easy to introduce a nonlinear spring with hysteretic damping to simulate the nonlinear soil behavior.

– *Finite element model.* In the finite element analysis, the subsoil profile is divided into a finite number of elements (see Fig. 4.24). This analysis is widely performed in various geotechnical engineering fields. A two or three-dimensional change in the layer thickness and physical properties of the soil can be considered easily in this analysis. The computer program FLUSH [39] is often used for equivalent linear analysis of two dimensional problems.

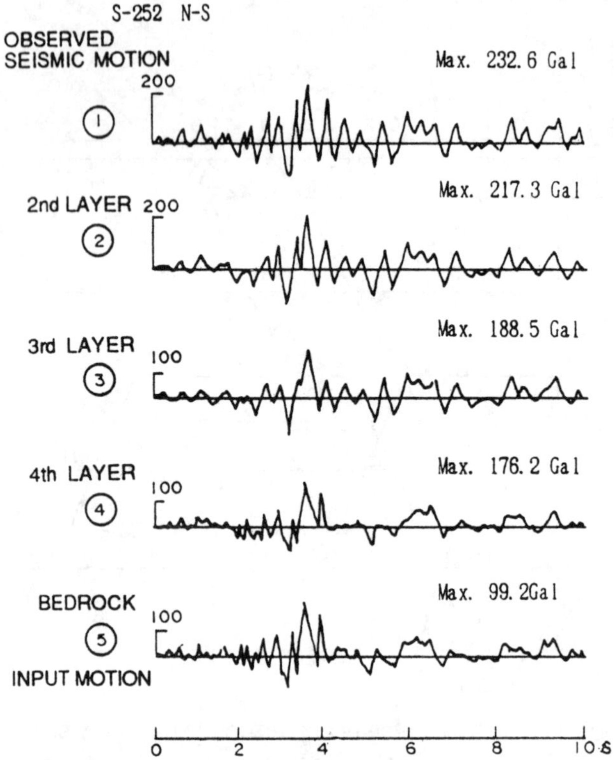

Figure 4.22. Incident and response waveforms at each layer boundary at Hachinohe site during the Tokachi-oki earthquake.

d) *Evaluation of the material properties.* For earthquake response analysis, it is necessary to determine the material nonlinearity of the soil constituting the subsoil profile. The stress-strain relationship is linear when the shear strain during earthquake is relatively low. For the stress-strain relationship for medium to large strain levels, nonlinearity of the soil is noticeable and hence should be adequately simulated in the analysis.

Various methods are used for modeling the nonlinearity. The equivalent linear analysis uses strain dependent material constants. Nonlinear response analysis idealizes the soil by consistently considering the stress-strain relationship of the soil during shaking. Various nonlinear models were discussed in Section 4.3. The difference in the results obtained by using these models is discussed below.

Figure 4.25 shows examples of one-dimensional earthquake response analyses with an input maximum acceleration of 100 Gals at a base layer using: 1) linear, 2) equivalent linear (SHAKE), 3) bi-linear, 4) tri-linear, 5) Hardin-Drnevich, and 6) Ramberg-Osgood models for the same subsoil profile [40]. $G/G_o - \gamma$ and $h - \gamma$ curves used for these models conform to the laboratory test results at a strain level of about 3×10^{-3} (0.3%). Figure 4.25 indicates that, when the nonlinearity of the material is fully taken into ac-

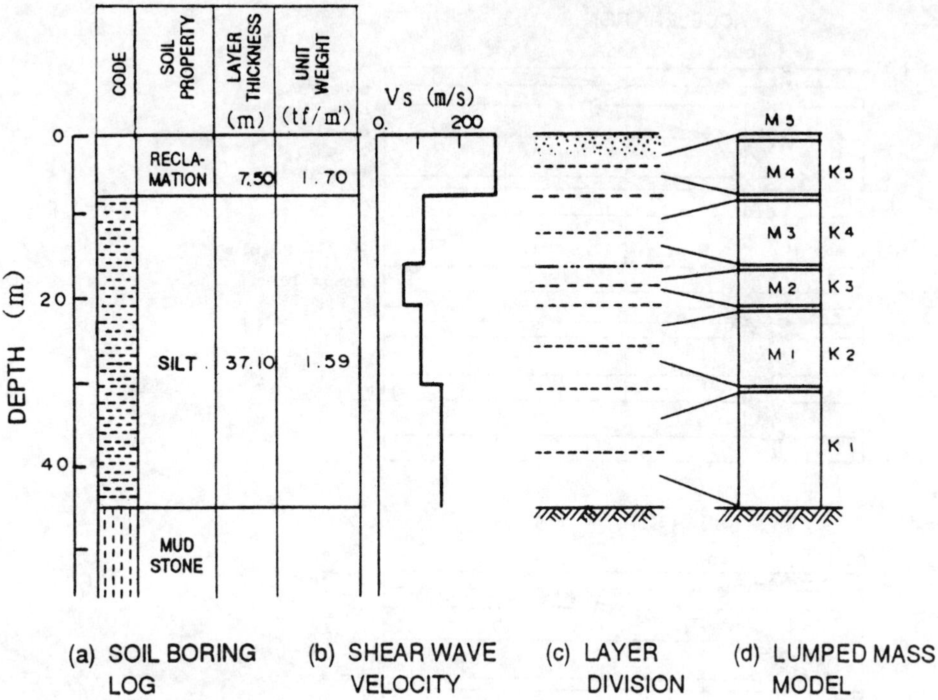

Figure 4.23. Lumped mass model.

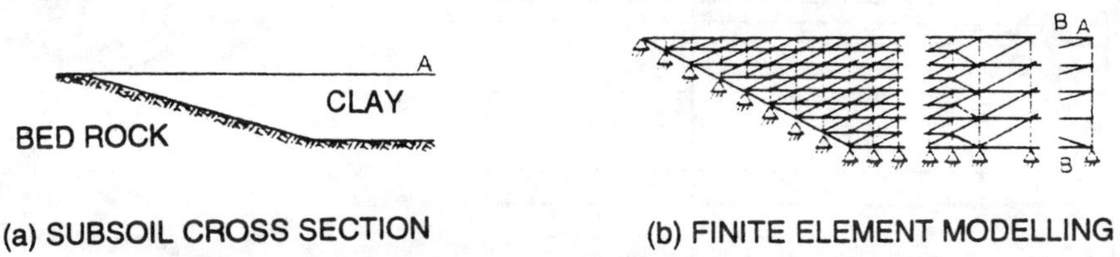

Figure 4.24. Finite element model.

count, the response acceleration and shear stress decrease, but there is a tendency for the shear strain to increase. A large difference, however, is not found in maximum shear strain γ_{max}.

For nonlinear response analysis, it is necessary to at least conform to the $G/G_o - \gamma$ and $h - \gamma$ curves obtained in the soil test at the anticipated strain level in the analysis. *Masing's rule* is often used for defining the stress-strain curve, but the hyperbolic model with Masing's rule gives an excessively large hysteretic damping factor at large strain levels [40, 41]. Modifications are underway to resolve this problem in modeling [42].

Figure 4.26 compares the results of: 1) SHAKE (equivalent linear model), 2) DESRA (hyperbolic model), and 3) CHARSOIL (Ramberg-Osgood mod-

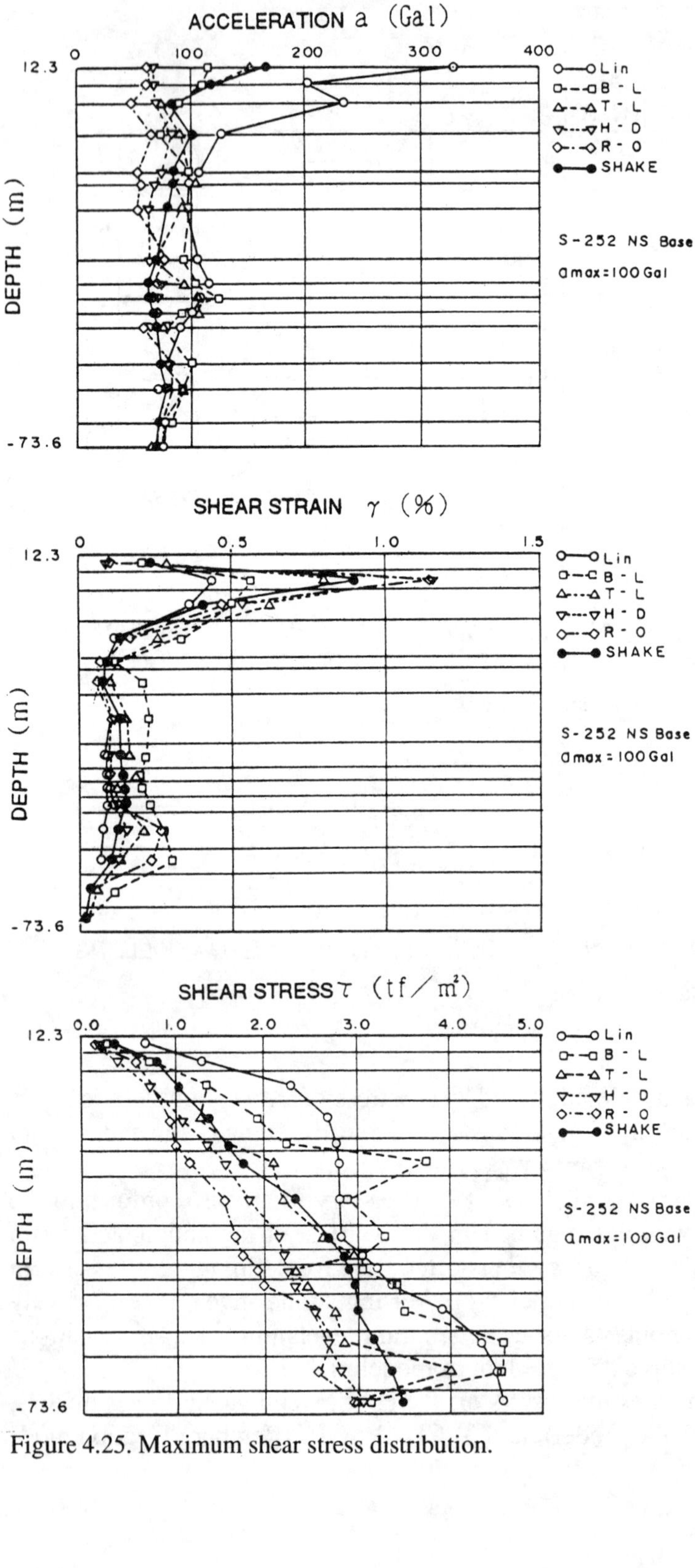

Figure 4.25. Maximum shear stress distribution.

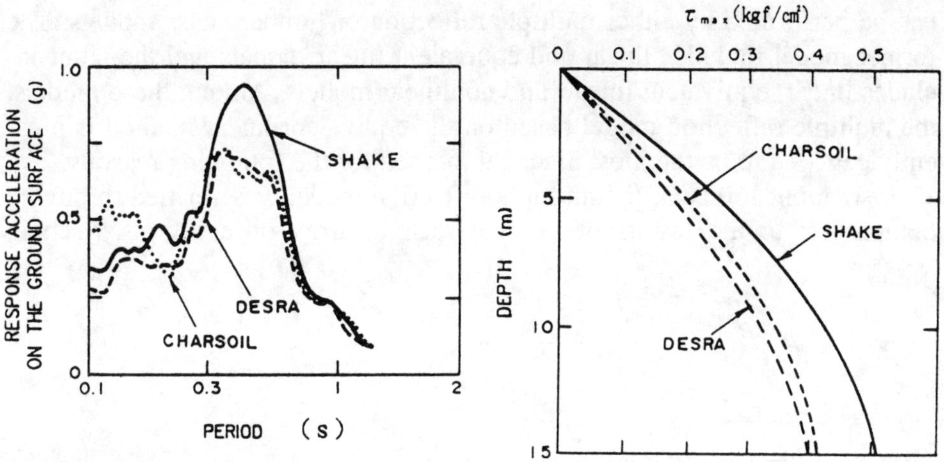

Figure 4.26. Comparison of earthquake response analyses (total and effective stress analyses) (Finn et al. [43]).

el) for the same subsoil condition (sandy soil with a layer thickness of 15 m) with the same earthquake input motion (El Centro 1940 NS scaled to a maximum acceleration of 0.1 g) [43]. In SHAKE, the maximum shear stress and response acceleration at the ground surface are larger compared to those obtained by other nonlinear response analyses. The difference in the results depends on many factors such as soil properties and earthquake motion characteristics. The equivalent linear method represented in SHAKE is generally believed to provide results that are conservative in liquefaction evaluation.

e) *Classification based on computation domain.* The earthquake response analysis can be classified into time domain and frequency domain analyses. The linear or equivalent linear analyses are performed both in the time and frequency domains. Nonlinear analyses are performed only in the time domain. Liquefaction is an unsteady phenomenon in which the excess pore water pressure in the soil changes during an earthquake. This phenomenon is more accurately simulated by analysis in the time domain which sequentially computes the phenomenon. In particular, the effective stress analysis is performed in the time domain. However, when the excess pore water pressure is not too large (less than 0.5 at effective stress ratio), it is possible to analyze in the frequency domain by equivalent linear analysis.

4.4.3.2 *Method used for response analysis in this handbook*

In coastal areas, stratified soils often prevail that can be assumed to be horizontal layers. In such cases, the results of a one-dimensional model can be sufficiently accurate for practical use.

To summarize the discussions presented earlier, one dimensional analysis can be performed by either multiple reflection or lumped mass models. The former model includes linear and equivalent linear models and the latter includes linear, equivalent linear, and nonlinear models. Among these models, the multiple reflection model based on the equivalent linear method is presently believed to be the most practical method for the following reasons.

First, applicability of the multiple reflection model was verified through a comparison to the results of vertical seismic array observations. Second,

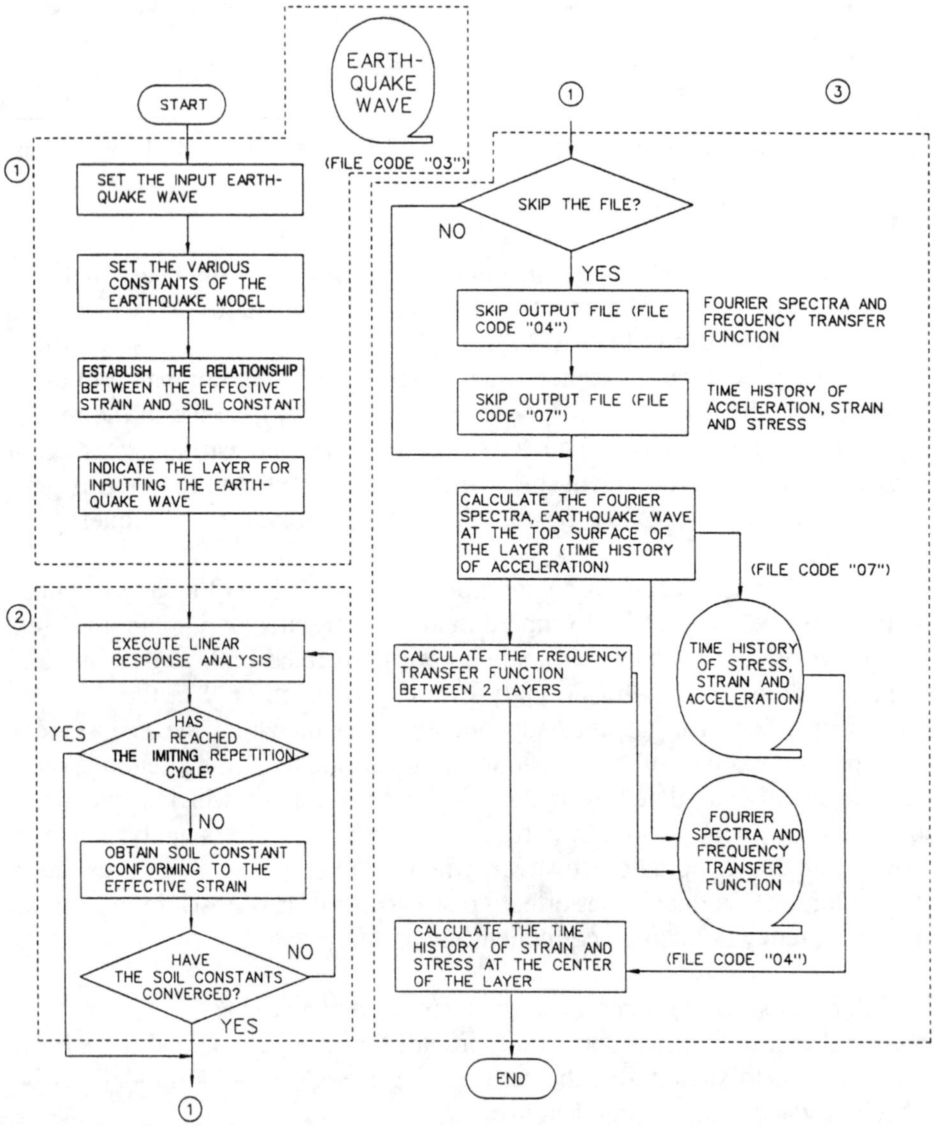

Figure 4.27. Flow chart for the SHAKE program.

from the practical point of view, the calculations based on the equivalent linear method represented in SHAKE are believed to provide conservative results as noted earlier. Third, equivalent linear analysis compares favorably with more rigorous non-linear analyses in acceleration response, and particularly in the vicinity of maximum response [45]. Finally, this computer program has been used for response analysis on the subsoil profile in the port and the harbour region in Japan since 1974 [46], and the computer programs are widely used in fields of activity in civil engineering practice other than ports and harbours.

Based on the above considerations, use of an earthquake response analysis for the soil using the multiple reflection model based on the one-dimensional equivalent linear method is recommended. For response analysis, the computer program SHAKE developed by Schnabel et al. is often used [38]. Figure 4.27 shows a flow chart for this program. Part 1 determines the strain dependence characteristics of the shear modulus and damping factor. Part 2 performs repetitive computation until the strain dependent shear modulus and damping factor are obtained. This is the main part of the equivalent linear analysis. Part 3 includes output for acceleration, stress, strain, Fourier spectra, and transfer function. In the earthquake response analysis based on this program, the earthquake motion provided in Section 4.4.2 is converted into the frequency domain in the computation process and the shear strain in each layer is computed based on the strain level dependent shear modulus and damping factor. Shear stress is obtained as a product of shear strain and shear modulus.

4.4.3.3 *Effective stress analysis*

In the process of soil liquefaction, excess pore water pressure is generated and the effective stress decreases. As a result, the shear modulus of the soil decreases and the response characteristics of the subsoil profile change accordingly as mentioned earlier. The effective stress analysis noted in Section 4.4.3.1 can capture these phenomena. Based on this effective stress computer code [47, 48], evaluation of liquefaction of port and harbour structures is being implemented [49-52]. Presently, however, specialized knowledge is required in the determination of the physical properties of the soil to be used in the effective stress analysis, and its use in general practice is limited.

4.4.4 *Earthquake response analysis*

The equivalent linear analysis as recommended in this handbook is performed based on the following procedure.

4.4.4.1 *Determination of base layer*

The base layer for the earthquake response analysis is established by referring to the soil survey results. The base layer refers to a bedrock or a relatively firm soil layer with a shear wave velocity of 300 m/sec or more as mentioned earlier. The base layer should be wide spread. Obviously, a soft layer should not appear below the base layer. When the shear wave velocity is not known, a sandy soil layer with an SPT *N*-value of 50 or more or a clay layer with an unconfined compressive strength (q_u-value) of 65 tf/m^2 or more can be considered the base layer. When the base layer cannot be verified due to a thick deposition of clayey soil, it is necessary to verify, by response analysis, that the earthquake response in the subsoil profile is not changed greatly by a change in the assumed position of the base layer. The base layer as referred to here is of engineering use only and does not always correspond to the bedrock defined in seismology (generally granite or hard rock having a shear wave velocity of about 3 km/sec).

4.4.4.2 *Layer division of the subsoil profile*

The subsoil profile is divided into layers based on variation in the soil properties, specified by soil type, shear wave velocity, SPT *N*-value, q_u-value, etc. When the subsoil profile is uniform in a thick deposit, layers 1 m thick are considered within 10 m below the ground surface and 2.5 m thick below 10 m. A thin soil layer of less than 50 cm is considered to have little possibility of being distributed continuously over a wide area within the soil profile in Japanese coastal areas and therefore it can be usually ignored in the modeling. As the number of soil layers increase, the computation time increases. It is generally preferable for the total number of layers to be less than 20. When modeling a cross section including a gravity type wall such as a caisson, cofferdam, etc., and the structural load is distributed in the horizontal direction, these structures can be modeled as artificial 'soils' having the wall height as the layer thickness. The influence of the layer division of the soil on the calculated results will be discussed in detail in Section 4.4.4.7.

4.4.4.3 *Soil classification*

The soil in each layer is classified into sandy soil, clayey soil, or gravel. Usually the actual soil consists of gravel, sand, silt, and clay mixture in various proportions. In this handbook, soil with a fines content (grain size of less than 74 μm) of less than 20% is considered sandy soil, and soils with other grain size distributions are considered clayey soil. Rubble used for mound and backfill are classified as gravel. In the response analysis, the strain dependent properties should be determined based on the soil classification discussed above.

4.4.4.4 *Unit weight of the soil*

If undisturbed samples are obtained and the unit weight of these samples is

Table 4.15. Standard values of unit weight.

Type of soil	Total unit weight γ_t (tf/m^3)	Submerged (buoyant) unit weight γ'_t (tf/m^3)
Clayey soil	1.5 (water content of 60% or more)	0.5
	1.7 (water content of less than 60%)	0.7
Sandy soil	1.8 (above the ground water level)	–
	2.0	1.0
Rubble backfill	2.0	1.0
Bedrock	2.2	1.2
Caisson	2.1	1.1

measured, the value measured is used for unit weight (γ_t) of the soil in each layer in the analysis. When the unit weight has not been measured, it is possible to use the values shown in Table 4.15.

4.4.4.5 *Shear modulus*

Shear modulus at a very small strain (shear strain of about 10^{-6}) in the soil used for the response analysis is obtained using the following equation based on the shear wave velocity obtained with velocity logging.

$$G_o = \frac{\gamma_t}{g} V_s^2 \quad (\text{tf/m}^2) \tag{4.15}$$

where: G_o = shear modulus at very small strain level (tf/m^2), V_s = shear wave velocity (m/sec), γ_t = unit weight of the soil (tf/m^3), g = acceleration of gravity (= 9.8 m/sec^2).

When the shear wave velocity is not obtained by velocity logging, the shear modulus is estimated from the SPT N-value with the equation shown in Section 4.3 and reproduced below.

$$G_o = 1440\, N^{0.68} \quad (\text{tf/m}^2) \tag{4.16}$$

This equation indicates an average value obtained from a large amount of measured data as indicated in Section 4.3. Caution is necessary due to the considerable variation in the original data [27].

When the unconfined compressive strength (q_u-value) of a clayey soil has been obtained, it is possible to estimate the shear modulus with the following equation, shown in Section 4.3 [17] and reproduced below.

$$G_o = 170\, q_u \quad (\text{tf/m}^2) \tag{4.17}$$

where: G_o = shear modulus at very small strain level (tf/m^2), q_u = unconfined compressive strength (tf/m^2).

When the SPT N-value of the soil can be obtained only before construction of a structure as in the case of soil directly under a caisson wall, it is possible to estimate the SPT N-value after construction by taking into consideration the influence of the increase in the effective overburden pressure from the wall structure. In this estimate, Figure 4.40 in Section 4.5 can be used. When the soil condition changes before and after the construction and velocity logging is performed only before construction, the shear wave velocity after construction can be estimated using the following equation.

$$V_s = V_{so} \left(\frac{\sigma'_v}{\sigma'_{vo}} \right)^B \tag{4.18}$$

where: V_s = shear wave velocity after construction (m/sec), V_{so} = shear wave velocity before construction (m/sec), σ'_v = effective overburden pressure after construction (tf/m^2), σ'_{vo} = effective overburden pressure before construction (tf/m^2), $B = 0.25$ in the case of sandy soil, 0.5 in the case of clayey soil.

Measurement of the shear wave velocity of rubble backfill and rubble mound is very difficult and the number of measurements are limited. In this handbook, the following values are used as shear wave velocity for rubble backfill and rubble mound for a large quay wall with a water depth of 10 m, based on the results [53] obtained from earthquake observation in a composite breakwater.

Shear wave velocity of rubble mound: $V_s = 300$ m/sec
Shear wave velocity of rubble backfill: $V_s = 225$ m/sec

When analyzing a subsoil profile including a caisson, the following value is used as the shear wave velocity of the caisson.

Shear wave velocity of caisson: $V_s = 2000$ m/sec

Pressure dependence of shear modulus may be represented by the following relationship based on existing laboratory test results [17, 54].

– Shear modulus for clayey soil with a plasticity index of $I_p = 30$ or more is proportional to the effective confining pressure;

– Shear modulus of sandy soil and clayey soil with a plasticity index I_p = less than 30 is proportional to the square root of the effective confining pressure.

Figure 4.28 shows the average shear wave velocity V_s of Toyoura sand obtained from a centrifuge test as a function of confining pressure σ'_c at the mid point of the sandy soil. The solid line in the figure is the curve of $V_s = k\,(\sigma'_c)^a$. The average value of the shear wave velocity of the sandy soil increases as the centrifugal acceleration increases, and the confining pressure dependence is noticeable. Figure 4.29 shows the distribution of shear wave

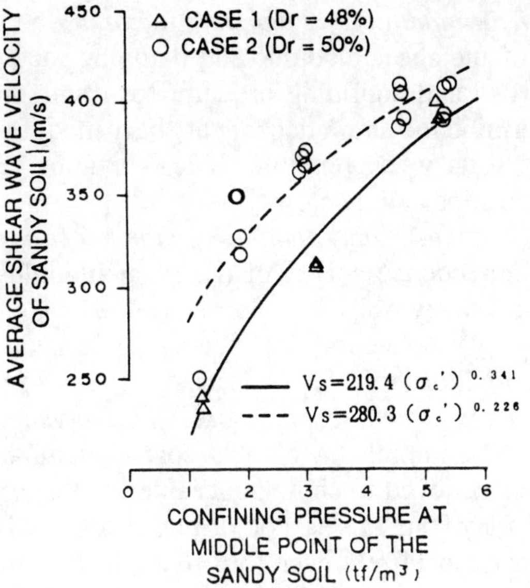

Figure 4.28. Relationship between the average effective confining pressure and average shear wave velocity of sandy soil.

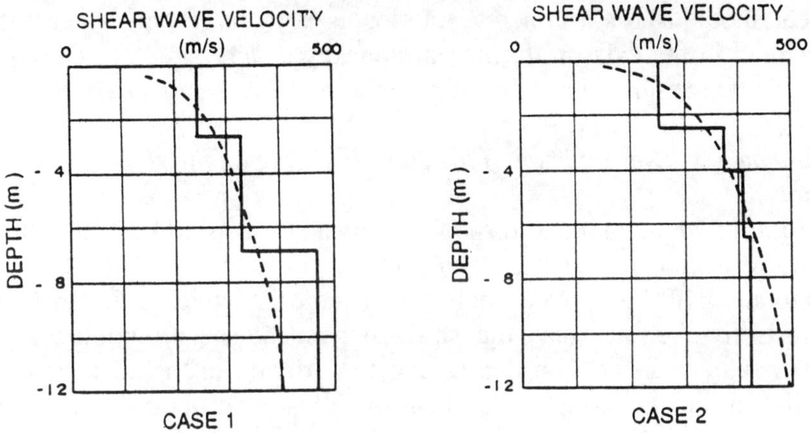

Figure 4.29. Distribution of shear wave velocity of sandy soil.

velocity with depth for the same model test. The broken line in the figure is proportional to the power of 0.25 of the confining pressure. This curve was obtained from the shear wave velocity of the mid point in the sandy soil. The shear wave velocity increases with depth for both cases in proportion to the power of 0.25 of the confining pressure. These results are based on measurements of the shear wave velocity when centrifugal forces ranging from 10 to 50 *G* are imposed on a 24 cm thick soil [55].

4.4.4.6 *Strain dependence of the shear modulus and damping factor*

Strain dependence characteristics of the shear modulus and damping factor differ according to the soil properties and confining pressure for each soil layer. It is better, therefore, to determine the strain dependent shear modulus and damping factor based on soil tests when possible. If tests cannot be performed, use the empirical relationships noted below.

– *Strain dependence of the shear modulus and damping factor for sandy and clayey soils* [17]. The strain dependence curves of the shear modulus and damping factor for sandy and clayey soils can be determined using Tables 4.10, 4.11, and Equation (4.14) in Section 4.3. For sandy soil, the curve obtained for the soils with I_p = NP or less than 9.4 is used.

– *Shear modulus and damping factor of concrete caisson, rubble mound, and backfill.* The shear modulus and damping factor of a coarse granular mass such as rubble and backfill are believed to show noticeable nonlinearity compared to that of sand and clay, but this is not firmly established. Existing research results are summarized in reference [56]. By referring to these results, it is possible to assume a strain dependence curve for the shear modulus and damping factor of the rubble and backfill.

The shear modulus for a concrete caisson is very large and hence the strain generated in the caisson during earthquakes is very small. A caisson can, therefore, be considered as a linear material in the response analysis.

4.4.4.7 *Influence of layer division of the subsoil profile on the response analysis*

When performing an earthquake response analysis of a subsoil profile with the multiple reflection model based on the equivalent linear method, it is necessary to subdivide the layer finely when there is a thick uniform soil layer. This is required because the strain dependent shear modulus and damping factor may differ with depth, due to variation in the shear strain generated by earthquake motion. The shear modulus of soil also depends on confining pressure, so that even when a uniform soil deposit is considered, shear modulus actually changes with depth. In an elastic wave exploration, it is preferable to consider the depth dependent properties in the division of the soil layers. This also relates to the measurement intervals in the elastic wave exploration method (refer to Section 4.2.4).

Figure 4.30 shows the influence of layer division of the subsoil profile on the response analysis [55]. The depth of the analyzed subsoil profile is 15 m; sandy and clayey soils are assumed. In the analysis, four types of subsoil profiles were assumed; uniform, 3 layers, 5 layers, and 15 layers profiles. A strong earthquake recording at Ofunato Port (1978 Miyagi-ken-oki earthquake) was used as input motion with a maximum acceleration of α_{max} = 100 Gal. The shear wave velocity of each layer was determined so that the propagation time Δt (= 0.1 sec) from the base layer to the ground surface be-

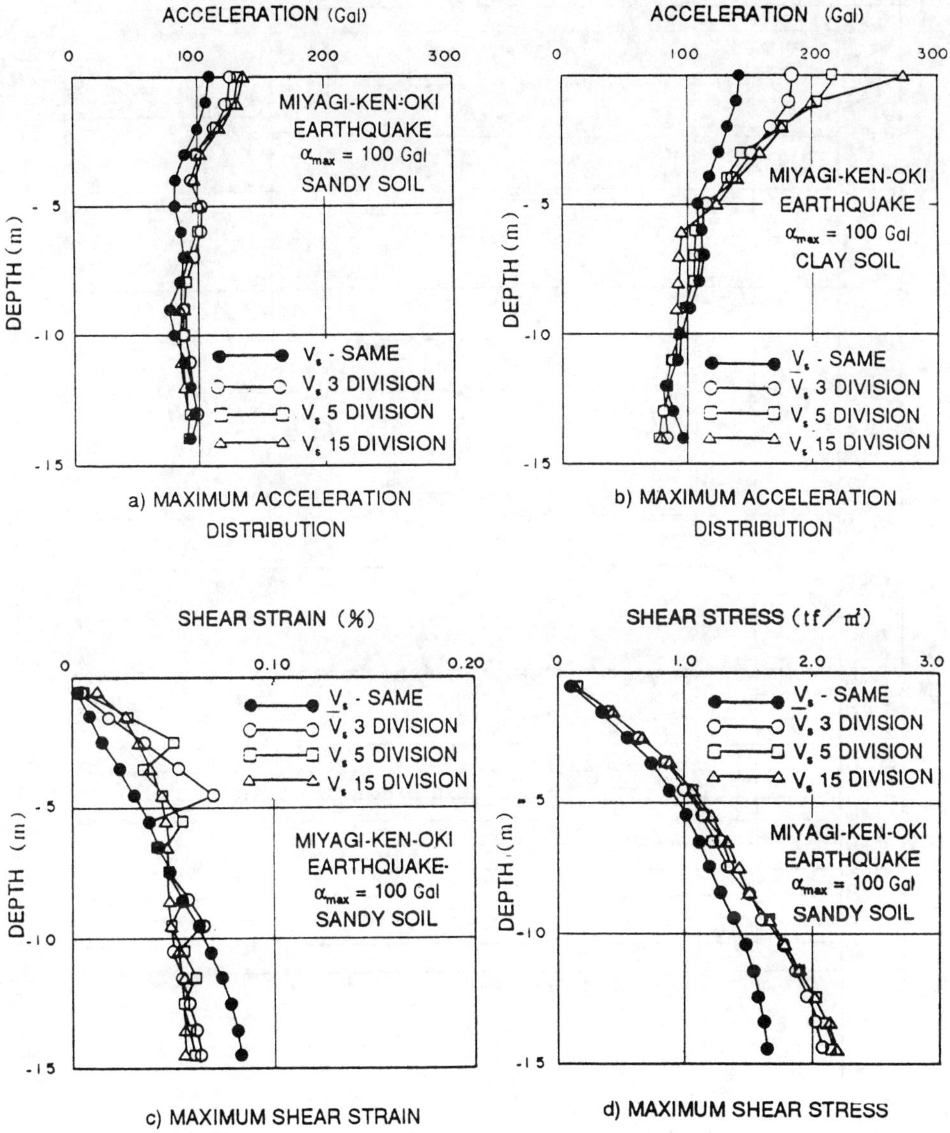

Figure 4.30. Influence of confining pressure dependence of shear wave velocity on the response analysis.

comes the same and satisfies the pressure dependence.

When the layer division was very small, the maximum acceleration of the ground surface tended to increase. This was due to the pressure dependent soil properties. For clayey soil, the response acceleration increased to about twice that of uniform soil. There was also a difference of about 20% in shear stress.

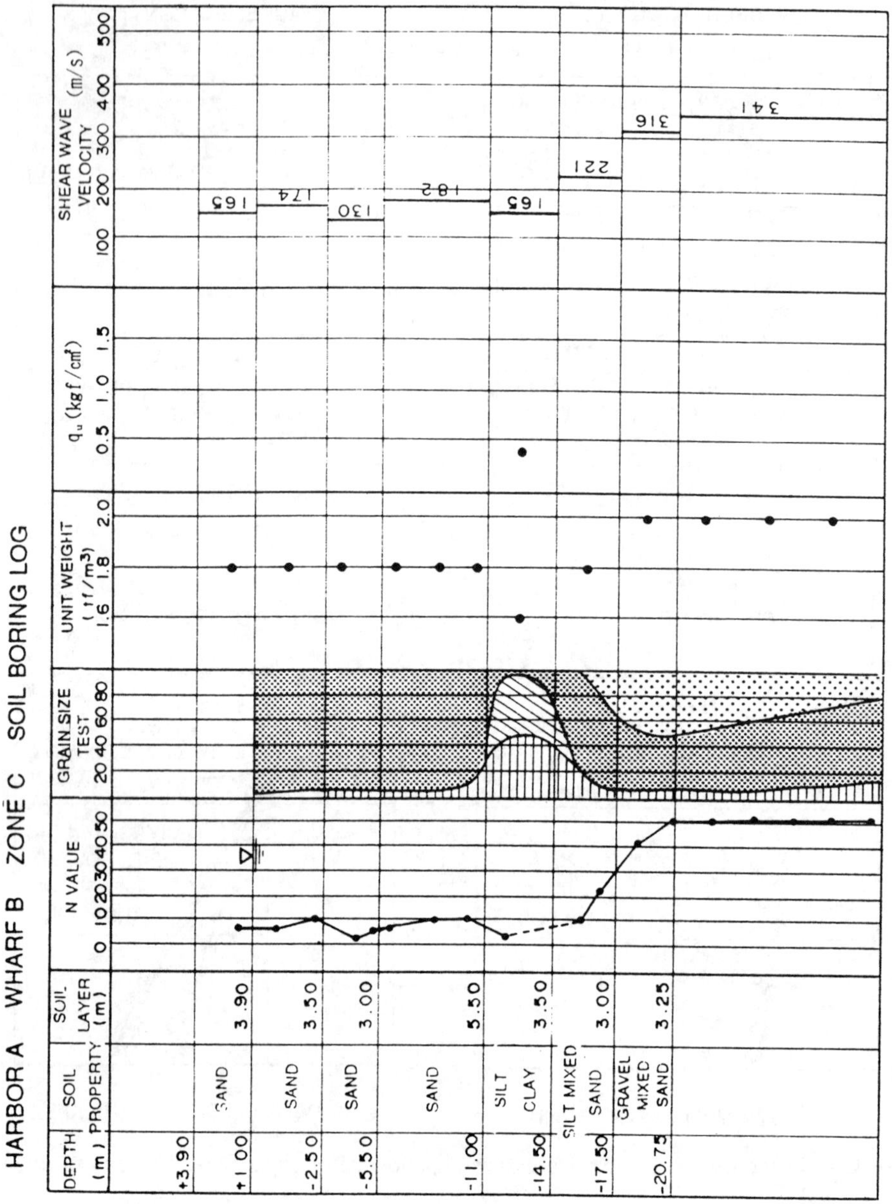

Figure 4.31. Example conditions for earthquake response analysis.

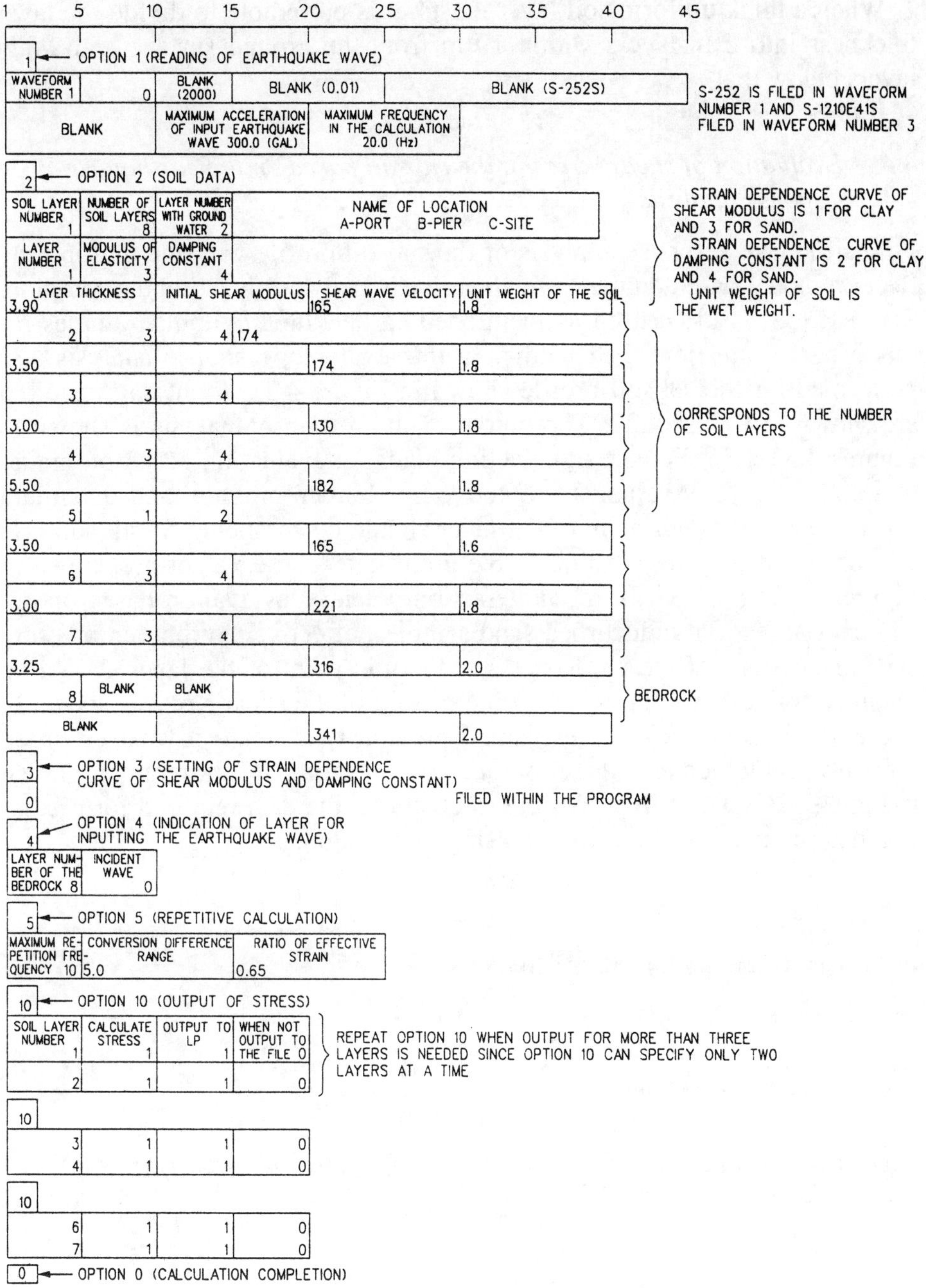

Figure 4.32. Example of input data.

When a thick uniform soil layer exists, it is preferable to divide the layer thickness into 2 m layers within 10 m from the ground surface and 2.5 m layers below that.

4.4.5 *Evaluation of shear stress in the subsoil based on the earthquake response analysis*

For earthquake response analysis of the subsoil profile with the multiple reflection model based on the equivalent linear analysis, the computer program SHAKE [38] is often used as mentioned earlier. Input/output examples for this program are described below. In this example, response analysis was performed for the subsoil profile shown in Figure 4.31. Input data required are shown in Figure 4.32. The output of the response analysis is shown in Figures 4.33-4.36. For the liquefaction potential evaluation based on gradation and the SPT *N*-value, the effective overburden pressure and maximum shear stress at the mid point of the layer should be computed for all soil layers composed of soil with liquefiable grain size. These values are shown in Figures 4.33 and 4.34. In the liquefaction potential evaluation based on cyclic triaxial test on undisturbed sand samples, effective overburden pressure and time history of the shear stress at the mid point of the layer should be obtained for each soil layer. The time history of the shear stress is shown in Figures 4.35 and 4.36. As explained previously, the effective overburden pressure is obtained as a direct output from the computer program as shown in Figure 4.33. Thus, it is possible to obtain all the data required for prediction/determination of liquefaction using these outputs.

```
* * * * * *    OPTION  2 * * *  READ SOIL PROFILE

NEW SOIL PROFILE NO.  1    IDENTIFICATION        A-PORT  B-PIER  C-SIT

NUMBER OF LAYERS                        8        DEPTH TO BEDROCK      25.65
NUMBER OF FIRST  SUBMERGED LAYER        2        DEPTH TO WATER LEVEL   3.90
```

LAYER	TYPE G B	THICKNESS (M)	DEPTH (M)	EFF. VER. PRESS. (T/SM)	MODULUS (T/SM)	DAMPING	UNIT WEIGHT	SHEAR VEL. (M/SEC)
1	3 4	3.90	1.95	3.51	5001.	0.	1.8000	165.
2	3 4	3.50	5.65	8.42	5561.	0.	1.8000	174.
3	3 4	3.00	8.90	11.02	3104.	0.	1.8000	130.
4	3 4	5.50	13.15	14.42	6084.	0.	1.8000	182.
5	1 2	3.50	17.65	17.67	4445.	0.	1.6000	165.
6	3 4	3.00	20.90	19.92	8971.	0.	1.8000	221.
7	3 4	3.25	24.03	22.75	20379.	0.	2.0000	316.
8	BASE				23731.	0.	2.0000	341.

Figure 4.33. Example of computed effective overburden pressure.

```
EARTHQUAKE    -  S-252  NS HACHINOHE
SOIL PROFILE  -         A-PORT  F-PIER  C-SIT


ITERATION NUMBER  9
 THE CALCULATION HAS BEEN CARRIED OUT IN THE TIME DOMAIN WITH EFF. STRAIN =  0.65 * MAX. STRAIN
```

LAYER	TYPE G B	DEPTH	EFF. STRAIN (%)	NEW DAMP (%)	DAMP USED (%)	ERROR (%)	G/GO	NEW G (T/SM)	G USED (T/SM)	ERROR (%)
1	3 4	1.95	0.01833	7.10	7.12	-0.4	0.653	3267.	3260.	0.2
2	3 4	5.65	0.08354	15.81	15.94	-0.8	0.346	1926.	1904.	1.1
3	3 4	8.90	1.08920	27.46	27.42	0.1	0.060	186.	189.	-1.6
4	3 4	13.15	0.16499	19.89	20.20	-1.6	0.231	1404.	1351.	3.8
5	1 2	17.65	0.10399	8.91	8.93	-0.2	0.638	2836.	2831.	0.1
6	3 4	20.90	0.13560	18.76	18.75	0.0	0.262	2352.	2355.	-0.1
7	3 4	24.03	0.03550	10.53	10.49	0.4	0.518	10557.	10591.	-0.3

VALUES IN TIME DOMAIN

LAYER	TYPE G B	THICKNESS (M)	DEPTH (M)	MAX. STRAIN (%)	MAX. STRESS (T/SM)	TIME (SEC)
1	3 4	3.90	1.95	0.02820	0.92	6.03
2	3 4	3.50	5.65	0.12852	2.47	6.06
3	3 4	3.00	8.90	1.67569	3.12	6.08
4	3 4	5.50	13.15	0.25382	3.56	6.01
5	1 2	3.50	17.65	0.15998	4.54	5.91
6	3 4	3.00	20.90	0.20862	4.91	5.41
7	3 4	3.25	24.03	0.05461	5.77	5.38

Figure 4.34. Example of computed maximum shear stress.

Δt=0.01 (EXAMPLE) SHEAR STRESS 6.01 SEC FROM THE START

① PEAK SHEAR STRESS: 657, 2.4925 (607, −1.9744)
② PEAK SHEAR STRESS: 610, −2.0453 (560)
③ PEAK SHEAR STRESS: 717, −1.2306 (667)

t(s)	σ,,(tf/m²)	t(s)	σ,,(tf/m²)	t(s)	σ,,(tf/m²)	t(s)	σ,,(tf/m²)	t(s)	σ,,(tf/m²)
501	0.0174	551	−1.4189	601	2.2487	651	−0.0904	701	−0.2921
502	0.1081	552	−1.5344	602	2.3282	652	−0.1888	702	−0.2912
503	0.1916	553	−1.6445	603	2.3913	653	−0.2895	703	−0.2880
504	0.2641	554	−1.7463	604	2.4385	654	−0.3916	704	−0.2822
505	0.3222	555	−1.8369	605	2.4703	655	−0.4937	705	−0.2733
506	0.3640	556	−1.9137	606	2.4879	656	−0.5943	706	−0.2609
507	0.3888	557	−1.9744	607	2.4925	657	−0.6918	707	−0.2445
508	0.3974	558	−2.0174	608	2.4852	658	−0.7846	708	−0.2235
509	0.3917	559	−2.0414	609	2.4671	659	−0.8712	709	−0.1979
510	0.3749	560	−2.0453	610	2.4383	660	−0.9504	710	−0.1674
511	0.3505	561	−2.0289	611	2.3990	661	−1.0210	711	−0.1325
512	0.3226	562	−1.9924	612	2.3488	662	−1.0822	712	−0.0936
513	0.2950	563	−1.9366	613	2.2872	663	−1.1334	713	−0.0518
514	0.2710	564	−1.8630	614	2.2139	664	−1.1741	714	−0.0084
515	0.2532	565	−1.7731	615	2.1289	665	−1.2041	715	0.0350
516	0.2429	566	−1.6689	616	2.0330	666	−1.2230	716	0.0768
517	0.2405	567	−1.5528	617	1.9274	667	−1.2306	717	0.1154
518	0.2456	568	−1.4272	618	1.8143	668	−1.2269	718	0.1494
519	0.2575	569	−1.2955	619	1.6962	669	−1.2121	719	0.1776
520	0.2749	570	−1.1618	620	1.5763	670	−1.1865	720	0.1994
521	0.2967	571	−1.0307	621	1.4577	671	−1.1510	721	0.2144
522	0.3217	572	−0.9069	622	1.3437	672	−1.1065	722	0.2227
523	0.3489	573	−0.7941	623	1.2376	673	−1.0546	723	0.2247
524	0.3774	574	−0.6946	624	1.1419	674	−0.9969	724	0.2209
525	0.4064	575	−0.6091	625	1.0586	675	−0.9352	725	0.2119
526	0.4348	576	−0.5365	626	0.9892	676	−0.8711	726	0.1983
527	0.4615	577	−0.4750	627	0.9342	677	−0.8064	727	0.1804
528	0.4844	578	−0.4216	628	0.8935	678	−0.7423	728	0.1585
529	0.5010	579	−0.3723	629	0.8663	679	−0.6799	729	0.1322
530	0.5082	580	−0.3225	630	0.8512	680	−0.6198	730	0.1012
531	0.5029	581	−0.2677	631	0.8460	681	−0.5627	731	0.0647
532	0.4827	582	−0.2038	632	0.8481	682	−0.5091	732	0.0219
533	0.4462	583	−0.1282	633	0.8543	683	−0.4595	733	−0.0279
534	0.3933	584	−0.0395	634	0.8613	684	−0.4143	734	−0.0852
535	0.3248	585	0.0625	635	0.8654	685	−0.3738	735	−0.1502
536	0.2426	586	0.1772	636	0.8636	686	−0.3385	736	−0.2226
537	0.1492	587	0.3032	637	0.8533	687	−0.3088	737	−0.3019
538	0.0475	588	0.4386	638	0.8331	688	−0.2851	738	−0.3867
539	−0.0594	589	0.5813	639	0.8022	689	−0.2675	739	−0.4755
540	−0.1689	590	0.7291	640	0.7607	690	−0.2560	740	−0.5668
541	−0.2792	591	0.8803	641	0.7095	691	−0.2500	741	−0.6587
542	−0.3892	592	1.0334	642	0.6499	692	−0.2488	742	−0.7495
543	−0.4990	593	1.1871	643	0.5832	693	−0.2513	743	−0.8376
544	−0.6089	594	1.3403	644	0.5110	694	−0.2564	744	−0.9214
545	−0.7197	595	1.4919	645	0.4346	695	−0.2631	745	−0.9993
546	−0.8322	596	1.6404	646	0.3548	696	−0.2703	746	−1.0700
547	−0.9466	597	1.7834	647	0.2719	697	−0.2772	747	−1.1323
548	−1.0632	598	1.9183	648	0.1861	698	−0.2833	748	−1.1856
549	−1.1815	599	2.0424	649	0.0972	699	−0.2879	749	−1.2289
550	−1.3006	600	2.1531	650	0.0050	700	−0.2910	750	−1.2617

Figure 4.35. Example of computed time history of shear stress in digital table.

In evaluating liquefaction potential, the most conservative results should be chosen from the results obtained using at least two earthquake waveform.

4.4.6 *Simplified evaluation of shear stress in the subsoil*

When performing an approximate evaluation for liquefaction potential of the soil at many locations, it is possible to obtain the shear stress using a simplified procedure [57].

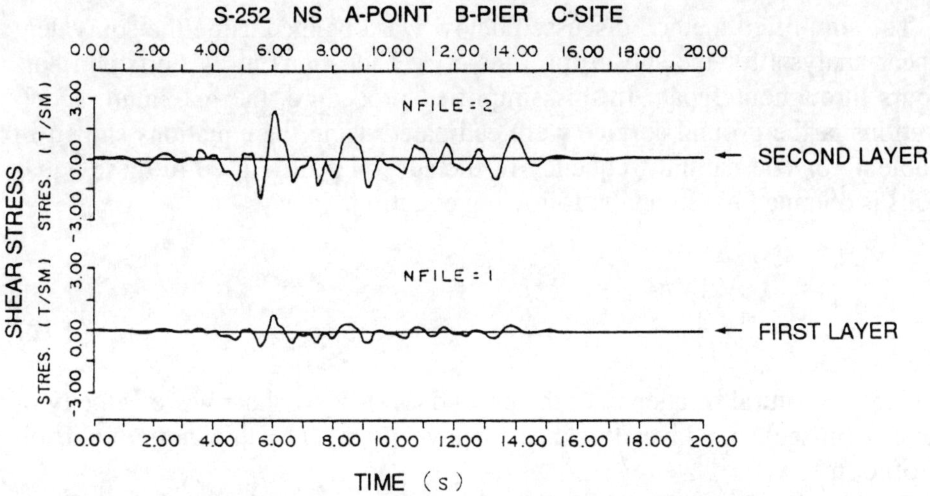

Figure 4.36. Example of computed time history waveform of shear stress.

Table 4.16. Estimation equation of maximum acceleration at the ground surface ($f > 0.5$ Hz).

Tokachi-oki earthquake Bedrock acceleration	S-252 NS Base Maximum acceleration at the ground surface α_{est} (Gal)	 (f in Hz)
200 Gal	$177 \times f$	$(0.5 \leq f \leq 1.3)$
	230	$(f > 1.3)$
250 Gal	$200 \times f$	$(0.5 \leq f \leq 1.5)$
	300	$(f > 1.5)$
300 Gal	$200 \times f$	$(0.5 \leq f \leq 1.6)$
	320	$(f > 1.6)$

Miyagi-ken-oki earthquake Bedrock acceleration	S-1210 E41S Maximum acceleration at the ground surface α_{est} (Gal)	 (f in Hz)
200 Gal	$190 \times f$	$(0.5 \leq f \leq 1.0)$
	$27.5 \times f + 162.5$	$(1.0 < f \leq 5.0)$
	300	$(f > 5.0)$
250 Gal	$190 \times f$	$(0.5 \leq f \leq 1.0)$
	$53.3 \times f + 136.7$	$(1.0 < f \leq 4.0)$
	350	$(f > 4.0)$
300 Gal	$190 \times f$	$(0.5 \leq f \leq 1.0)$
	$52.5 \times f + 137.5$	$(1.0 < f \leq 5.0)$
	400	$(f > 5.0)$

The simplified method discussed below was obtained using the equivalent linear analysis for the subsoil profiles at 357 locations in 24 ports and harbours throughout Japan. In this simplified procedure, the maximum accelerations at the ground surface were estimated using the equations shown in Table 4.16. The natural frequency of the subsoil profile used in these equations is obtained based on the following equation.

$$f = \frac{\sum_{i=1}^{n} V_{si} \times h_i}{4H^2} \tag{4.19}$$

where: f = natural frequency of the ground (Hz), V_{si} = shear wave velocity of layer i (m/sec), h_i = layer thickness of layer i (m), H = thickness of subsoil profile (m).

A reduction factor for estimating the shear stress in the subsoil profile is obtained as follows:

For 'S-252 NS Base'

$$r_d\,(z) = 1.0 - 0.019z \quad (z \leq 25\text{ m}) \tag{4.20}$$

For 'S-1210 E41S'

$$r_d\,(z) = 1.0 - 0.026z \quad (z \leq 25\text{ m}) \tag{4.21}$$

where: $r_d(z)$ = reduction factor, z = depth from the ground surface (m).

The maximum shear stress in the soil is obtained using the reduction factor given in the equation above and the estimated maximum acceleration at the ground surface based on the following equation.

$$\tau_{\text{est}}\,(z) = r_d\,(z)\left(\frac{\alpha_{\text{est}}}{g}\right)\sigma_v\,(z) \tag{4.22}$$

where: $\tau_{\text{est}}(z)$ = estimated value of shear stress in the ground (tf/m^2), $r_d(z)$ = reduction factor, α_{est} = estimated maximum acceleration at the ground surface (Gal), g = acceleration of gravity (980 Gal), $\sigma_v\,(z)$ = total overburden pressure (tf/m^2).

The simplified analysis for earthquake response of the subsoil profile discussed herein is simply to obtain the rough determination of liquefaction as discussed in Section 4.5. When this simplified analysis is used for other objectives, caution is advised. The range of use for this simplified analysis includes subsoil profiles close to horizontal stratification, and subsoil depth less than 25 m having a natural frequency greater than 0.5 Hz.

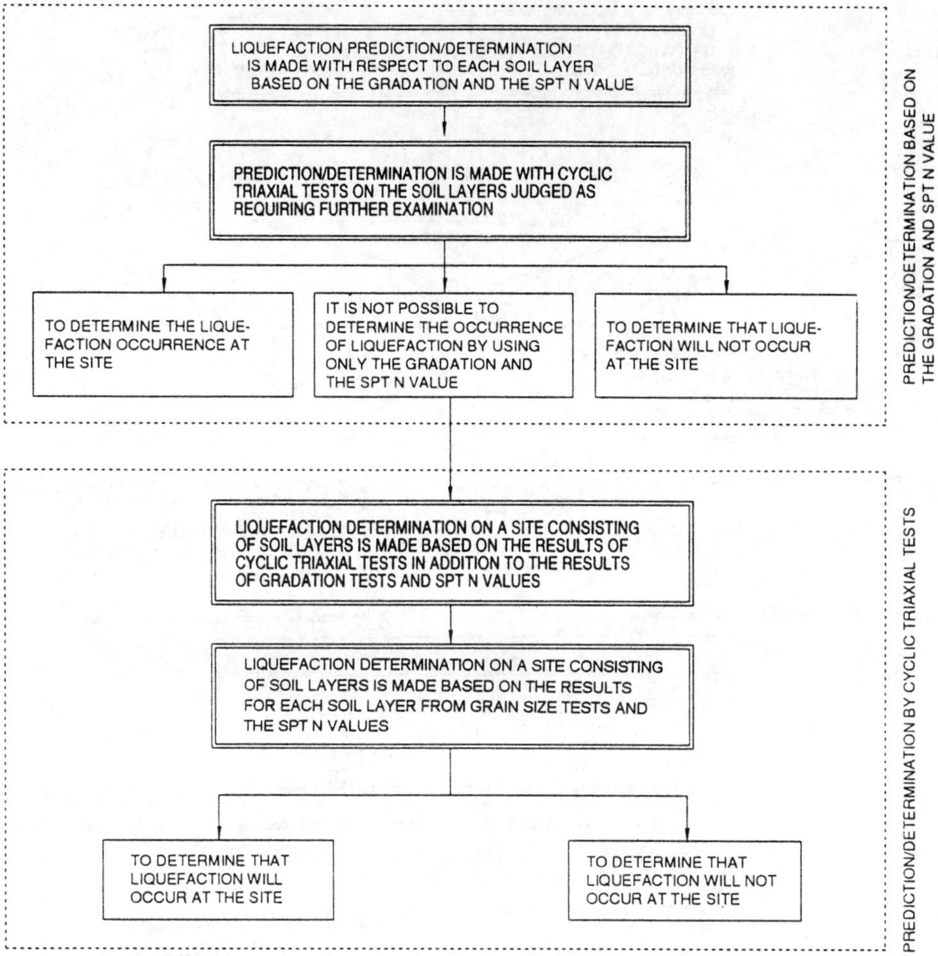

Figure 4.37. Flow chart of liquefaction prediction/determination.

4.5 LIQUEFACTION PREDICTION/DETERMINATION

4.5.1 *Procedure for liquefaction prediction/determination*

Liquefaction prediction/determination basically follows the procedure shown in Figure 4.37. The first step of this procedure uses the grain size distribution (gradation) of soil and standard penetration test blow counts (SPT *N*-value) for a soil layer. If this first step of the procedure is difficult, i.e. the results are close to a borderline between liquefaction and nonliquefaction conditions, then the procedure can be supplemented with cyclic triaxial tests.

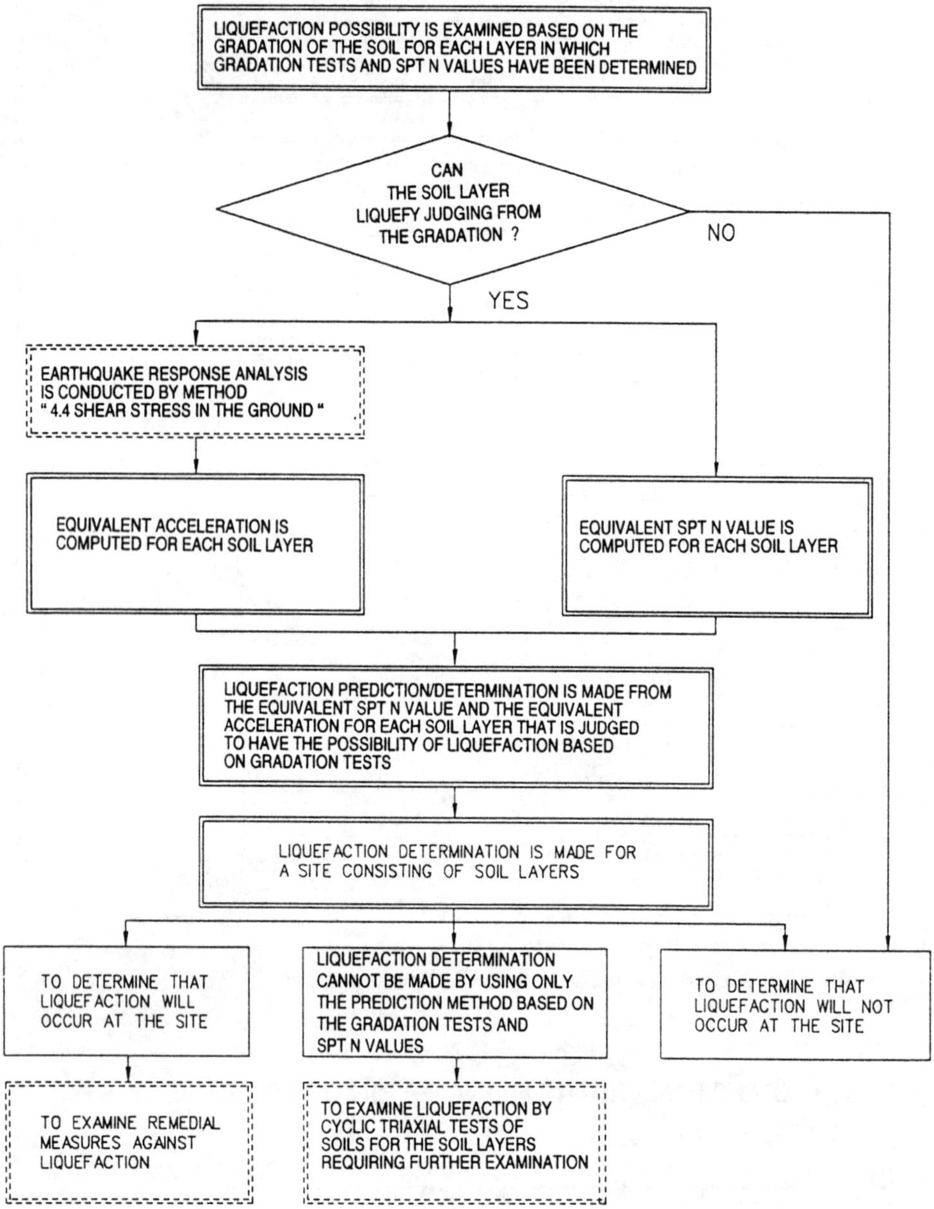

Figure 4.38. Prediction/determination procedure based on gradation and SPT *N*-values.

4.5.2 *Prediction/determination based on gradation and SPT N-value*

Prediction/determination based on gradation and SPT *N*-value is made utilizing the flow chart shown in Figure 4.38.

4.5.2.1 *Standard procedure*

– *Step 1*. Initially, soil is classified according to the gradation shown in Fig-

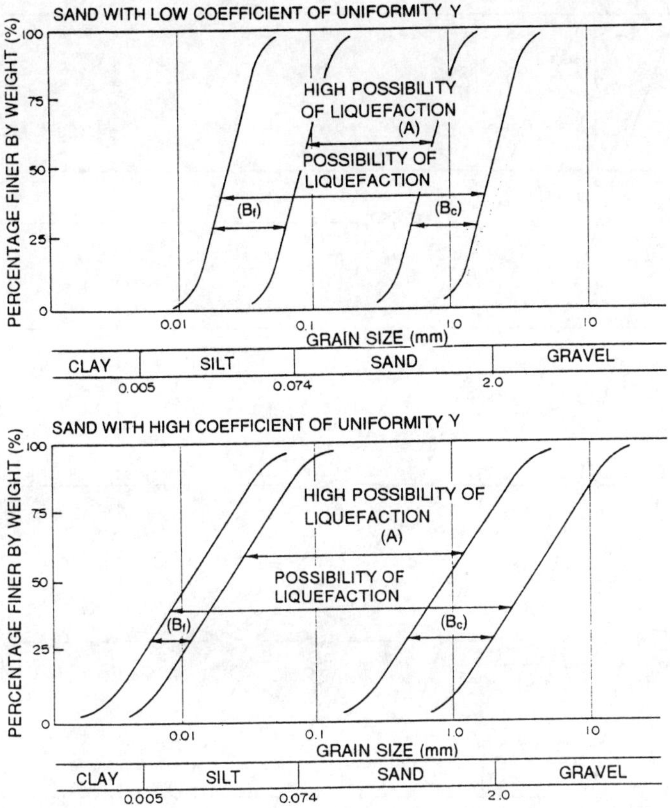

Figure 4.39. Gradation of soil having the possibility of liquefaction.

ure 4.39. One of the figures is chosen based on the coefficient of uniformity of the soil. The coefficient $U_c = D_{60}/D_{10} = 3.5$ may be used as a criterion in this choice, where U_c = coefficient of uniformity, D_{60} = 60% diameter, and D_{10} = effective diameter (10% diameter). The soils for which gradation falls in zones other than A, B_f, and B_c are considered nonliquefiable.

For difficult classification, such as when the grain size distribution covers two zones of gradation, leading to a noticeable difference in the results of the liquefaction prediction/determination depending on the zonal choice, then suitable engineering judgement becomes necessary such as using a prediction/determination method based on cyclic triaxial tests.

Local soils in Japan such as Shirasu (i.e. volcanic ash in Kyushu, Japan) are known to have special properties differing from generic sand. Before adopting the prediction/determination method shown for generic sand, it is necessary to evaluate the applicability of the method by comparing cyclic triaxial test results at a representative test site.

– *Step 2*. For a soil layer with gradation falling in zones A and B_c shown in Figure 4.39, an equivalent N-value is computed by the following equation.

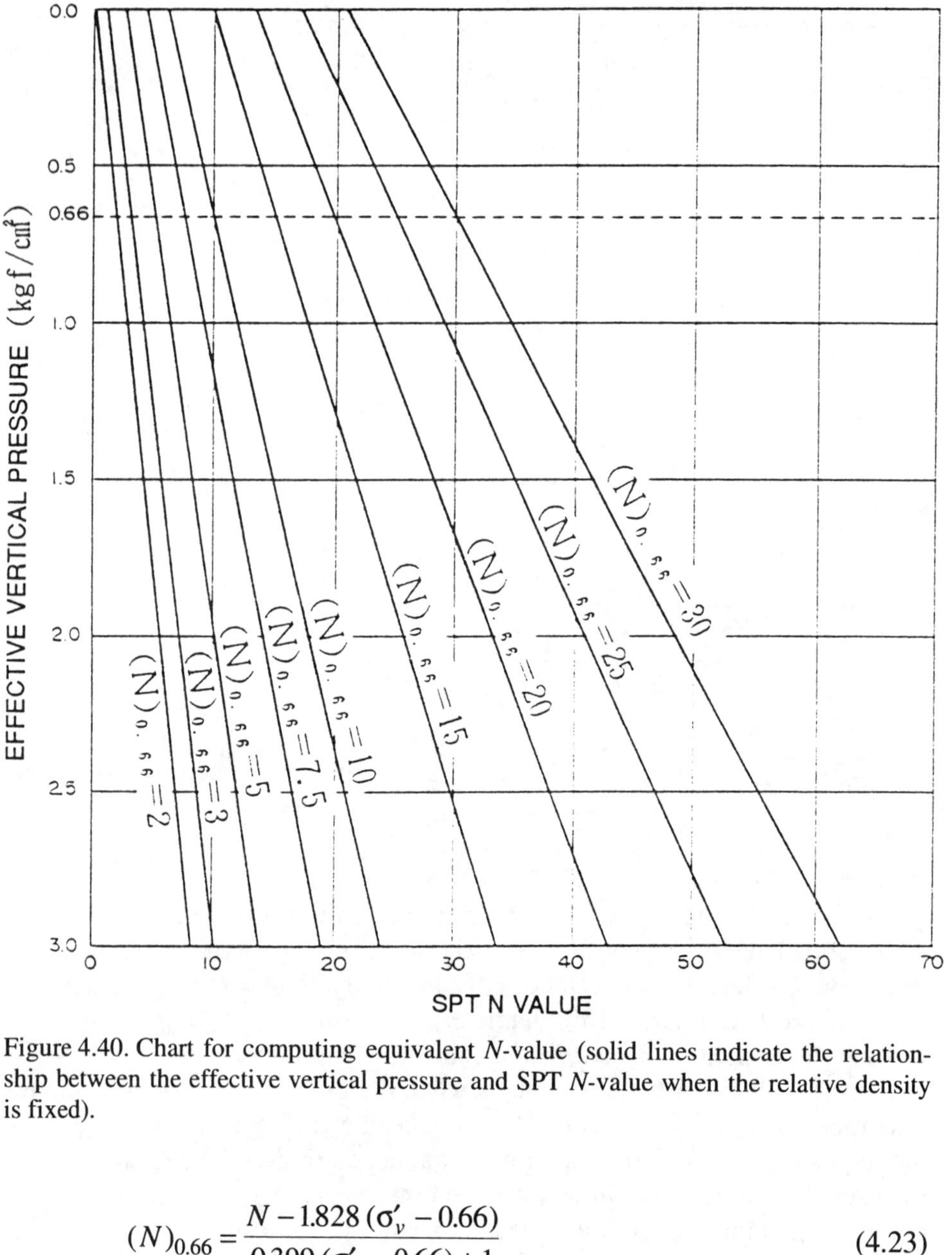

Figure 4.40. Chart for computing equivalent *N*-value (solid lines indicate the relationship between the effective vertical pressure and SPT *N*-value when the relative density is fixed).

$$(N)_{0.66} = \frac{N - 1.828\,(\sigma'_v - 0.66)}{0.399\,(\sigma'_v - 0.66) + 1} \tag{4.23}$$

where: $(N)_{0.66}$ = equivalent *N*-value, N = SPT *N*-value of a soil layer, σ'_v = effective vertical pressure of a soil layer (kgf/cm^2) (effective vertical pressure used here should be computed with respect to the ground surface elevation at the time of the standard penetration test).

Equation (4.23) is applicable for the ranges $2 \leq (N)_{0.66} \leq 40$, $0 \leq \sigma'_v$ 3 kgf/cm^2. For conditions outside of these ranges, it is preferable to evaluate the applicability by referring to reference [12]. The relationship for the above equation is shown in Figure 4.40.

The equivalent *N*-value refers to the SPT *N*-value corrected for an effective vertical pressure of 0.66 kgf/cm^2. This conversion reflects the practice that, in the former 'Technical standards for port and harbour facilities and commentaries' [58] before the 1989 revision, liquefaction prediction was made on the basis of an SPT *N*-value for a soil layer in the vicinity of the ground water table. The derivation of Equation (4.23) is explained in detail in reference [12].

For a layer with a grain size distribution in zone B_f shown in Figure 4.39, the equivalent *N*-value is the same as the SPT *N*-value of the soil layer without correction. The reason for this is explained in detail in reference [13].

– *Step 3*. Equivalent acceleration is computed for the soil layers with a grain size distribution in zones *A*, B_f, and B_c. It is computed for each layer using the following equation based on the maximum shear stress obtained from earthquake response analysis (analysis by the method shown in Section 4.4).

$$\alpha_{eq} = 0.7 \times \frac{\tau_{max}}{\sigma'_v} \times g \tag{4.24}$$

where: α_{eq} = equivalent acceleration (Gal), τ_{max} = maximum shear stress (kgf/cm^2), σ'_v = effective vertical pressure (kgf/cm^2) (The effective vertical pressure used here is computed with respect to the ground surface elevation at the time of earthquake), g = acceleration of gravity (980 Gal).

The equivalent acceleration is a maximum shear stress ratio based on the right hand side of the equation. The factor used here to convert the stress ratio to equivalent acceleration reflects the practice outlined in the former 'Technical standards for port and harbour facilities and commentaries' [58] prior to the 1989 revision in which the maximum acceleration of earthquake motion rather than the maximum shear stress ratio was used as an index of intensity of shaking.

– *Step 4*. The zone in Figure 4.41 to which a soil layer belongs to is determined from the equivalent *N*-value and the equivalent acceleration. A soil layer with a grain size distribution in zone *A* in Figure 4.39 is checked by Figure 4.41a and a soil layer with a gradation in zone B_f or B_c is checked by Figure 4.41b. For those soils which include more than 5% fines content (fine grained soils of less than 74 μm) of the zone *A* soils, the boundary line for each area shown in Figure 4.41a is lowered by multiplying the reduction factor shown in Figure 4.42 to the equivalent *N*-value. Alternatively, instead of lowering the boundary line, conversion can be applied to the equivalent *N*-value by dividing it by the reduction coefficient shown in Figure 4.42. The converted equivalent *N*-value can then be directly checked with respect to Figure 4.41a.

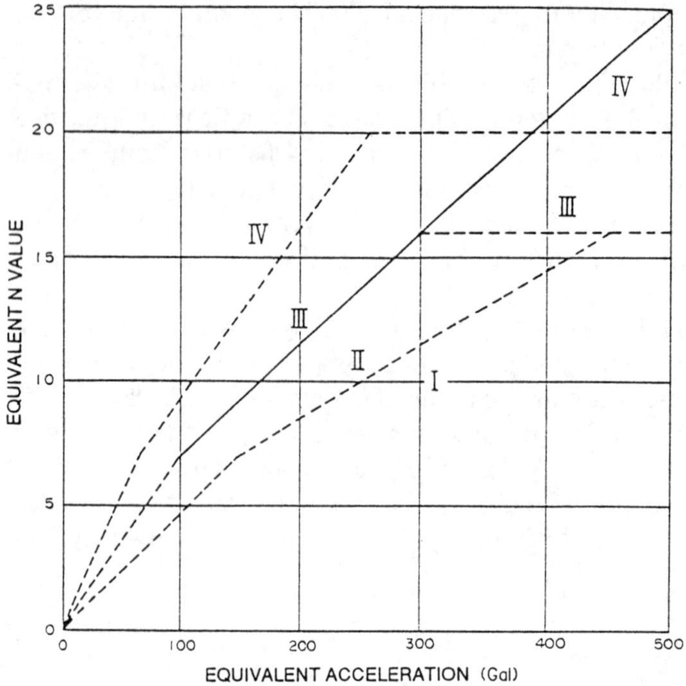

(a) FOR SOIL LAYER WITH GRADATION IN ZONE A

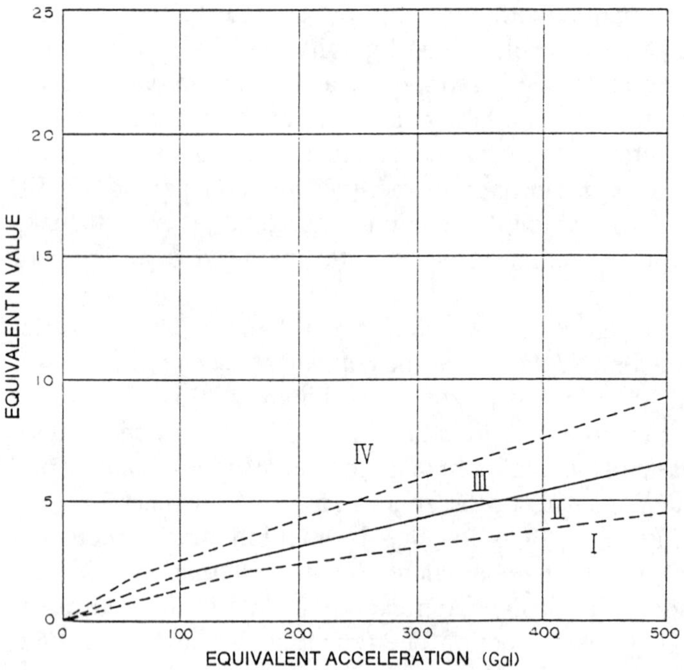

(b) FOR SOIL LAYER WITH GRADATION IN ZONES B_f AND B_c

Figure 4.41. Classification of soil layer for liquefaction prediction based on equivalent acceleration and equivalent *N*-values.

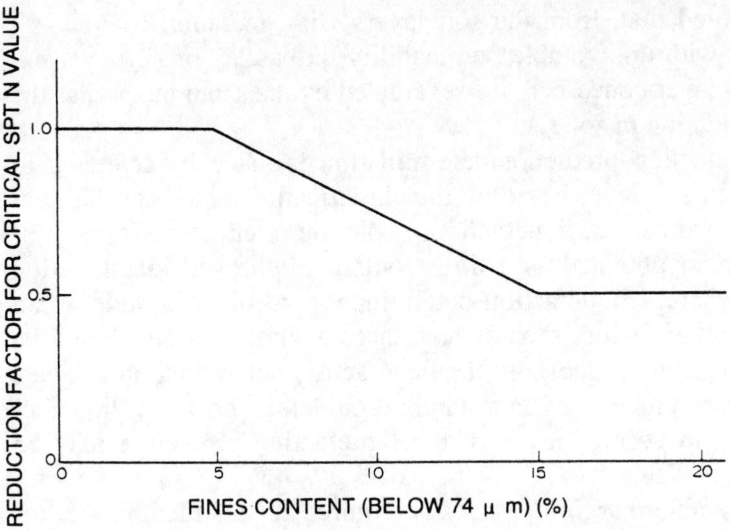

Figure 4.42. Reduction factor for critical SPT *N*-value based on the fines content (for a soil layer of zone A gradation only).

Table 4.17. Liquefaction prediction/determination for each soil layer based on grain size distribution and SPT *N*-value.

Zones indicated in Figure 4.41	Liquefaction prediction based on gradation and SPT *N*-value	Liquefaction determination based on gradation and SPT *N*-value
I	Possibility of liquefaction occurrence is very high	Liquefaction will occur
II	Possibility of liquefaction occurrence is high	Decide either to determine that liquefaction will occur, or to conduct further evaluation based on cyclic triaxial tests
III	Possibility of liquefaction occurrence is low	Decide either to determine that liquefaction will *not* occur, or to conduct further evaluation based on cyclic triaxial tests When it is necessary to allow a significant safety margin for a structure, decide either to determine that liquefaction will occur, or to conduct further evaluation based on cyclic triaxial tests
IV	Possibility of liquefaction occurrence is very low	Liquefaction will not occur

It should be noted that, from the soil layers with gradation in zone B_c, when a soil layer with unfavorable permeability like a clay or silt layer exists above the soil layer concerned, it is evaluated by the same method as the soil layer with gradation in zone A.

– *Step 5.* Liquefaction prediction/determination is made by referring to Table 4.17 for each soil layer based on the classification of the soil layer in Step 4. As mentioned earlier, liquefaction prediction refers to assessment of liquefaction potential obtained as a direct output of the soil classification made in Step 4 whereas liquefaction determination requires in addition an understanding of other factors such as how large a safety margin should be allowed for the structure in question; the liquefaction determination is hence not a straightforward procedure. As a standard guideline, however, liquefaction determination in accordance with the liquefaction prediction may be made by referring to Table 4.17.

When a considerable margin of safety is required for a structure, the liquefaction determination with respect to zone III in Table 4.17 should read *to conduct further evaluation with cyclic triaxial test.* In this case, if the evaluation by cyclic triaxial test is difficult due to restrictive conditions at the site, the liquefaction determination should read *to determine that liquefaction will occur* with respect to the zone III.

– *Step 6.* Overall liquefaction determination at a site consisting of soil layers is made by considering the thickness and depth of the liquefiable soil layers determined in Step 5, and permeability of soils.

Normally, whether a site is subject to liquefaction or not during an earthquake is judged by such phenomena as sand boil, water spouting, settlement, etc. The liquefaction determination for a site in this handbook refers to the occurrence of liquefaction involving these phenomena.

4.5.2.2 *Simplified procedure for rough evaluation of liquefaction potential*

When making an initial approximate evaluation by adopting the simplified procedure shown in Section 4.4.6 for computing earthquake induced shear stress in subsoil in Step 3, the following procedure is followed for liquefaction prediction /determination.

Initially, estimates of the earthquake induced shear stress are obtained as follows. As mentioned earlier, the regression analysis resulted in the shear stress estimate τ_{est} given by Equation (4.22) in Section 4.4.6. By considering the variation in the data in the regression analysis, the upper and lower bound estimates of the earthquake induced shear stress are obtained by 1.35 τ_{est} and 0.65 τ_{est}, respectively. Next, the upper and lower bound estimates of the equivalent acceleration are obtained by substituting into τ_{max} of Equation (4.24). Finally, the Steps 4 and 5 are followed for liquefaction prediction/determination; if it is determined that liquefaction will not occur by using the upper bound estimate of the equivalent acceleration or if it is determined that liquefaction will occur by using the lower bound estimate of the equivalent acceleration, this liquefaction determination through the simplified procedure is considered the final result of the liquefaction determination. Otherwise, the simplified procedure is considered insufficient to produce the final results and further evaluation based on the earthquake response analysis must be made by returning to Step 3.

4.5.2.3 *Commentaries on overall liquefaction determination*

The procedures mentioned above describe the liquefaction prediction/determination based on the grain size distribution and SPT *N*-values. Using these procedures, overall liquefaction determination for a site as designated in Step 6 may require a high degree of engineering judgement.

In the overall liquefaction determination for a site consisting of soil layers, it is preferable to refer to past examples of damage during earthquakes. When small light wooden structures are concerned, Figure A2.1 in Appendix 2 is a good reference. In this figure, when the ground water level for a reclaimed land area is at a depth of 2 m, for example, layers shallower than this water table are considered nonliquefiable and hence, if for example the maximum acceleration is 200 Gal, liquefaction at a site will not occur when the thickness of the liquefiable layer is less than 2 m.

It is important in the overall liquefaction determination that the cause and mode of damage be thoroughly understood. A site located at the back of a large gravity bulkhead at a water depth of about 10 m, for example, may be judged nonliquefiable based on the overall liquefaction determination for the site if most of the soil layers are judged nonliquefiable except for one or two points of liquefaction determined from the SPT *N*-values obtained at 1 m depths. A site or a subsoil profile which significantly contributes to the stability of a structure, such as the soil directly below a gravity bulkhead, may be judged liquefiable based on the overall liquefaction determination for the site even if the liquefiable layer thickness is only 1 m. There is one study for evaluating effects of liquefaction on a sand mound type breakwater by effective stress analysis as shown in Technical note 1 [59].

4.5.3 *Prediction/determination based on cyclic triaxial test results*

4.5.3.1 *Prediction procedure*

When the liquefaction potential cannot be determined from the gradation and SPT *N*-value, liquefaction prediction is made by performing cyclic triaxial tests on undisturbed samples. This method evaluates liquefaction potential by comparing the liquefaction strength of soil and the cyclic shear stress generated in the soil layers during earthquakes. This is fundamentally based on the procedure proposed by Seed et al. [60].

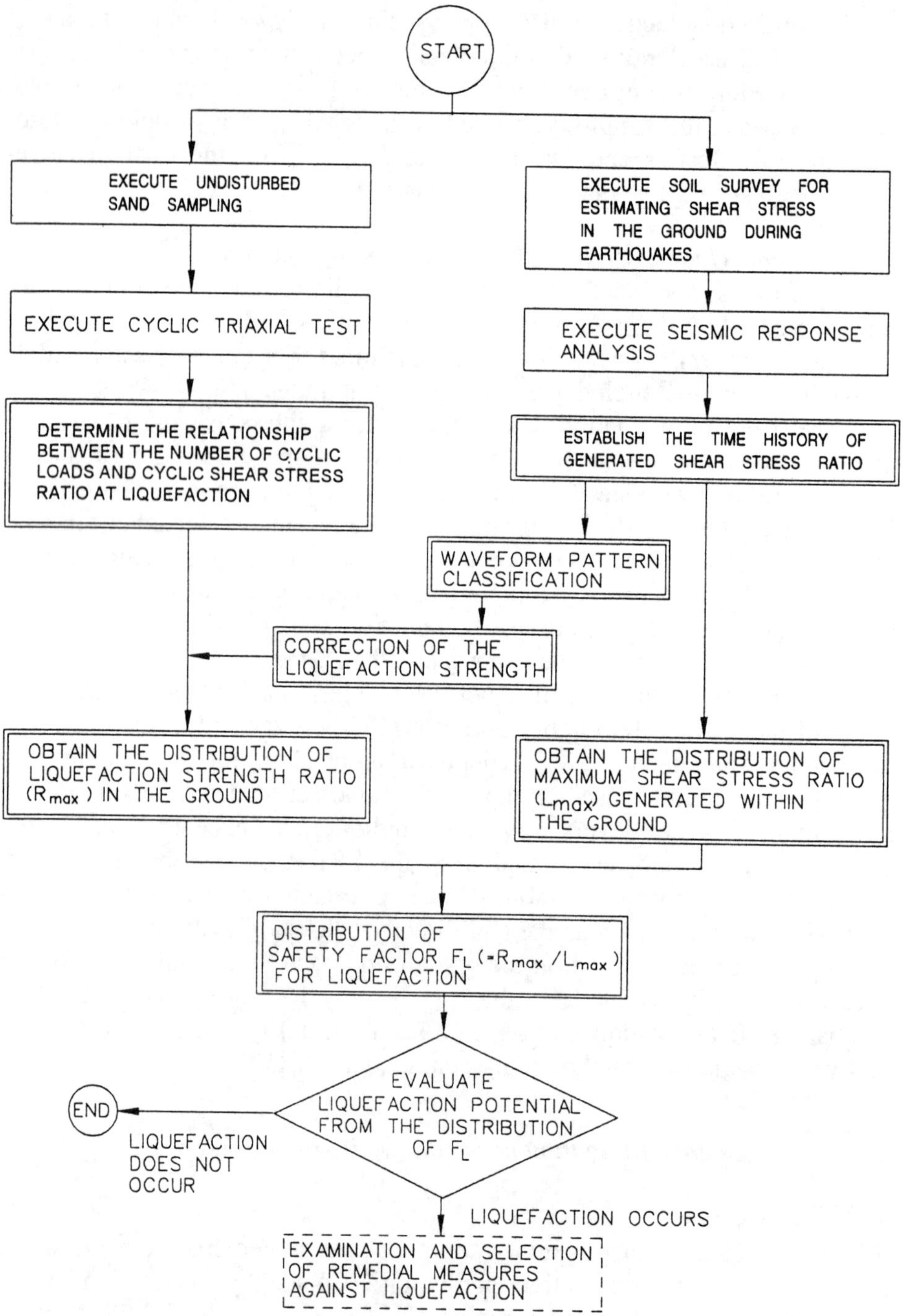

Figure 4.43. Procedure for liquefaction prediction based on the cyclic triaxial method.

The shear stresses generated by an earthquake fluctuate with time in a complex way. In evaluation of liquefaction potential, the shear stress generated by earthquakes is represented by a maximum value of shear stress (maximum shear stress τ_{dmax}) generated in each layer. Liquefaction strength τ_{lmax} of the soil is evaluated based on cyclic triaxial test results. The evaluation of liquefaction potential is made by comparing the two. The quantities τ_{dmax} and τ_{lmax} are made dimensionless by normalization with respect to effective vertical pressure σ'_v. These dimensionless quantities are respectively called maximum shear stress ratio L_{max} and liquefaction strength ratio R_{max}, and are defined as $L_{max} = \tau_{dmax}/\sigma'_v$ and $R_{max} = \tau_{lmax}/\sigma'_v$.

Liquefaction prediction is made by using the following procedure as shown in Figure 4.43.

1. Time history of the shear stress generated within the soil layer is computed from an earthquake response analysis;
2. The maximum shear stress ratio L_{max} at the depth of a layer is determined from the effective vertical stress and time history computed in 1.
3. The cyclic triaxial test is performed on undisturbed samples and the relationship between the cyclic shear stress ratio and the number of load cycles is obtained. The cyclic shear stress ratio τ_l/σ'_v is obtained by dividing cyclic shear stress τ_l necessary for generation of liquefaction at the number of load cycles N_l by the initial effective confining pressure σ'_v of the test.
4. From the relationship between the cyclic shear stress ratio and the number of load cycles obtained in 3 , the liquefaction strength ratio R_{max} is evaluated by making the corrections discussed in Section 4.5.3.3.
5. The safety factor against liquefaction F_L is calculated using Equation (4.25) from L_{max} and R_{max} obtained in 2 and 4 as

$$F_L = \frac{R_{max}}{L_{max}} \tag{4.25}$$

4.5.3.2 *Maximum shear stress ratio*

There are two methods to evaluate the maximum shear stress ratio L_{max} in the subsoil layer; one is a simplified method in which an approximate value is obtained based on the maximum acceleration of the ground surface, and the other is an analytical method in which L_{max} is computed by performing a seismic response analysis. In the liquefaction prediction/determination using the cyclic triaxial test, the seismic response analysis (refer to Section 4.4) must be performed to obtain L_{max}.

Based on the time histories of shear stresses computed at an arbitrary depth, L_{max} can be obtained using Equation (4.26) considering the maximum value of shear stress τ_{dmax} during an earthquake.

$$L_{max} = \frac{\tau_{d\,max}}{\sigma'_v} \tag{4.26}$$

L_{max} is calculated for each representative depth of the layer in order to evaluate the liquefaction potential, and its distribution is plotted with depth. L_{max} does not always appear at the same time for each depth investigated.

4.5.3.3 *Liquefaction strength ratio*

a) *Representation and correction of cyclic triaxial test results.* The liquefaction strength ratio R_{max} of the layer is obtained from a cyclic triaxial test. The results of the cyclic triaxial tests provide a relationship between the cyclic shear stress ratio and the number of load cycles as shown in Figure 4.44.

There are two methods used to define the number of load cycles causing liquefaction in the cyclic triaxial test; one focuses on excess pore water pressure and the other focuses on axial strain. Regarding the intrinsic phenomenon of liquefaction, the method which focuses on excess pore water pressure to define initial liquefaction is preferable. For dense sand, however, this method has the disadvantage of not being precise, and hence definition with the axial strain amplitude is considered. When strain amplitude is used for defining liquefaction, a double amplitude of 5% strain is often selected as a reference for liquefaction. It is not established, however, what strain amplitude percentage should be used as the reference for liquefaction. The state of 5% strain generally corresponds to initial liquefaction except for dense sand

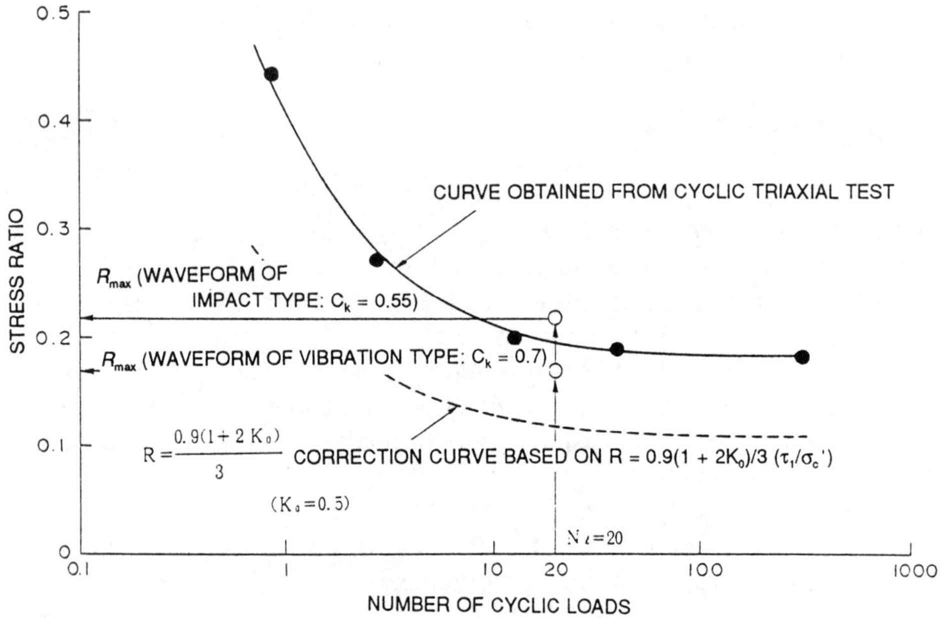

Figure 4.44. Relationship of $\tau_l / \sigma'_c - N_l$.

with relative density of $D_r \geq 70\%$. For these dense sands, 10% strain is often used for defining liquefaction.

The cyclic triaxial test is performed by simplifying actual field conditions. While the actual soil layer is in the K_o state, the cyclic triaxial test is performed in the isotropic stress state. While the principal stress direction for the actual earthquake changes continuously, the principal stress direction cannot change for the triaxial cyclic test. While the actual earthquake wave is irregular, a regular waveform such as a sine wave is applied in the cyclic triaxial test. Therefore, the cyclic triaxial test does not exactly simulate the in-situ liquefaction condition. In order to obtain the in-situ liquefaction strength, it is necessary to correct the cyclic triaxial test results by considering the differences between those conditions.

b) *Correction for in-situ stress conditions.* The cyclic triaxial test is performed in an isotropic stress condition. The subsoil, however, is consolidated in the K_o condition. When $(1 + 2K_o)/3$ is multiplied by the cyclic shear stress ratio obtained from the results of the liquefaction test under isotropic stress conditions, a liquefaction strength corresponding to the stress condition in the subsoil profile is determined [62, 63]. Therefore, in order to correct for the in-situ stress conditions, τ_l/σ'_c should be multiplied by $(1 + 2K_o)/3$.

c) *Correction for multidirectional earthquake motion.* In a cyclic triaxial test, shear stress is applied in one direction alone. Actual earthquake motion constantly changes the direction of shear. The strength is reduced by multidirectional shearing. The reduction factor is found to be about 0.9 by laboratory studies [32]. Therefore, by considering the reduction in strength with respect to multidirectional shearing, τ_l/σ'_c should be multiplied by 0.9.

d) *Correction for irregularity in the waveform.* In the dynamic analysis of the shear stress generated in the soil, irregular waveforms representing earthquake motions are used, and a time history of shear stress is computed for each depth. Applying the shear stress thus computed to the undisturbed samples obtained from each depth is, however, very complicated and difficult in the actual cyclic test. Predicting the actual earthquake waveform in advance is also very difficult at the present time.

The torsion shear test results by Ishihara [9, 64] indicated that the irregularity of the waveform does not have much influence on pore water pressure generation if the effective dynamic shear stress is the same. For practical use, the shear stress-waveform can be classified, for evaluation of liquefaction potential, into impact and vibration types as shown in Figure 4.45 and Table 4.18. This means that even complex earthquake motion can be simplified into two vibrational patterns regardless of the number of waves and duration of motion in the evaluation of liquefaction potential. When the stress generated by earthquake motion is represented by the maximum shear stress $\tau_{d\max}$ (or the maximum shear stress ratio $\tau_{d\max}/\sigma'_v$) and vibrational patterns

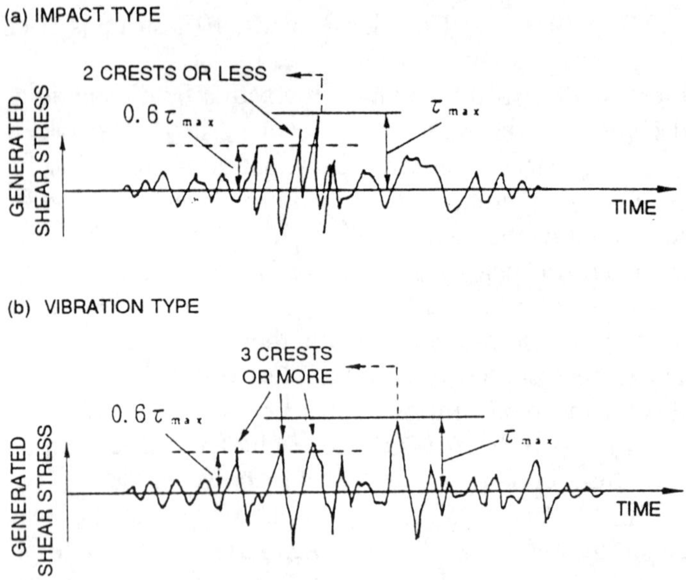

Figure 4.45. Classification of waveform patterns by Ishihara [9, 64].

Table 4.18. Classification of waveforms and conversion coefficient.

Waveform pattern	Definition	Conversion coefficient C_k
Impact type	The case where only 2 or fewer crests having an amplitude of over 60% of τ_{max} precede τ_{max}	0.55
Vibration type	The case where 3 or more crests having an amplitude of over 60% of τ_{max} precede τ_{max}	0.7

are known, it is possible to evaluate the liquefaction potential by performing a torsion shear test under those conditions.

It is, however, not easy to perform the torsion shear test using an irregular waveform as a general practice. Since the cyclic triaxial test using a regular waveform is used most widely, it is useful if an equivalent liquefaction strength can be mutually converted between the two tests. For the cyclic triaxial test, the cyclic shear stress ratio at an arbitrary number of load cycles can be used as the index of liquefaction strength. For convenience, however, the cyclic shear stress ratio $(\tau_l/\sigma'_c)_{N_l=20}$ at the number of load cycles for 20 waves is used. When the cyclic shear stress ratio τ_{dmax}/σ'_c necessary to cause liquefaction is obtained for both impact and vibration type waveforms, and the cyclic shear stress ratio $(\tau_l/\sigma'_c)_{N_l=20}$ is obtained from the same undisturbed samples, the conversion coefficient C_k is obtained as shown in Table 4.18. By using the conversion coefficient in Table 4.18, correction with re-

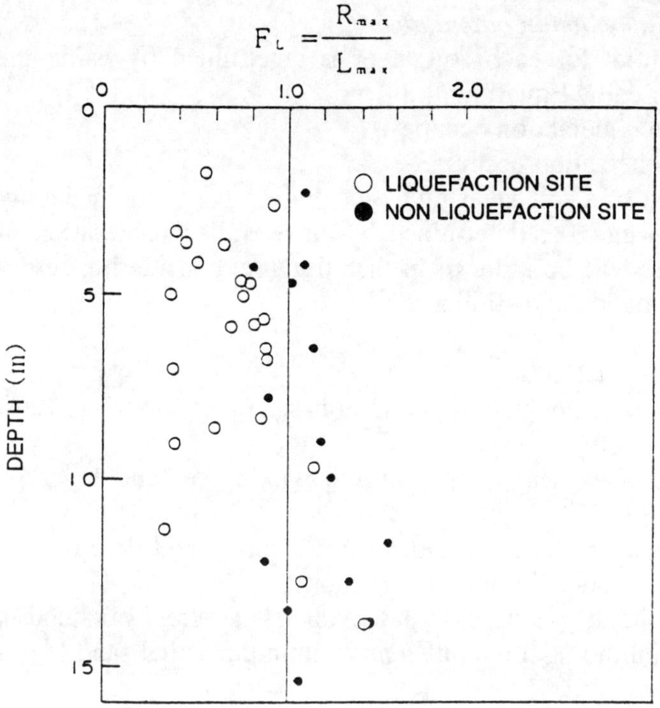

Figure 4.46. Value of F_L.

spect to irregular waves is possible. The in-situ liquefaction strength for the impact type earthquake motion becomes 1/0.55 times $(\tau_l/\sigma'_c)_{N_l=20}$ and the liquefaction strength for the vibration type earthquake motion becomes 1/0.7 times $(\tau_l/\sigma'_c)_{N_l=20}$. Even if the same earthquake waveform is used as an input motion for response analysis, the appearance of either impact or vibration type depends on the layer composition. The waveform pattern should be determined following the result of seismic response analysis for each layer.

On the basis of the corrections noted in a to d, the in-situ liquefaction strength ratio R_{max} which is used in the evaluation of liquefaction potential is given by the following equation:

$$R_{max} = \frac{0.9}{C_k} \frac{(1+2K_o)}{3} \left(\frac{\tau_l}{\sigma'_c} \right)_{N_l=20} \tag{4.27}$$

The coefficient of earth pressure at rest K_o can ordinarily be considered as 0.5 when a horizontal layer is assumed. The method for calculating R_{max} from the cyclic triaxial test results is shown schematically in Figure 4.44. If L_{max} and R_{max} are calculated respectively using Equations (4.26) and (4.27), the F_L value of each depth can be obtained by substituting into Equation (4.25). Figure 4.46 shows examples of F_L values of the layer computed for the port and harbour areas [65].

4.5.3.4 *Prediction of liquefaction potential*

The liquefaction potential for each soil layer is determined by using the safety factor F_L expressed by Equation (4.25):

– When $F_L \geq 1.0$, no liquefaction occurs;

– When $F_L < 1.0$, liquefaction occurs.

If only some portion of a soil layer undergoes liquefaction, the necessity for remedial measures must be determined by an overall consideration of whether or not damage will be induced in the structure, considering settlement, horizontal deformation and sliding failure.

4.5.3.5 *Other prediction methods*

The prediction method for liquefaction potential based on cyclic triaxial test results basically consists of:

1. Computing the cyclic shear stress induced within the sand layer by seismic response analysis;
2. Performing liquefaction tests on undisturbed samples and determining the liquefaction strength under simplified conditions;
3. Correcting the liquefaction strength and cyclic shear stress obtained in 1 and 2 to convert them into in-situ equivalents within the actual sand layers during earthquakes; and
4. Performing liquefaction prediction by comparing the converted in-situ liquefaction strength and in-situ cyclic shear stress.

The method noted in this handbook discussed earlier is based on the method proposed by Ishihara [9]. In addition, there are the method proposed by Seed et al. [60, 66], the method proposed by Iwasaki et al. [67], and the method based on cumulative damage theory [30, 65]. The main differences among these methods are in the concepts related to corrections. These methods are discussed in summary below.

a) *Method by Seed et al.*

– Cyclic shear stress during earthquakes. The waveform of cyclic shear stress generated within the layer during earthquakes is obtained from seismic response analysis. From the maximum value $\tau_{d\max}$ of the computed cyclic shear stress, the cyclic shear stress τ_e of the uniform amplitude is defined as follows:

$$\tau_e = 0.65\ \tau_{d\max} \tag{4.28}$$

– Liquefaction strength in the layer. The relationship between cyclic shear stress ratio τ_l/σ'_c and a number of load cycles N_l as shown in Figure 4.44 is drawn from cyclic triaxial test results. The correction coefficient C_r ($C_r = 0.57$ when the coefficient of earth pressure at rest $K_o = 0.4$; $C_r = 0.9$ to 1.0 when $K_o = 1.0$) is introduced in order to consider the in-situ stress condition. The in-situ liquefaction strength τ_r corresponding to τ_e is obtained using Equation (4.29).

$$\tau_r = C_r \sigma'_v \left(\frac{\tau_l}{\sigma'_c}\right)_{N=N_e} \tag{4.29}$$

Because the (τ_l/σ'_c) in Equation (4.29) is a function of the number of load cycles, an equivalent number of load cycles N_e is newly defined as the number of load cycles when the external earthquake force was represented by shear stress τ_e of uniform amplitude. The (τ_l/σ'_c) corresponding to N_e is determined from the relationship between the (τ_l/σ'_c) and N_l. N_e has been provided with a practical correlation diagram for the magnitude of the earthquake. For example, when the magnitude M is 7, it becomes $N_e = 10$ and when M is 8, $N_e = 30$. The value of (τ_l/σ'_c) corresponding to N_e is obtained from the relationship of $(\tau_l/\sigma'_c) - N_l$, then substituted into Equation (4.29). When M increases, N_e also increases, and as shown in Figure 4.44, the liquefaction strength τ_r decreases.

– Prediction of liquefaction potential. The τ_e and τ_r respectively obtained from Equations (4.28) and (4.29) are compared with each other. If the following condition

$$\tau_e > \tau_r \tag{4.30}$$

is satisfied, then it is considered that there is a large potential for liquefaction.

b) *Method by Iwasaki et al.*

– Cyclic shear stress ratio during earthquakes. Maximum shear stress ratio L_{max} is obtained using Equation (4.26).

– Liquefaction strength ratio of the layer. The in-situ liquefaction strength ratio R_{max} is calculated as follows:

$$R_{max} = c_1 c_2 c_3 c_4 c_5 \left(\frac{\tau_l}{\sigma'_c}\right) \tag{4.31}$$

where: c_1 = the correction coefficient related to isotropic state of stress $(= (1 + 2K_o)/3)$, c_2 = the correction coefficient related to irregularity of earthquake waveform (= 1.67 as an average of 1/0.55 to 1/0.70), c_3 and c_4 = the correction coefficients related to disturbance and densification of the sample, generally $c_3 \times c_4 = 1$, c_5 = the correction coefficient related to multi-directional motion of earthquake, being about 0.9.

From these correction coefficients,

$$R_{max} = \left(\frac{\tau_l}{\sigma'_c}\right)_{N_l=20} \tag{4.32}$$

is obtained as $c_1 c_2 c_3 c_4 c_5 = 1$. The value of (τ_l/σ'_c) in this equation corresponds to $N_l = 20$.

– Prediction of liquefaction potential. The safety factor F_L with respect to liquefaction is calculated using Equation (4.25) from L_{max} and R_{max} obtained respectively using Equations (4.31) and (4.32).

c) *Method based on cumulative damage theory.*

– Cyclic shear stress ratio during earthquakes. Similar to the method by Seed et al. [60, 66], the maximum value τ_{dmax} of cyclic shear stress obtained from the seismic response analysis is converted into a cyclic shear stress of uniform amplitude. The shear stress ratio L is calculated as follows:

$$L = \left(\frac{\tau_e}{\sigma'_v}\right) = 0.65\left(\frac{\tau_{d\max}}{\sigma'_v}\right) = 0.65\, L_{\max} \tag{4.33}$$

– Liquefaction strength ratio of the layer. The liquefaction strength ratio R is obtained with Equation (4.34) based on the same correction method noted in Section 4.5.3.3 to obtain the relationship of R and N_l by

$$R = \frac{0.9\,(1+2K_o)}{3}\left(\frac{\tau_l}{\sigma'_c}\right) \tag{4.34}$$

The (τ_l/σ'_c) in Equation (4.34) uses a value corresponding to the equivalent number of load cycles N_{eq} which is determined by the method noted in Appendix 5 based on cumulative damage theory.

– Prediction of liquefaction potential. The value of R corresponding to N_{eq} is determined from the $R - N_l$ relationship, designated as $R(N_{eq})$, defined as follows:

$$R(N_{eq}) = \frac{0.9\,(1+2K_o)}{3}\left(\frac{\tau_l}{\sigma'_c}\right)_{N=N_{eq}} \tag{4.35}$$

The safety factor F_L for liquefaction is obtained using Equations (4.33) and (4.35) as follows:

$$F_L = \frac{R(N_{eq})}{L} \tag{4.36}$$

d) *Comparison of the correction procedures.*

The methods noted above can all be expressed in the form of Equation (4.37).

$$F_L = a\,\frac{(\tau_l/\sigma'_c)_{N=b}}{L_{\max}} \tag{4.37}$$

where: a and b are summarized in Table 4.19 according to the correction method. In the method by Ishihara et al. and the method by Iwasaki et al., the number of load cycles of 20 is used as the value of b. In the method by

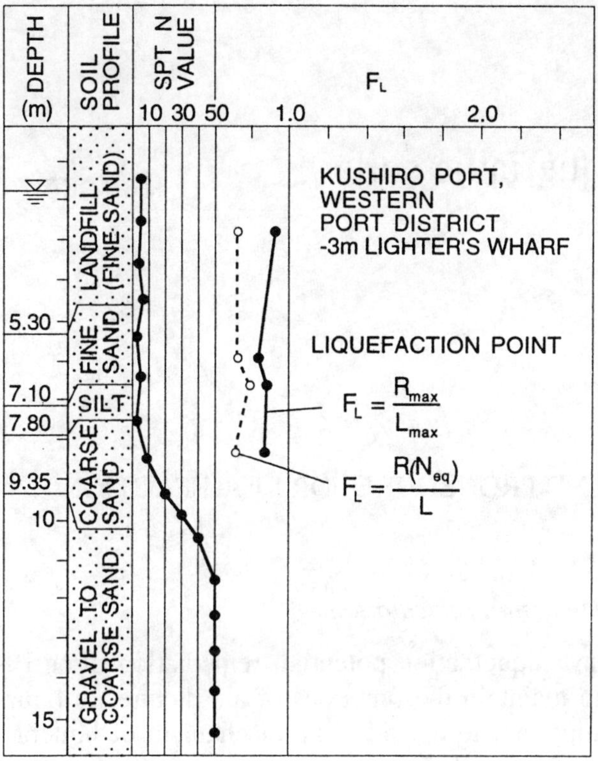

Figure 4.47. Depth distribution of liquefaction safety factor.

Table 4.19. Comparison of correction factor.

	Correction factor	
	a	b
Method proposed by Seed [66]	$C_r / 0.65$	N_e
Method proposed by Ishihara [9]	$0.9 / C_k \times (1 + 2 K_o) / 3$	20
Method proposed by Iwasaki et al. [67]	$0.9 / 0.63 \times (1 + 2 K_o) / 3$	20
Cumulative damage theory by Annaki & Lee [30]	$0.9 \times (1 + 2 K_o) / 3$	N_{eq}

Seed et al. and the method based on cumulative damage theory, the value of b is not fixed but rather depends on the magnitude and waveform of shear stress generated by earthquakes.

Figure 4.47 shows the results of case studies of the liquefaction potential proposed by Ishihara et al.'s method and the method based on cumulative damage theory. In the latter result, the F_L value becomes slightly less than in the former. The method based on cumulative damage theory is influenced by the $(\tau_l/\sigma'_c) - N_l$ relationship at a small number of load cycles at about $N_l = 1$, and hence it is necessary to obtain very accurate laboratory results at about $N_l = 1$ in the cyclic triaxial test.

CHAPTER 5

Remediation of liquefiable soils

5.1 BASIC STRATEGY AND PROCEDURE FOR LIQUEFACTION REMEDIATION

5.1.1 *Basic strategy for liquefaction remediation*

When a site is found to have liquefaction potential, remediation often becomes necessary in order to maintain the functions of a structure. In liquefaction remediation, it is important to consider the mechanism of liquefaction as well as the cause and mode of damage to structures.

Liquefaction remediation measures can be classified into the following two categories:

1. Soil improvement so that liquefaction will not occur (liquefaction remediation by soil improvement);

2. Structural design to minimize damage even when liquefaction occurs (liquefaction remediation incorporated into structural design).

In practice, a combination of these two measures is also often adopted.

The first approach, adopting soil improvement, can be classified further into two categories based on the mechanism of liquefaction explained in Section 2.1.

a) Soil Improvement so that the soil skeleton will not collapse under earthquake loading;

b) Achievement of speedy dissipation of excess pore water pressure.

It is possible to combine these two in practice.

Similarly, the second approach, based on structural design, can be classified further into two categories:

c) Reinforcement of the structure and maintenance of structural stability with strength and rigidity;

d) Relief of external force (unbalanced force caused by gravity, external force generated by excessive displacement of liquefied soil) by attaching flexible joints or modifying a structure.

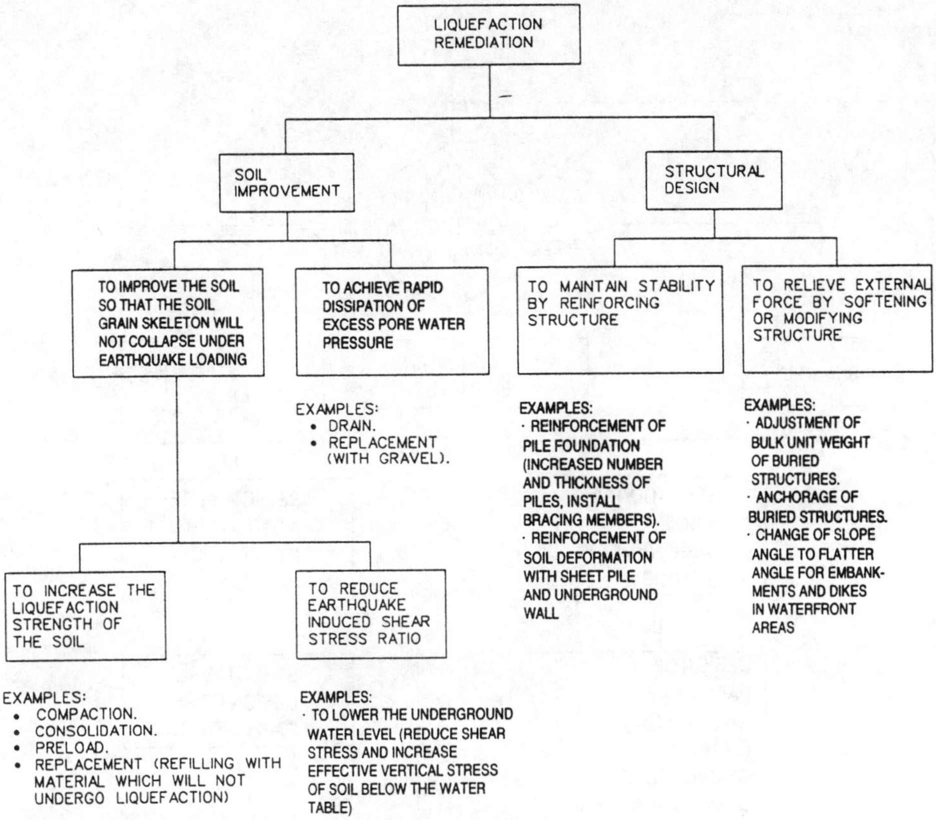

Figure 5.1. Basic strategy for liquefaction remediation.

It is also possible to combine these two in practice.

The basic strategies for liquefaction remediation are summarized in Figure 5.1. Specifics with respect to various remediation techniques are explained in Section 5.2.

5.1.2 *Standard procedure for liquefaction remediation*

A standard procedure for liquefaction remediation is shown in Figure 5.2. As shown in this figure, it is general practice to design the soil improvement area to protect against liquefaction after selecting a remediation method. This handbook is written considering this sequence of design procedure.

In practice, it often becomes necessary to re-select a remediation method due to construction site restrictions for implementing soil improvement as indicated by the broken line in the figure. It also often becomes necessary to combine a number of engineering methods to achieve efficiency. Consequently, the procedure shown in Figure 5.2 should not be complied with me-

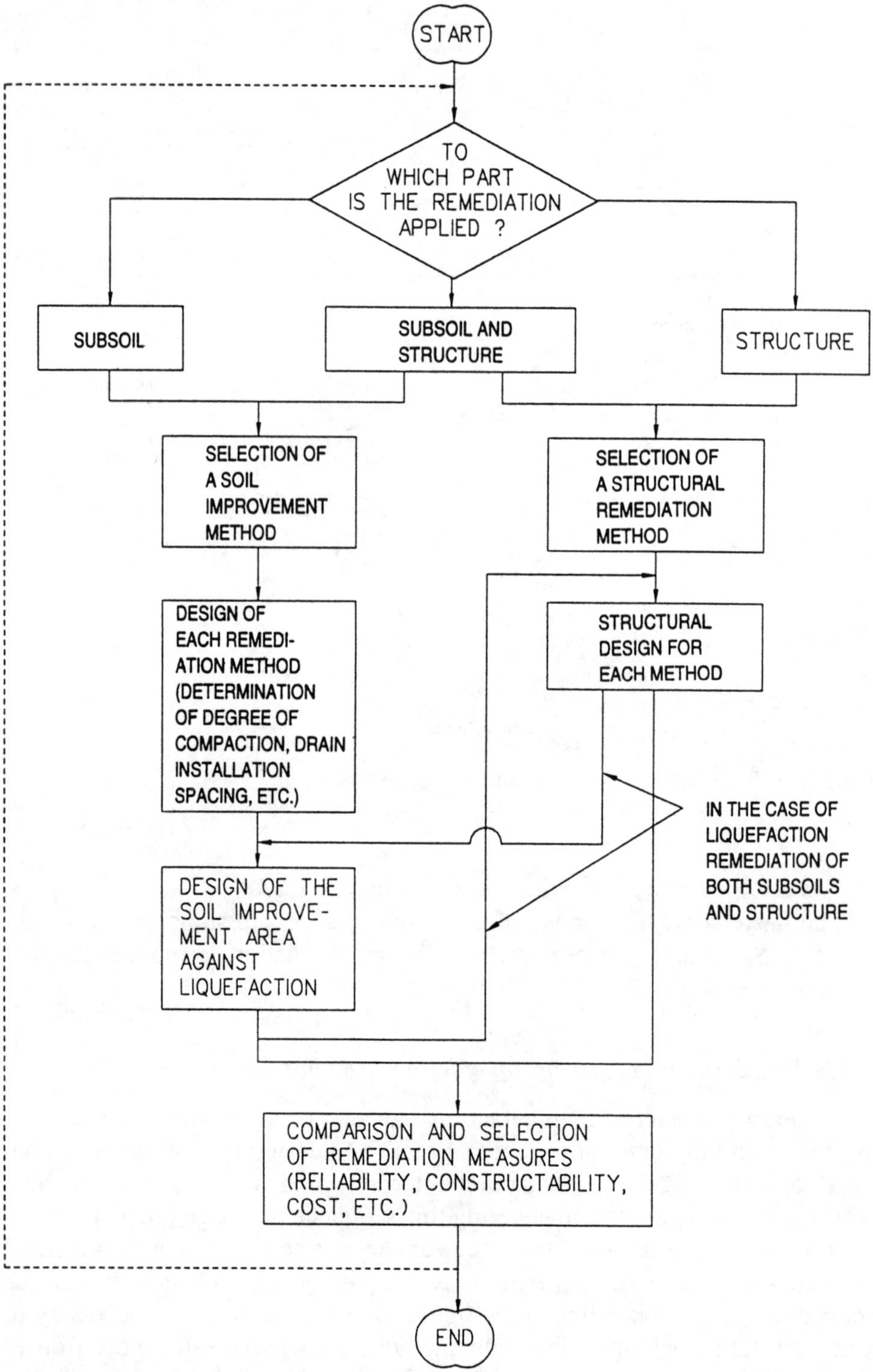

Figure 5.2. Standard procedure for liquefaction remediation.

chanically. In practice, a remediation method may be chosen efficiently with rough estimates of the soil improvement area necessary for each remediation method and a rough grasp of restrictions anticipated at the construction site, thereby screening out the remediation methods which are clearly unsuitable.

A combination of two or more remediation methods is very effective in many cases. An example is the combination of a low noise/low vibration method, such as the drain method and, a remediation method which assures ductile resistance of the improved soil, such as compaction around the improved area with the drain method. This combination is already being used in liquefaction remediation for existing structures in reclaimed land. A procedure for determining which combination of methods is optimal has not yet been established. Engineers' creative minds are required for each site specific condition.

5.2 OUTLINE OF REMEDIAL MEASURES AGAINST LIQUEFACTION [1,2]

Many methods have been developed for liquefaction remediation based on Section 5.1. The representative remediation measures are shown in Table 5.1. An outline of each remediation measure is discussed below.

5.2.1 *Compaction*

Methods based on compaction are used to increase liquefaction resistance of soil by densifying sandy soil with vibration and impact. Compaction methods have been used extensively, and there are many case histories. The sand compaction pile method [3], vibration rod method [4], vibroflotation method [5], dynamic compaction method [6] are representative examples. These methods usually increase earth pressure and cause vibration and noise during the compaction and thus, the influence of compaction on the surrounding soil and adjacent structures during its installation should be considered. In addition, when the fines content of the soil is large, compaction is difficult.

Certain modifications have recently been made for a vibroflotation method to reduce these negative effects (refer to Appendix 4).

5.2.2 *Pore water pressure dissipation*

Liquefaction remediation based on pore water pressure dissipation enhances dissipation of excess pore water pressure generated in sandy soil during earthquakes and reduces excess pore water pressure by installing permeable drain piles. The gravel drain and plastic drain methods [7] are representative examples. The drain method causes less vibration and noise than the com-

Table 5.1. Remedial measures against liquefaction.

Principle	Methods designation	Depth	Summary	Influence on the surrounding areas	Remarks
Compaction	Sand compaction pile method	About GL-35 m	Insert steel pipe casing underground, install sand compaction pile by forcing out sand during extraction of casing, and compact the sand pile and the natural soil in a horizontal direction simultaneously	This method produces high levels of noise and vibration. The extent of noise and vibration differs according to the type of construction equipment used	Compaction efficiency is high for soil with fines content of less than 25 to 30%. SPT *N*-value increases to about 25 to 30
	Vibro-rod method	About GL-20 m	Compaction of soil is conducted using vibratory penetration of a rod and filling with additional sand from the ground surface	This method produces noise and vibration that are slightly less than those produced by the sand compaction pile method	Compaction efficiency is high with respect to soil having fines content of less than 15 to 20%. SPT *N*-value increases to about 15 to 20.
	Vibroflotation method	About GL-20 m	The surrounding soil is compacted with water sprayed from nozzle tip end and with horizontal vibration from vibrator with a built-in eccentric load. Additional sands are filled into the voids in the subsoil profile	This method produces little noise and vibration compared to other methods based on compaction	Compaction efficiency is high with respect to soil having fines content of less than 15 to 20%. SPT *N*-value increases to about 15 to 20
	Dynamic compaction method	About GL-10 m	The soil is compacted with an impact load by dropping a weight of 10 to 30 tf from a height of about 25 to 40 m above the ground	This method produces high levels of vibration and large impact	Compaction is difficult when the fines content is high
Pore water pressure dissipation	Gravel drain method	About GL-20 m	The gravel pile is installed by inserting gravel into a casing placed at prescribed position and then extracting the casing. Excess pore water pressure is dissipated through the gravel pile during earthquakes	Influence on the surrounding areas is minimal	This method is often used when compaction is difficult. If fines content of the subsoil is high and permeability is low, application of this method is difficult

Table 5.1. Continued.

Principle	Methods designation	Depth	Summary	Influence on the surrounding areas	Remarks
Pore water pressure dissipation	Attachment of drainage device for steel piles or sheet piles	—	Piles with a drainage device are inserted into the subsoil as liquefaction remedial measures	Noise and vibration depend on the installation method used	The pile prevents uplift and settlement of upper-structures
Cementation and solidification	Deep mix method	About GL-30 m	Stabilizing material such as cement is mixed and solidified in the soil. There are two approaches: one is total improvement which solidifies all the soil, the other is partial improvement which makes a solidified wall in the subsoil profile	Influence on the surrounding areas is minimal	Liquefaction remediation is possible even when there is a possibility of liquefaction in the soil under the existing structure, such as for an embankment, by installing an improvement wall at the periphery
	Premix method	—	Landfill is deposited by adding and mixing a stabilizing material such as cement in advance in the soil. Liquefaction remediation following the landfill becomes unnecessary	Water quality control is necessary since sand mixed with a stabilizing material is dropped into water	This method can reduce the construction period because liquefaction remediation can be accomplished simultaneously with landfill, and treatment can be performed in a large volume
Replacement	Replacement method	About GL-5 m	The liquefiable soil is replaced with a non-liquefiable material	Handling of the excavated soil is necessary	The construction is reliable since the liquefiable soil is replaced with non-liquefiable material
Lowering of ground water level	Deep well method	Water level lowering of about 15-20 m	The perimeter of the liquefiable soil area is surrounded with a cutoff retaining wall constructed with sheet piles, etc., and the ground water level is lowered using deep wells	Caution should be exercised with regard to consolidation and settlement of the soil and influence on the surrounding area from the lowering of the ground water level	This method can be applied to the subsoils below existing structures. The operating cost should be considered

Table 5.1. Continued.

Principle	Methods designation	Depth	Summary	Influence on the surrounding areas	Remarks
Shear strain restraint	Underground wall method	—	Liquefaction remediation can be achieved by surrounding the liquefiable soil using underground walls and restraining the shear deformation of subsoil during earthquakes	Influence on the surrounding areas is minimal	This method can also prevent the propagation of excess pore water pressure from the surrounding liquefied soil
Preload	Soil overlay	Depending on embankment width	Preload is applied by overlaying soils in order to overconsolidate the subsoil	Influence on the surrounding areas is minimal	The stress propagation in the soil should be considered since there is load dispersion
	Deep well method	Water level lowering of about 15-20 m	The ground water level of liquefiable soil is lowered temporarily and thereafter restored	Caution should be exercised with regard to consolidation and settlement of the soil and influence on the surrounding area from the lowering of the ground water level	This method can be applied to the subsoils below existing structures
Structural strengthening	Pile foundation	—	The strength of the pile increases resistance even if liquefaction occurs	Influence on the surrounding areas is minimal	—
	Sheet pile	—	Shear deformation after liquefaction is restrained by surrounding the soil with sheet piles	There is noise and vibration during sheet pile installation	This method can also prevent propagation of excess pore water pressure from the surrounding liquefied soil
	Top shaped foundation	—	Used for spread foundation by laying top-shaped concrete blocks on the soil	Influence on the surrounding areas is minimal	This method is applied to small buildings such as residential housing

paction method and hence has the advantage that the influence on the surrounding area is minimized.

The methods based on this principle require complex calculations and measurement of many parameters in establishing the pile installation spacing in the design. Since the natural soil itself is not improved, this method allows some generation of excess pore water pressure. In addition, there is a possibility that large deformations may suddenly be induced when the excess pore water pressure ratio approaches 1.0 and therefore it may lack *ductility* [8]. This method, therefore, is often used when application of other methods such as compaction are difficult. As a variation of the method, a liquefaction remediation method which attaches a water drainage device to a pile and combines the effect of excess pore water pressure dissipation with the strength of the pile itself has been developed [9]. In soils with low permeability and a high fines content, the application of the drain method is difficult.

5.2.3 *Cementation and solidification*

Liquefaction remedial measures based on solidification include the deep mix method [10] and premix method [11]. The deep mix method achieves liquefaction remediation by mixing a stabilizing material such as cement in sandy soil and solidifying the soil. This method is considered applicable even for liquefaction remediation directly below existing structures. By surrounding the soil at the periphery of the structure with an underground wall made using the deep mix method, liquefaction remediation may be achieved by the shear deformation restraining effect noted in Sections 5.2.6 and 5.2.8.

The premix method is used for constructing new reclaimed land by adding stabilizing material to the landfill soil and sand in advance and placing the treated soil into open water with a tremie chute or bottom dump barge. Liquefaction remediation after the landfill placement such as compaction becomes unnecessary and the construction period can be reduced.

Factors which require attention for cementation and solidification methods are quality control of the treated subsoil profile, mix control of the stabilizing material, and water quality control of the surrounding area.

5.2.4 *Replacement [12]*

The method which replaces soil with a material that does not undergo liquefaction is called replacement. It is apparent from Figure 4.39 in Section 4.5.2 that liquefiable soil is limited to a material of which grain size distribution is in a given range. Gravel and soil mixed with cement are non-liquefiable material and used in the replacement for liquefaction remediation.

In the implementation of this method, it is necessary to evaluate the stability of the excavated cross section during the replacement. It is also necessary to consider the handling of the excavated soil and sand.

5.2.5 *Lowering of ground water level [13]*

Liquefaction remedial measures based on lowering the ground water level achieve liquefaction remediation by making the soil unsaturated above the lowered ground water table and increasing the effective stress for soils below the ground water table. For this method, it is always necessary to maintain the lower ground water level, thereby resulting in an operating cost in maintenance for pumping from the deep well. It is therefore necessary to keep the cost of operation in mind.

5.2.6 *Shear strain restraint [14]*

In remedial measure based on shear strain restraint, a continuous underground wall installed below the ground surface restrains the shear deformation in the soil during earthquakes. This method simply surrounds the liquefiable soil with a continuous underground wall, and thus it can be used for the soil at the bottom of existing structures where liquefaction remediation is difficult with other methods.

It is, however, difficult to evaluate the shear strain restraint effect on liquefaction prevention during earthquakes.

5.2.7 *Preload [15]*

Preload works by overconsolidating the soil with a surcharge embankment and thus increasing the liquefaction resistance. This method is applicable to soil with a large fines content and can also be applied even below existing structures since the soil can be overconsolidated by lowering the ground water level.

Two factors require attention in this method; first, a detailed evaluation of soil properties is necessary in advance since the case history record of the this method is limited and a design procedure has not yet been fully established; second, a large overconsolidation ratio cannot be obtained when the existing vertical load is very large.

5.2.8 *Structural remedial measures*

The measures in Sections 5.2.1-5.2.7 implement liquefaction remediation mainly by improving the soil characteristics. Structural remedial measures, however, accomplish the same by strengthening the foundation with sheet piles [16], piles, etc., so that damage is not induced in the structure by liquefaction.

In this section, an outline of liquefaction remedial measures was given in Sections 5.2.1-5.2.8 and Table 5.1. In the next and subsequent sections, the most effective liquefaction remedial measures for the structure/subsurface conditions often found in coastal regions are explained with design/installation examples.

5.3 DESIGN/INSTALLATION OF COMPACTION METHOD

Compaction of the subsoils as a remedial measure against liquefaction is accomplished by using a steel rod or a steel casing for forcing sands or gravel into liquefiable subsoils and applying strong vibrations to these subsoils. This chapter begins with the design of the compaction degree common to various compaction methods, followed by the design/installation of the sand compaction pile method, the rod compaction method, and the vibroflotation method.

5.3.1 *Design of compaction degree*

5.3.1.1 *Determination of the SPT N-value to be achieved*

The SPT *N*-value of the subsoils can be determined according to the flow chart shown in Figure 5.3. This flow chart describes the procedure for taking into account the effect of compaction on earthquake response characteristics of the subsoil profile.

Normally, even when a wide area of soil is predicted to undergo liquefaction, the soil improvement area necessary for maintaining the stability and the function of a structure is limited to the perimeter of the structure (refer to Section 5.7). Thus, improved soil around the structure will be adjacent to unimproved soil outside of the perimeter. In this case, it is necessary to ensure that the degree of compaction for the improved soil is adequate and liquefaction of the improved soil is fully prevented. The equivalent *N*-value to be achieved in order to satisfy this condition is 16 or higher, which corresponds to relative density of 80% or higher at an effective vertical pressure at a depth less than about 20 m. The required SPT *N*-value is near the upper bound on compaction degree feasible with the compaction method.

When there is no possibility of liquefaction triggered by liquefaction of adjacent soil, such as when an impermeable underground wall prevents pore water flow into the improved subsoil profile, the equivalent *N*-value to be reached by compaction may be determined in accordance with Section 4.5. In particular, it can be determined from the boundary of areas II and III in Figure 4.41. If the fines content of the soil layers with grain size distribution in zone A is 5% or more, or if the soil has grain size distribution in zone B_c

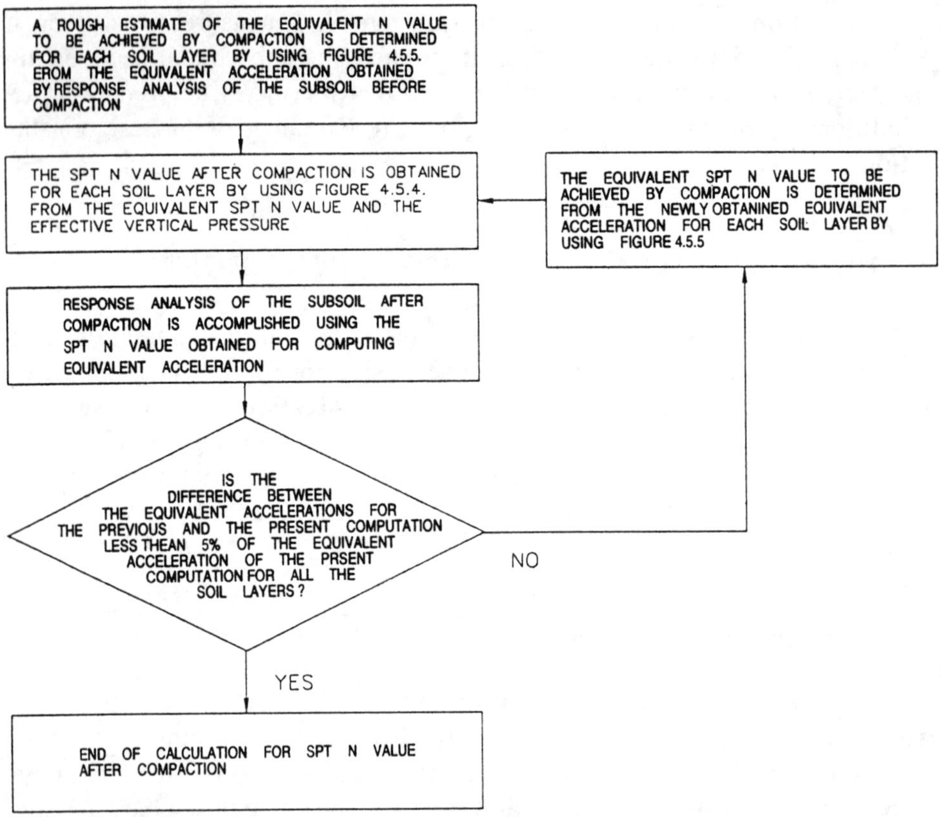

Figure 5.3. Determination procedure of SPT N-value to be achieved by compaction.

overlain by a soil layer with small permeability, the SPT N-value to be achieved by compaction is determined in accordance with the notes shown in Section 4.5.

For important structures requiring a considerable margin of safety, the equivalent N-value to be achieved by compaction is determined from the boundary of zones III and IV in Figure 4.41. As mentioned earlier, it is important to prevent liquefaction triggered by adjacent soil conditions.

There may be cases when it is difficult to achieve the required SPT N-value by compaction. Then, the liquefaction resistance necessary for the soil may be evaluated by cyclic triaxial testing as noted in Section 4.5. Alternatively, the use of other remediation methods such as preload or cementation and solidification methods can be considered.

5.3.1.2 *Designing compaction*

The degree of compaction attainable by compaction methods is affected by many factors as listed below [18]:

– Characteristics of soil (grain size distribution and fines content (i.e. the amount of fine grains smaller than 74 μm));
– Degree of saturation and level of ground water table;
– Relative density of soil before compaction;
– Initial stresses in soil layer before compaction (effective vertical stress, etc.);
– Skeleton structure and cementation of soil before compaction;
– Distance from a source of vibration during compaction;
– Properties of fill sand;
– The characteristics of compaction methods (types of installation equipment, vibration capacity of the equipment, installation method, expertise of the operating engineers, etc.).

Thus, a simple theoretical approach for the design of compaction is not easy to supply. The compaction design should be based on case histories and recent test compaction programs performed at a test site chosen with regard to the characteristics of the installation method and the characteristics of the soil. Test compaction considerably enhances reliability of the design.

5.3.2 *Sand compaction pile method*

5.3.2.1 *Types of installation*

The procedure for subsoil improvement using the installation of sand compaction piles is shown in Figure 5.4. In this compaction method, a steel casing is inserted to the design depth below the ground surface in the underlying subsoil profile with vibration induced by a vibrator attached to the head of the installation rig. Sand fill is then placed into the casing and a portion of the sand column is installed in the subsoil profile by gradually raising the casing. The casing is then vibrated and re-inserted in the subsoil profile,

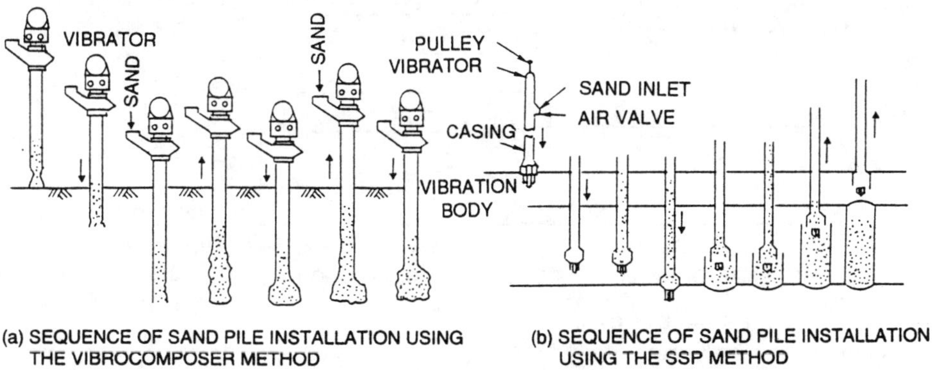

Figure 5.4. Schematic diagram for compaction of subsoil profile by installation of sand compaction piles.

compacting the sand column and increasing the diameter of the sand column below the ground surface. In this way, loose sandy subsoils are compacted by forcing sand columns below the ground surface and vibrating the natural soil around the columns. The methods included in this category are the vibrocomposer, SSP, NTK compaction, and OFB methods. In comparison to the rod compaction and vibroflotation methods to be described later, the sand compaction pile method generally produces more compaction. The effects on the environment in the vicinity of the installation site, however, are more significant, requiring careful consideration. The effects of the compaction methods are discussed in Section 5.8.

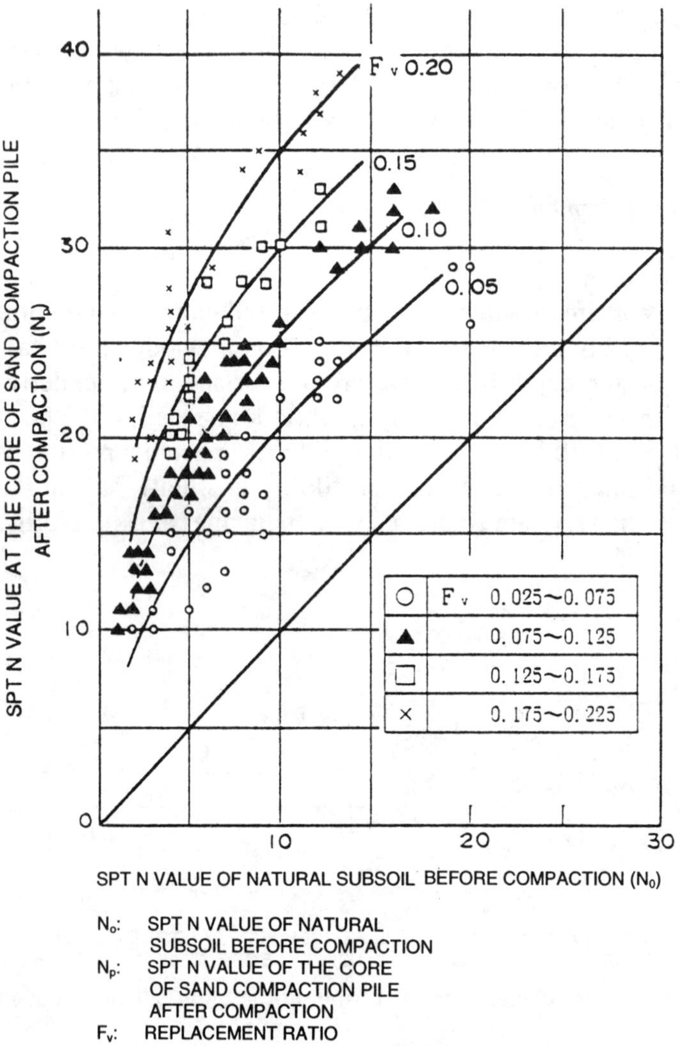

Figure 5.5. Relationship between the SPT *N*-value in the pile core after compaction and that of natural subsoils before compaction (fines content ≤ 20%) [3].

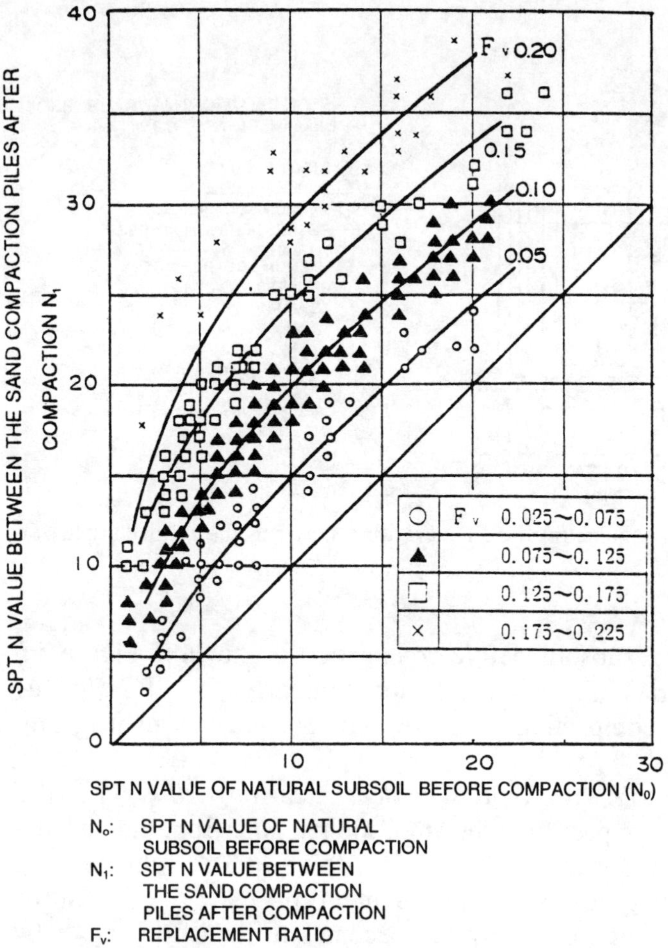

Figure 5.6. Relationship for the SPT *N*-value between piles after compaction and that of natural subsoil before compaction (fines content ≤ 20%) [3].

5.3.2.2 *Design based on installation case history data*

Soil compaction can be designed based on case history data if reliable data are available on the replacement ratio of sand piles, characteristics of the soil and the SPT *N*-value of the subsoil profile before and after compaction.

The vibrocomposer method has been used extensively and installation case history data are readily available for its design as shown below [19]. Other compaction methods may have similar case history data and, if so, a procedure similar to the one described below may be used.

– Design for sandy soil with a fines content less than 20%. For loose sandy soil with fines content less than 20%, the replacement ratio F_v of the sand compaction pile is determined from the SPT *N*-values before and after compaction based on the case history data. The relationship between the SPT *N*-value (N_p) at the sand pile core after compaction and the SPT *N*-

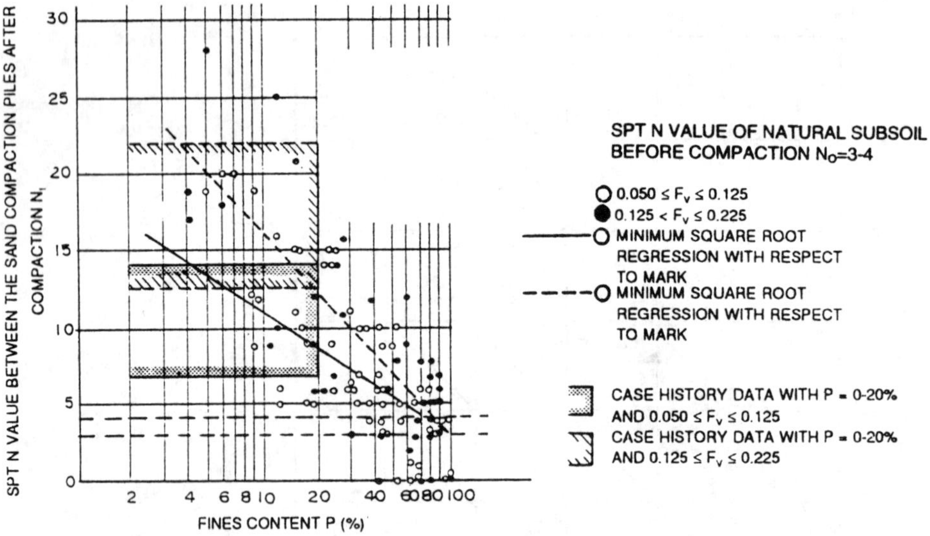

Figure 5.7. Effect of fines content on the SPT *N*-value between the sand compaction piles after compaction.

value (N_o) of the natural subsoils before compaction is shown in Figure 5.5 with the replacement ratio F_v as a parameter. Similarly, the SPT *N*-value (N_1) between the sand compaction piles after compaction is shown in Figure 5.6.

– *Design for sandy soil with a fines content exceeding 20%.* The SPT *N*-value between sand compaction piles after compaction, from the installation case history records for loose sandy soil with an SPT *N*-value ranging from 3 to 4 [3], is plotted versus the fines content in Figure 5.7. When the fines content increases, high SPT *N*-value becomes difficult to achieve by compaction, and variation in the SPT *N*-value increases. For soils with a high fines content, therefore, it is preferable to use other methods such as preload or follow a design based on a test installation shown below to improve the soil.

5.3.2.3 *Design based on test installation*

For soil containing a large amount of silt or when no reliable case history data is available for the chosen compaction method, a test installation should be planned based on the relationship between void ratio and degree of compaction with the case history data.

a) *Significance of test installation.* For soil containing a large amount of silt, or when no reliable case history data are available for the concerned compaction method, a design based on the method noted in Section 5.3.2.2 is difficult to apply. It is, therefore, preferable to design by implementing a test installation.

b) *Estimating amount of sand fill for test installation.* A rough estimate of

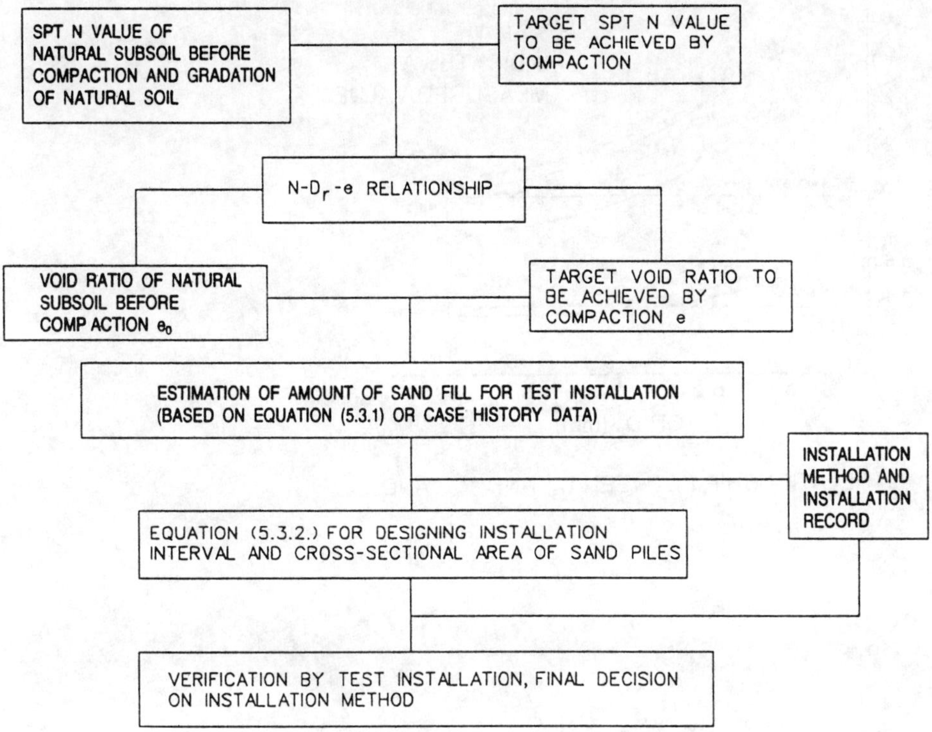

Figure 5.8. Flow chart for evaluating compaction based on the void ratio.

the amount of sand fill for a test installation using compaction methods is determined from the original SPT *N*-values and those to be achieved by compaction, provided that a reliable relationship between the SPT *N*-value and the void ratio can be established. The flow chart for this evaluation is shown in Figure 5.8.

c) *Relationship between void ratio and SPT N-value of sand.* In the design of the sand compaction pile method, the relationship between void ratio e and the SPT *N*-value is necessary. A relationship with the effective vertical pressure, relative density D_r, 60% grain size D_{60}, and coefficient of uniformity U_c $(= D_{60}/D_{10})$ of sand is often used [20] (refer to Figure 5.9).

d) *Estimation of void ratio.* When the compaction method is designed, the grain size distribution of the sand and the SPT *N*-value (N_o) of natural subsoils before compaction are at least known from the subsurface investigation. The void ratio e_o of the natural subsoil profile before compaction is estimated using Figure 5.8. Similarly, the void ratio e to be achieved by compaction is estimated from the SPT *N*-value to be achieved.

e) *A rough estimate of the amount of sand fill required for the test installation.* A rough estimate of the amount of sand fill can be determined using Equation (5.1) [21].

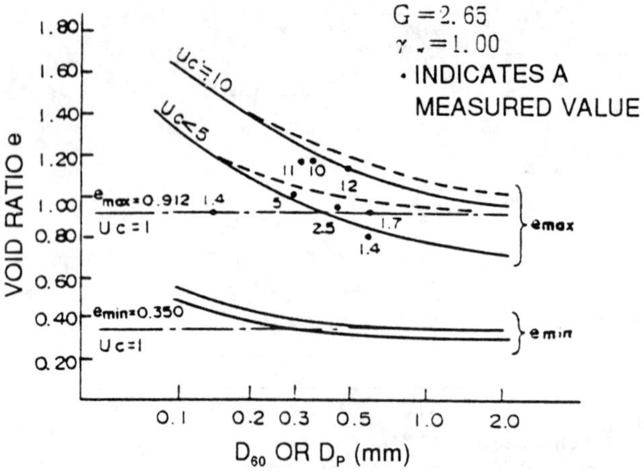

(a) RELATIONSHIP BETWEEN D_{60} AND e_{min} AND e_{max}

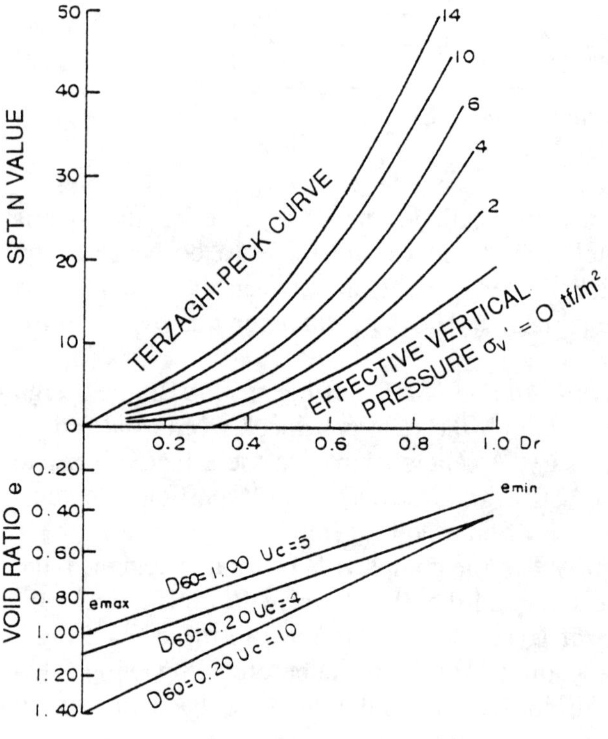

(b) RELATIONSHIP BETWEEN SPT N VALUE, D_r AND e

Figure 5.9. *N-D_r-e* relationship [20].

$$v = \frac{e_o - e}{1 + e_o} \tag{5.1}$$

where: v = amount of sand fill per unit volume of natural soil before compaction: replacement ratio (m^3/m^3), e_o = void ratio of natural soil before compaction, e = void ratio to be achieved for compacted soil between the piles.

Equation (5.1) indicates that, provided the total volume of the soil does not change before and after compaction, the volume of void is decreased by forcing fill material of volume v into 1 m^3 of subsoil profile, resulting in the average void ratio e between the piles. In this computation, only the effect of volume decrease related to the volume of sand pile is taken into account; the compaction effect caused by vibration during installation is not considered. In reality, bulging or settlement of the ground surface is induced during the compaction, resulting in volume changes. The amount of sand fill based on Equation (5.1), therefore, is merely a rough estimate. It is preferable to design the compaction method by performing a test installation.

f) *Estimation for installation spacing.* The spacing of the sand pile installation can be estimated using Equation(5.2). It is preferable to determine the spacing, however, by referring to installation records.

$$d = \alpha\sqrt{\frac{S}{v}} \tag{5.2}$$

where: d = sand pile installation spacing (m), $\alpha = 1$ for a square arrangement, $\alpha = 1.075$ for a equilateral triangular arrangement, S = amount of sand fill per unit length of sand pile (m^3/m), v = replacement ratio (m^3/m^3) of sand, based on Equation (5.1).

For a given replacement ratio v of sand, many combinations of installation spacing and diameter of sand pile exist. There is a limit, however, to the quantity of sand which can be placed in the subsoil profile with one sand pile. There is also a limit in the compaction effects on the sand between the piles. These limits vary according to the capacity of installation methods and equipment. It is necessary to establish suitable ranges for S and d based on installation records.

In the vibrocomposer method, the diameter of the sand pile is usually 70 cm ($S = 0.385$ m^3/m) [22]. When there are no obstacles in the subsoil profile, about 1.5 to 3.0 m is considered suitable for the sand pile spacing [21], and in most of actual cases, 1.8 to 2.2 m [23]. When obstacles such as tie rods exist behind sheet pile type quay walls, adjustment is usually necessary.

g) *Design for soils with high fines content.* In estimating the pile arrangement, a new method has been proposed as shown in Figure 5.7 for soils with high fines content [3]. This method basically follows the steps of the design method shown in Figure 5.9. The e-D_r relationship for soils with high fines

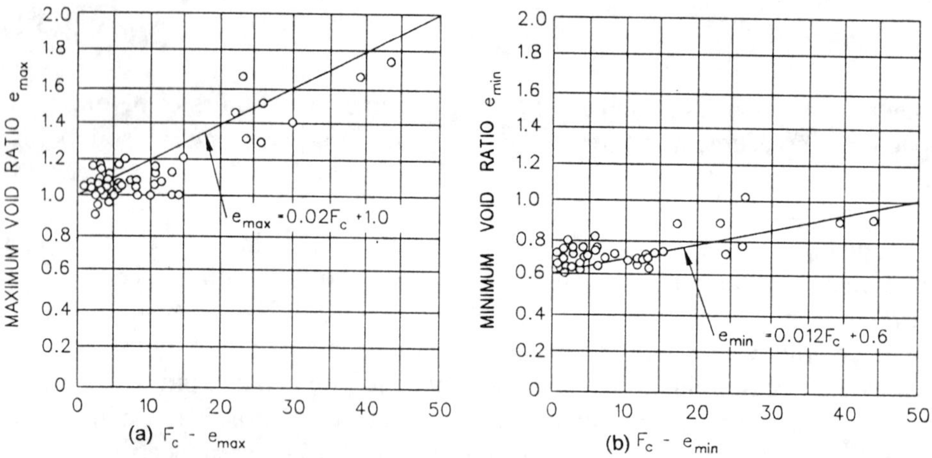

Figure 5.10. Relationship between fines content F_c and e_{max}, e_{min}.

content shown in Figure 5.9a, however, is replaced by the new relationship shown in Figure 5.10 [24]. Instead of the relationship of N-D_r-σ'_v shown in Figure 5.9b, the equation presented by Meyerhoff [25], shown below, is used.

$$D_r(\%) = 21\sqrt{\frac{N}{\sigma'_v + 0.7}} \tag{5.3}$$

where: σ'_v = the effective vertical pressure (kgf/cm^2).

If the relationship shown above is used instead of Figure 5.9, the replacement ratio corresponding to the designated SPT N-value can then be calculated. Since the N-D_r-σ'_v relationship presented by Meyerhoff does not take into account the effects of fines content, an empirical reduction coefficient is introduced by comparing the estimated SPT N-value using this method and the case history data shown in Figure 5.7. For the soils with very high fines content, the case history data is limited and therefore it is preferable to perform a test. In the test installation, it is convenient to make a plan using Figure 5.7.

5.3.2.4 *Evaluation of compacted subsoil profile*

a) The subsoils after compaction are evaluated using a standard penetration test in Japan.

In evaluating compacted subsoil, several approaches are used as follows:

– One approach evaluates the compaction effects conservatively by using only the SPT N-values between the piles, where the effect of compaction is considered minimum. In this approach, a representative value of the SPT N-values between the piles should exceed the SPT N-value to be achieved, and the representative value is defined as an average minus a standard deviation.

– The other approach uses an average of the SPT *N*-value (N_p) at the pile core and the SPT *N*-value (N_1) between the piles. The average is given as a weighted average with the weight factors given by the replacement ratio F_v as

$$N = (1 - F_v) N_1 + F_v N_p$$

– Another approach is based on random data sampling, in which a survey range is arbitrarily chosen in the subsoil profile concerned, a reference mesh is defined in the range, and the boring sites are determined by random number. The average SPT *N*-value is obtained from the boring data. This approach is effective when a uniform installation is expected because only a portion of the subsoil is used for the survey. This approach is often used to verify the replacement ratio for the sand compaction pile method in clayey subsoils.

b) The plate load bearing test may be adopted for evaluation of the improved soil. It is necessary to note, however, that evaluation of only the surface layer can be made.

c) In sandy soil with a high fines content, excess pore water pressure induced by pile installation is not dissipated immediately after the installation and hence there is a possibility to underestimate the effect of compaction. In such a case, it is preferable to perform standard penetration testing when the excess pore water pressure is completely dissipated.

d) Liquefaction criteria based on SPT *N*-values are widely adopted in Japan so that the standard penetration test is often used in Japan for evaluating the soil strength after compaction. There are many examples in the USA and Europe, however, of cone penetration tests which provide continuous data with depth [26].

e) In recent years, seismic wave surveys have been incorporated for evaluating the compaction effect [26]. When the objective of compaction is to reduce the settlement of subsoils, then the evaluation based on a minimum SPT *N*-value between the piles may be too conservative and hence a macroscopic seismic wave survey may be more appropriate.

5.3.2.5 *Remarks on design/installation*

a) *For improving clay layers intermixed with sand layers.* When the subsoil profile consists of mixed layers of sand and clay, sand compaction may be accomplished for both the sand and clay layers simultaneously.

b) *Quality of sand fill.* Well graded (i.e. having a large coefficient of uniformity) sand with low fines content with hard grains is the preferred sand fill material.

When installing sand compaction piles in soil consisting of mixed layers of sand and clay, compaction in the sand layer and drainage in the clay layer can be produced by using sand fill of high permeability, similar to that used for sand drain material.

c) *Additional amount of sand fill.* The sand fill quantity is theoretically computed using Equation (5.1). A reference exists, however, that considering the volume decrease during installation, it is necessary to add 10 to 20% or 20 to 30% [27] of the calculated volume. Verification during a field test is necessary with regard to this point.

d) *Depth of installation.* With the installation equipment presently used in Japan, it is possible to compact to a maximum depth of about 35 m. The surface layer of 1 to 2 m from the ground surface has a low overburden pressure and hence an increase in the SPT *N*-value is small. In compaction near the ground surface, it is necessary either to perform extra banking or consider another compaction method.

e) *Influence on adjacent structures.* When there are structures adjacent to the compaction installation site, careful evaluation of the existing case history record is necessary because there is a possibility of a negative influence on the adjacent structures during subsoil compaction. Some of the points to be considered are 1) changing the installation method to that with a smaller influence, 2) changing to the drain or other methods for liquefaction remediation, and 3) installing sand piles from the structure perimeter outward.

f) *Monitoring and control of installation.* During installation, it is necessary to monitor and control the quantity of sand fill, the position of the tip of the casing for sand piles, the level of sand fill in the casing, and the position of sand piles over the construction site. It is important to achieve uniformity of compaction and to keep a careful record of the installation. When the natural subsoil profile consists of sand with a high fines content, monitor the vertical displacement of the ground surface.

5.3.3 *Rod compaction method*

5.3.3.1 *Types of installation*

The principle of the rod compaction method is to drive a rod-shaped or pile-shaped vibrating body into the soil and thereby cause sand to undergo local and temporary liquefaction during vibration, and densify sand by placing new sand along the rod or pile from the ground surface. The methods classified in this category include the vibro-rod method using a steel pipe, the soil-vibro-stabilizing method (refer to Figure 5.11), the KF compaction method using a H-shaped steel beam and the cross shaped vibro method using a steel rod with thorn-shaped protrusions.

The rod compaction method (RCP) and sand compaction pile method

NAME	VIBRO ROD (V·R) METHOD	SOIL-VIBRO-STABILIZING (S·V·S) METHOD
MAXIMUM DEPTH OF COMPACTION	−15m	−18m
VIBRATION DIRECTION	VERTICAL	VERTICAL (AT THE ROD) AND HORIZONTAL (AT THE TIP END)
CRAWLER CRANE	35 TF LOADING	40 TF LOADING
VIBRO	75kW	90kW
ROD	DOUBLE CYLINDER TYPE ϕ 318.5 mm	ϕ 318.5mm
ROD TIP SHAPE	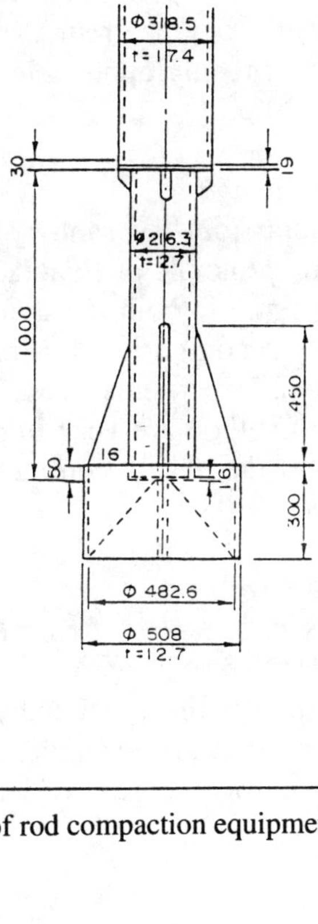	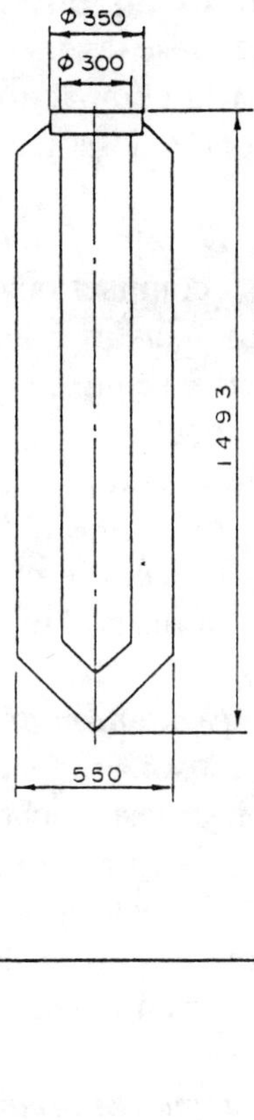

Figure 5.11. Examples of rod compaction equipment.

(SCP) differ in various aspects of the compaction mechanism. The main difference is the fact that sand fill is placed from the ground surface into the pores created in the soil with vibration in the RCP whereas, in the SCP, the amount of sand fill placed is controlled by forcing the sand into the subsoil profile by vibrating a casing. For RCP, therefore, the quantity of sand fill varies according to the method of installation, specification, and capacity of the vibrator. The effective range of compaction also varies. It is, therefore, necessary to revise the design method by obtaining new installation data if, for example, the capacity of the vibrator changes.

The rod compaction method is often used for liquefaction remediation in the vicinity of existing structures (refer to Section 5.8). It is necessary, however, to consider the influence of vibration on the existing structures depending on the type of structure. The improvement efficiency of this method is greatly affected by the stratification, grain size, and layer thickness of the natural subsoil profile before compaction. It is necessary to perform a field test in advance and understand the relationship among compaction effect, sand fill quantity, spacing and procedure for installation. Efficiency of compaction is relatively low for compaction near the ground surface with a small overburden pressure. Efficiency is also low for compaction of subsoils overlain by a sandy layer containing a large amount of silt or a clay layer because these layers prevent sand fill from the ground surface.

The rod compaction method is generally applicable to sandy soil with a fines content less than about 15 to 20%.

5.3.3.2 *Design of rod compaction method*

The rod compaction method differs from the sand compaction method in many respects as noted above. The principle of liquefaction remediation by compacting sand, however, is common to both. Design of the rod compaction method, therefore, is basically accomplished by referring to the design of the sand compaction pile method. In particular, case history data and test installation of the rod compaction method are very important because they directly reflect the characteristics of the installation method and the characteristics of the soil to be compacted.

5.3.3.3 *Evaluation of compacted subsoils*

Evaluation of compacted subsoils in Japan is often based on the standard penetration test between sand piles. To enhance reliability, efficiency and economy, it may be necessary to modify this practice by considering the areas and characteristics of subsoil profile. For example, the cone penetration test is often used in the USA and Europe because of its reliability and efficiency as mentioned earlier.

5.3.3.4 *Remarks on installation*

a) *Installation depth.* With the installation equipment presently used in Japan, it is possible to compact to a depth of about 15 to 20 m.

b) *Influence on adjacent structures.* Refer to Section 5.8 and Section 5.3.2.5e.

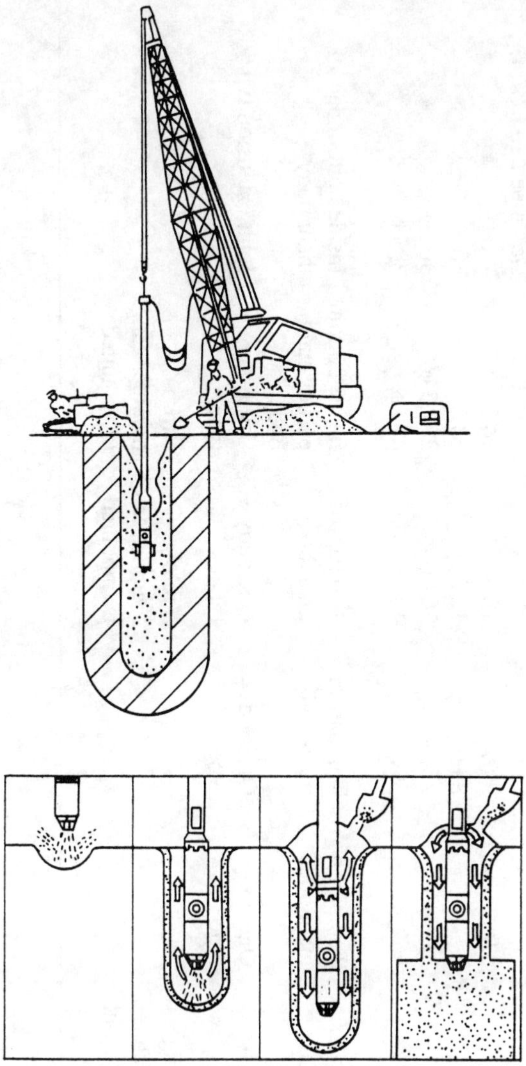

Figure 5.12. Schematic diagram of vibroflotation method.

5.3.4 *Vibroflotation method*

5.3.4.1 *Types of installation*

Vibroflotation compacts loose sandy soil by inserting a horizontally vibrating vibroflot into the subsoil profile while jetting water from the bottom end of the vibroflot, then gradually retracting it while forcing sand or gravel from the ground surface into the voids created at the perimeter of the vibroflot. A schematic diagram is shown in Figure 5.12 [28].

The vibroflotation method (VF) and the sand compaction pile method (SCP) differ in various aspects of the compaction mechanism. The main

Table 5.2. Types of vibroflots (supplements made to reference [31]).

	7.5 HP model	15 HP model	20 HP model	20 HP model	30 HP model	150 HP model
	50Hz / 60Hz	50Hz / 60Hz	50Hz / 60Hz	50Hz / 60Hz	50Hz / 60Hz	
Motor (kW)	5.5 kW (4p)	11 kW (4p)	15 kW (4p)	15 kW (4p)	22 kW (4p)	118 kW (hydraulic type)
Frequency (rpm)	1,450/1,740	1,450/1,740	1,450/1,740	1,450/1,740	1,450/1,740	1,500~3,000
Eccentric moment (kgf cm)	49.8/37.1	110.2	144.5	120	200	–
Excitation force (tf)	1.1/1.3	2.5/3.7	3.4/4.8	2.7/4.0	4.5/6.7	25
Excitation displacement (mm)	3~4	4~5	5~7	4~5	6~7	14
Excitation acceleration (g)	6.8/13.5	9.0/16.9	12.8/18.2	9.0/16.9	13.5/23.6	
Total weight (kgf)	1,200	1,700	2,185	3,525	3,200	Vibration part + isolator 1,300 kgf (extension pipe 220 kgf/m)
Total equipment length (m)	8.5	8.9	9.67	13.9	14.2	–
Effective length (m)	7	7	8 (10)	12.5	13.5	30
Diameter of casing (mm)	216 ϕ	319 ϕ	267 ϕ	267 ϕ	298.5 ϕ	300 ϕ
Outer diameter of casing including fin (mm)	346 ϕ	469 ϕ	397 ϕ	407 ϕ	470 ϕ	325 ϕ
Motor mounting position	Top part of main body	Top part of main body	Top part of main body	Top part of main body	Inside bottom part of main body	Inside bottom part of main body
Structure	One body type	One body type	One body type	One body type	One body type	Joint extension type
Remarks	Not being manufactured Presently out of use	–	–	Vibro-hammer can be mounted	Vibro-hammer can be mounted	–

difference is the same as the difference between the rod compaction method and the sand compaction method noted in the previous section of this handbook. Namely, sand fill is placed into the voids created in the soil profile by the vibration in VF whereas, in the SCP, the quantity of sand fill is determined in advance, and compaction is accomplished by placing the sand fill below the ground surface by vibrating the casing. For VF, therefore, the quantity of sand fill varies according to the installation method and capacity of the vibroflot. The effective range of compaction also varies. It is, therefore, necessary to revise the design method by obtaining new installation data if, for example, the capacity of the vibroflot changes.

The VF method was invented in Germany in the 1930's, and developed further in the late 1940's in the USA and Europe (particularly in the USA). This method was introduced in Japan with research and development by Mogami [29] in the 1950's. Since detailed information could not be obtained in Japan, a new vibroflot was developed based on in-situ tests and research by Watanabe [30]. The vibroflots presently used in Japan are shown in Table 5.2 [31] with a capacity ranging from 7.5 to 30 HP. The vibroflots used in the USA and Europe since the 1950's are the 30 HP type [32]. Recently those of 100 to 150 HP type have been used in Japan (refer to Appendix 4).

5.3.4.2 *Design based on installation case history data*

Soil compaction can be designed based on reliable case histories if data are available on the SPT *N*-values before and after compaction, capacity of vibroflot, installation spacing and characteristics of the soil.

a) *Installation case history data.* Descriptions of vibroflotation case history data here pertain to those used in Japan, with a vibroflot of 7.5 HP or 10 HP unless stated otherwise.

b) *Limits on applicability.* Limits on applicability of the vibroflotation method are shown in Figure 5.13 [30]. Figure 5.13 was obtained from eleven case histories with an equilateral triangular installation and a pile spacing ranging from 1.2 to 1.5 m.

– *Applicability limit according to gradation of natural soil.* According to Figure 5.14, the vibroflotation method is applicable to soils with a fines content less than 40% although the effectiveness decreases with increasing fines content. The case histories in the USA and Europe using a powerful vibroflot indicate that there is no compaction when the fines content exceeds 25% (Figure 5.14) [28]. The limit of applicability generally depends on the SPT *N*-value to be achieved in the design, and this may be the reason for the difference in the limit of applicability.

In Figure 5.14, three main points are illustrated: 1) the vibroflotation method is most effective in loose sand with gradation in zone B, 2) compaction by vibroflotation becomes very difficult in sand with gradation in zone C, and 3) compaction is possible for soils with a portion of the gradation

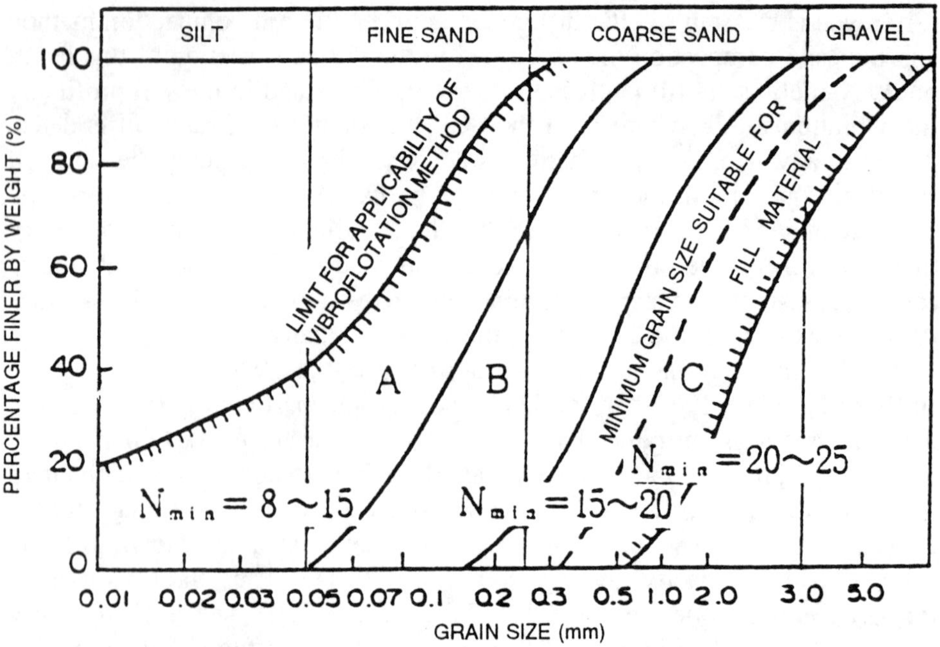

Figure 5.13. Relationship between the minimum SPT *N*-value achievable by compaction and gradation of natural soil (examples in Japan) [30].

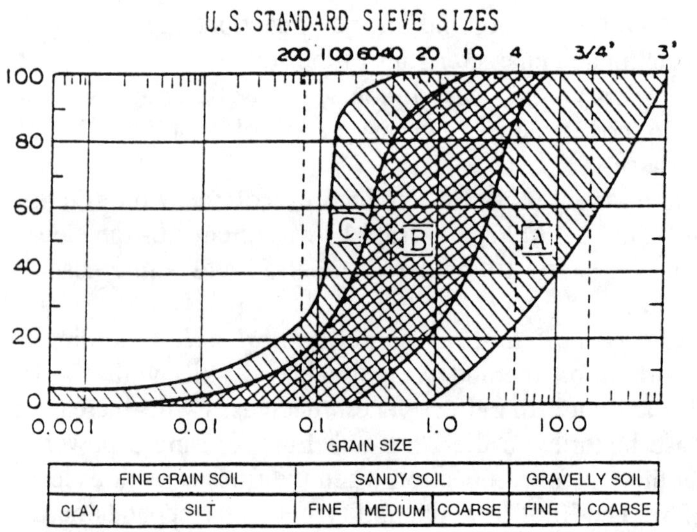

Figure 5.14. Gradation appropriate for vibroflotation method [28].

curve passing through zone C. Penetration of the vibroflot becomes difficult in relatively dense sand, cemented sand, and gravel in zone A.

– *Gradation limit for filling materials*. Gravel, coarse sand, slag, and on-site sand are used in many cases as the fill material. The compaction is greater for larger grain size. When the maximum grain size exceeds 5 cm, however, compaction is reduced [33].

If the grain size is small, compaction cannot be performed smoothly because of the scatter of fill material due to the rising mud flow and increasing viscous resistance to the penetration of the vibroflot. A favorable minimum grain size for a fill material is indicated with the broken line in Figure 5.13.

– *Expected SPT N-value after compaction.* The SPT *N*-values shown in Figure 5.13 indicate typical values obtained at the center of a triangular installation of vibro piles. It is possible to give a rough estimate of the SPT *N*-value which can be achieved under the conditions of installation noted in a and b.

The highest relative density which can be achieved using the vibroflotation method is considered to be between 70 and 90% [34].

– *Attenuation of the compaction effect from the vibro pile.* The compaction effect of vibroflotation decreases exponentially according to the distance from the vibro pile [32]. The effective range of compaction is affected greatly by the fines content. It has been reported that the effective range of compaction of sand having more than 20% of fines content is about half of that of clean sand [35].

c) *Installation spacing.* An arrangement of vibroflot in an equilateral triangle is considered most effective and many of the installation spacings in the conventional vibro pile range from 1.2 to 1.5 m [33].

d) *Amount of sand fill.* In the case histories using 7.5 HP and 10 HP vibroflots, the amount of sand fill was about 0.14 to 0.22 m^3 per 1 m^3 of natural soil. Alternatively, the amount of sand fill per 1.0 m vibro pile is over 3 times the volume of the vibroflot [36]. The volume of the vibroflot presently used in Japan can be calculated assuming a cylindrical shape with a diameter of 216 to 319 mm (refer to Table 5.2).

These estimates of sand fill can be used as a reference. It is, however, necessary to confirm them with in-situ testing because they are affected by installation and soil conditions.

5.3.4.3 *Design based on test installation*

For a soil containing a large amount of silt or a subsoil profile consisting of mixed layers of sand and clay, or when no reliable case history data is available for the vibroflotation method concerned, it is preferable to design based on a test installation. A test installation is planned based on the relationship between the degree of compaction and void ratio and case history records.

The design of a test installation using the vibroflotation method is similar to the sand compaction pile method. It is, however, necessary to refer to the case history records of the vibroflotation method in order to determine a suitable installation spacing and the amount of sand fill.

5.3.4.4 *Evaluation of compacted subsoil*

The subsoil after compaction is evaluated in a similar manner as noted in Section 5.3.2.4.

5.3.4.5 *Remarks on the installation*

a) *Extra amount of sand fill.* The amount of sand fill is theoretically computed with Equation (5.1). Considering any volume decrease during installation, an extra amount of sand fill is necessary. There is a reference which states that it is necessary to add 20 to 40% considering the loss during transport in the vibroflotation method [31].

b) *Installation depth.* With the 7.5 HP and 10 HP vibroflots, the maximum depth of installation was found to be about 8 m, but with the installation equipment used in Japan, compaction up to a maximum depth of 13.5 m is possible [31]. With the 150 HP equipment introduced recently, compaction to a maximum depth of 30 m is possible (refer to Appendix 4).

The effect of compaction is minimum near the ground surface and the bottom of the treated area (about 1.5 m from the both ends based on the case history data in the USA and Europe [36]). For compaction of the surface layer, it is necessary to consider other compaction methods.

c) *Influence on adjacent structures.* Refer to Section 5.3.2.5.

d) *For a clay layer interposed in the sand.* If a clay layer exists within the desired depth of improvement, large voids will not be created in the clay layer by vibration. This makes compaction difficult due to increased resistance to sand fill, which should be sunk to a greater depth. In such a case, protrusions in the vibroflot should be provided.

The water circulated for the vibroflotation method may become dense and viscous due to fine grained soil in the mixture, requiring extra time for the sand fill to drop into the subsoil. It may be necessary to use coarse grained material or allow a long waiting period for washing out of fine grained soil.

e) *Monitoring and controlling installation.* Installation of the vibroflotation depends significantly on the expertise of the engineers. In order to achieve expected compaction results, it is necessary to monitor the following items and compare them with the previous case history records: 1) arrangement and spacing of the installation locations, 2) penetration depth, 3) change in the speed of the vibroflot depending on stratification of subsoil profile, 4) extraction speed of the vibroflot, 5) amount of sand fill for each installation location, 6) gradation of sand fill, 7) filling water pressure and quantity during installation, 8) elevation of the ground surface before and after the installation, and 9) electric current consumed during the installation [28].

It is important to establish the items to be monitored during installation and to establish the standard time and schedule based on the test installation. In order to avoid re-installation due to unsatisfactory compaction, sounding may be made at specified time intervals to reevaluate the method of installation.

The change in the average void ratio may be computed with the following equation from the measurements of settlement or uplift of the ground surface before and after the installation and the amount of sand fill placed during installation.

$$e' = \frac{\beta d^2 (H \pm h)}{\{(\beta d^2 H)/(1+e_o)\} + \{(V)/(1+e_1)\}} - 1 \quad (5.4)$$

where: e' = average void ratio after improvement, β = area coefficient, $\beta = 1.0$ for square arrangement, $\beta = 0.866$ for equilateral triangular arrangement, d = vibro pile installation spacing (m), βd^2 = installation area per pile (m^2), H = improvement layer thickness (m), h = uplift (+) or settlement (–) of the subsoils, V= amount of sand fill per pile (m^3/pile), e_o = void ratio of natural subsoil, e_1= void ratio of sand fill.

5.4 DESIGN/INSTALLATION OF DRAIN METHOD

The purpose of the drain method for liquefaction remediation is to enhance the overall drainage capacity of the subsoil profile by installing drains made of very permeable materials below the ground surface. Such a drain is generally installed in a pile configuration [37]. Other forms of drains, however, have been proposed as well, such as a wall shaped drain, drain surrounding a structure [38] and attachment of a drainage device to steel piles or sheet piles [39]. Rubble or gravel used as backfill behind quay walls may also be considered as a type of drain if a permeable sheet is used to prevent the migration of sand. Gravel and crushed stone are generally used as material for drains. In recent years, however, pipes made of synthetic resin (i.e. plastic) with many small holes are also used as discussed later. As noted above, a variety of drainage methods exist for liquefaction remediation. In this section of the handbook, design/installation of a pile shaped drain will be discussed.

The method using a pile shaped drain with gravel and crushed stone also has variations. One variant is the gravel compaction method [40] which combines drainage and compaction. In this method, however, the natural subsoils adjacent to the drain may locally experience liquefaction due to strong vibration during drain installation. Careful attention to clogging during the drain installation is, therefore, necessary when using this method. In recent years, a method has been developed which increases soil density by pushing gravel into the soil with a steel rod (refer to Appendix 4). The various drain methods noted above can be summarized as shown in Figure 5.15.

For the vibroflotation method, gravel and crushed stone may be used as

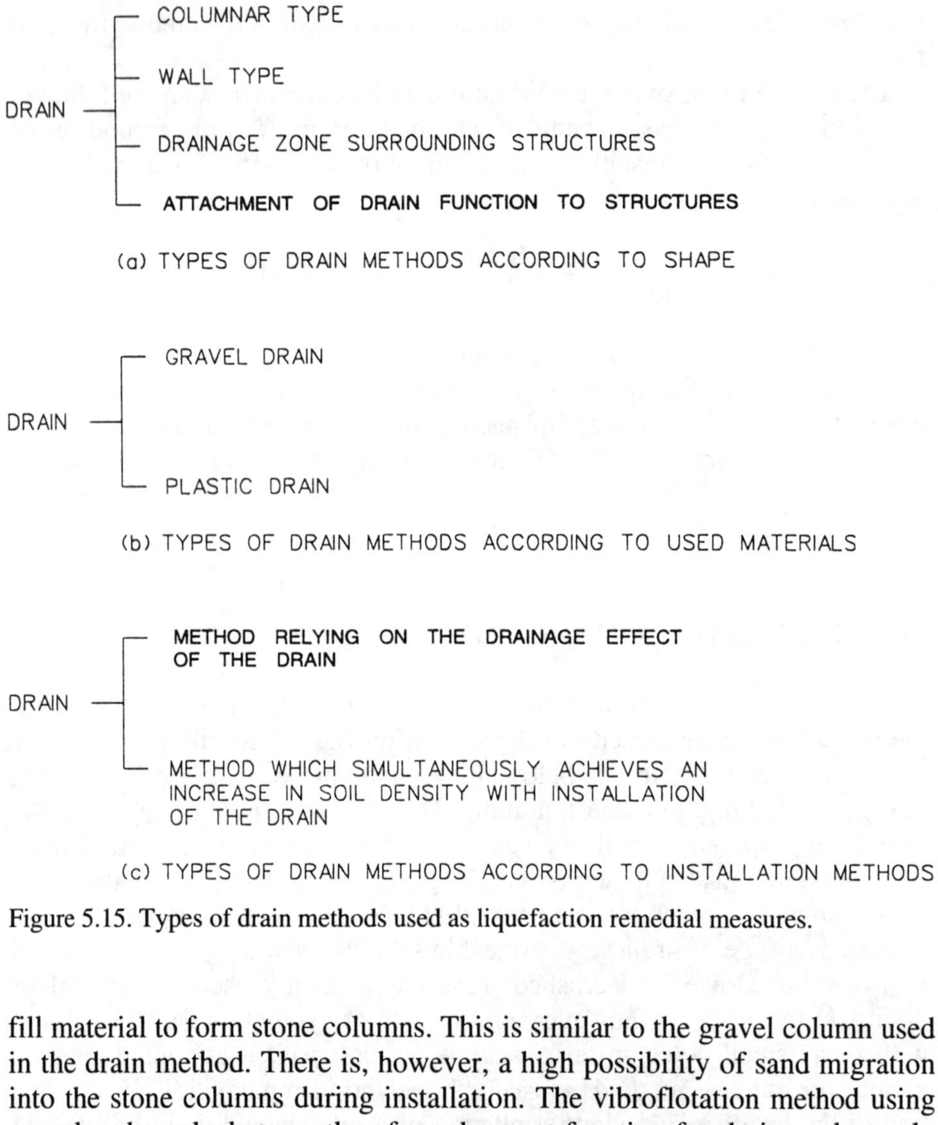

Figure 5.15. Types of drain methods used as liquefaction remedial measures.

fill material to form stone columns. This is similar to the gravel column used in the drain method. There is, however, a high possibility of sand migration into the stone columns during installation. The vibroflotation method using gravel and crushed stone, therefore, does not function for drainage but only compacts the subsoils.

In the drain method, slight settlement accompanying water drainage and slight increase in excess pore water pressures are inevitable. Compared to the compaction method, the drain method is a recent development and there are not sufficient case histories during earthquakes. In addition, the deformation characteristics of loose sandy soil improved by the drain method are not always favorable in view of ductility [41]. *Ductility* in this context refers to the fact that the performance, such as required resistance against deformation, is not significantly affected by a slight change in the level of shaking caused by earthquakes. For newly constructed structures with minimal restriction on the installation or in existing structures for which a large safety factor should be allowed, other methods such as replacement with gravel, and stone backfill are preferable to the drain method.

It should be noted that the design method mentioned below is based on cumulative damage theory. Thus, procedure developed according to this theory is often conservative, as described in Section 4.5.3. At times, therefore, the design for the drain method may become excessively conservative. In the actual design, it is important to qualify the design with good overall engineering judgement by considering the condition of the existing structure and other factors.

5.4.1 *Outline of design for drain method*

Design of drain method is accomplished using the procedure shown in Figure 5.16. As shown in this figure, (1) the material of the drain is determined, then (2) a soil survey and in-situ tests are conducted in the natural subsoil profile, and finally (3) the installation spacing of drains is established.

The design method is based on a balance between the rate of excess pore water pressure increase due to earthquake shaking and the rate of dissipation of pore water pressure due to drainage. During earthquakes, pore water is drained simultaneously with generation of excess pore water pressure. The rate of excess pore water pressure increase is increased by high levels of earthquake motion, sudden and intense shaking, and in loose subsoils. The rate of dissipation of pore water pressure increases if the coefficient of volume compressibility of the natural soil is small and the coefficient of permeability is high. The rate of dissipation increases more if the overall drainage capacity of the subsoil profile is increased by installing closely spaced drains and using higher permeability drain material.

Based on the above considerations, the design conditions which cause some difficulty for the drain method are as follows:

- Permeability of the natural subsoil profile is low;
- Coefficient of volume compressibility of the natural soil is high;
- Liquefaction strength of the natural soil is low;
- Earthquake motion level is high;
- Earthquake motion is an impact type;
- Liquefiable soil layer is thick.

In designing the drain, it is necessary to select a suitable drain material and spacing of the installation considering the conditions above. The design of a gravel drain is explained in Section 5.4.2 and that of a plastic drain in Section 5.4.3.

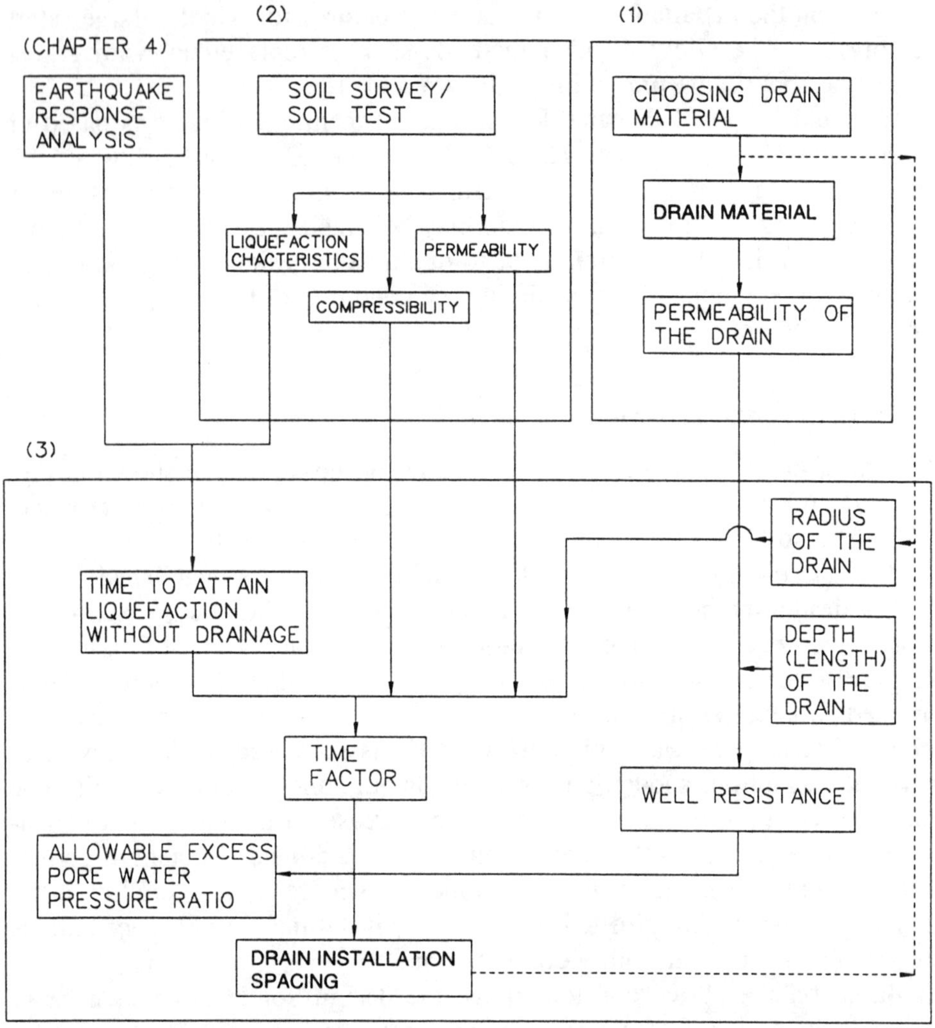

Figure 5.16. Design procedure for drain method.

5.4.2 *Design of gravel drain*

5.4.2.1 *Material for gravel drain*

Gravel/crushed stone used for a gravel drain must have a sufficiently high permeability to be adequate for liquefaction remediation and a high resistance against liquefaction. In selecting the gravel/crushed stone, it is preferable to evaluate the possibility of clogging.

In designing the gravel drain, the permeability of the gravel/crushed stone should be identified by a suitable survey/test. The results of the survey/test will not only be used for selecting suitable gravel/crushed stone but will also

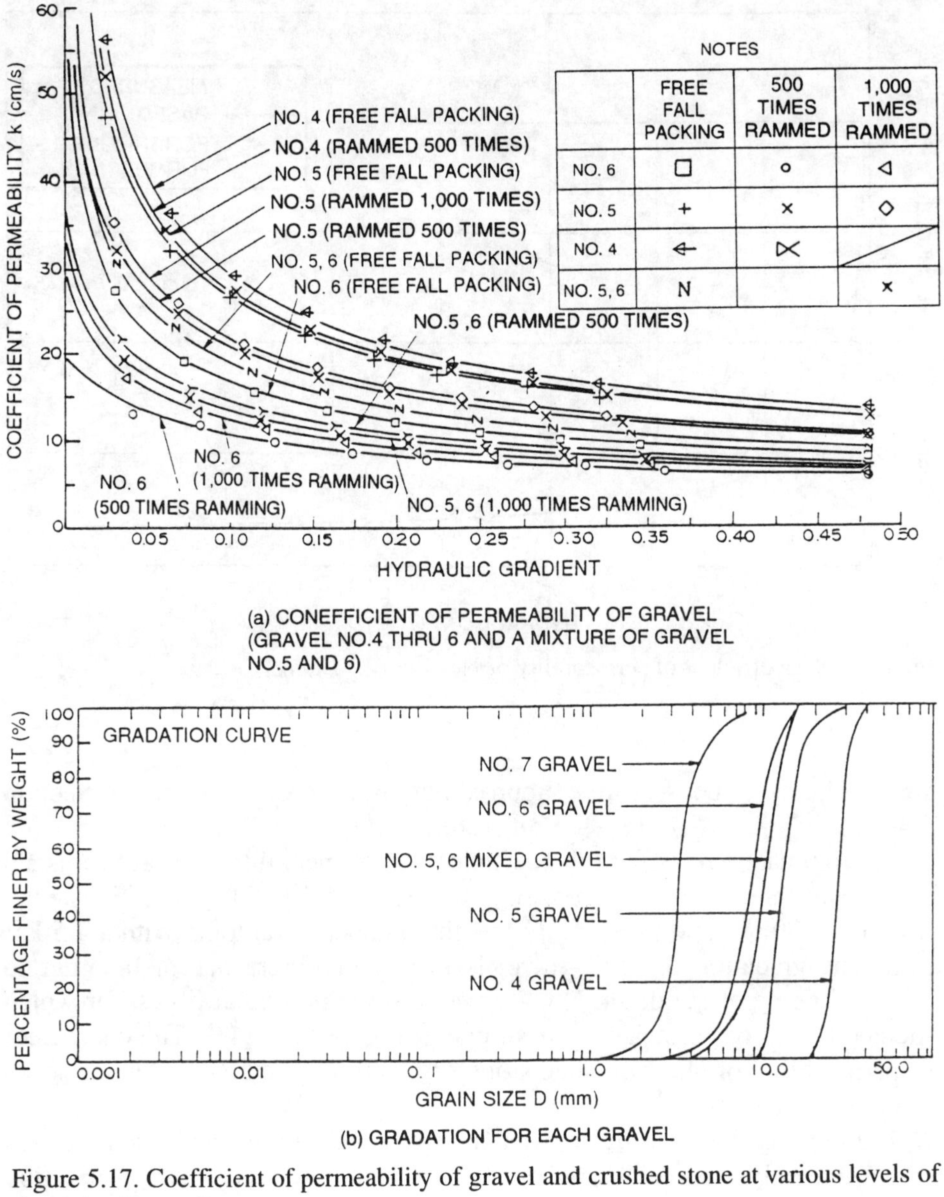

Figure 5.17. Coefficient of permeability of gravel and crushed stone at various levels of the hydraulic gradient.

be used in the design procedure presented in Section 5.4.2.3.

Seepage flow to a drain made of gravel/crushed stone is in many cases turbulent flow. The coefficient of permeability of the gravel/crushed stone, therefore, becomes dependent on the hydraulic gradient. The hydraulic gradient has a close relationship with the excess pore water pressure ratio, considered approximately to be equal to the allowable excess pore water pres-

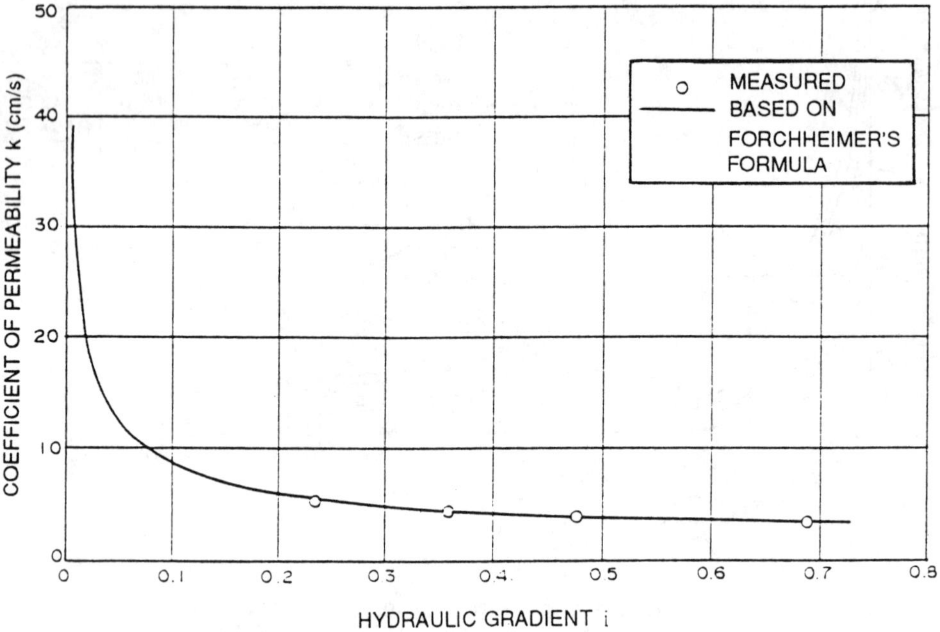

Figure 5.18. Coefficient of permeability of No. 7 crushed stone.

sure ratio as a conservative approximation in design (refer to Section 5.4.2.3).

Approximate values for the coefficient of permeability of gravel/crushed stone can be obtained by referring to Figure 5.17a [42]. In this figure, 'rammed 500 or 1000 times' indicates the number of tamping with a 4.5 kgf ram. The gradation of each gravel is shown in Figure 5.17b. In order to avoid clogging of the drain, No. 7 crushed stone is used at times, for which the coefficient of permeability is shown in Figure 5.18 [43]. The coefficient of permeability of No. 7 crushed stone shown in Figure 5.18 is a value when ramming is not done.

The required level of the coefficient of permeability of the gravel/crushed stone generally changes based on various conditions as shown in Section 5.4.2.3. A condition necessary for the drain to function as an empty well (i.e. to have the same drain function as when the coefficient of permeability of gravel/crushed stone is infinite) is one criterion in selecting the gravel/crushed stone. Based on Section 5.4.2.3, this condition roughly corresponds to a well resistance coefficient R being less than 1.0. This condition can be rewritten as follows:

$$k_d \geq k_s \left(\frac{h}{a}\right)^2 \tag{5.5}$$

where: k_d = coefficient of permeability of gravel/crushed stone used as drain material (cm/sec), k_s = coefficient of permeability in the horizontal direction of the natural subsoil profile (cm/sec), h = depth of the drain installation (cm) (refer to the first paragraph of Section 5.4.2.3), a = radius of the drain (cm).

When the installation depth of the drain is 10 m and the diameter of the drain is 60 cm for example, Equation (5.5) requires that the coefficient of permeability of gravel/crushed stone for the drain should be about 1000 times that of the natural subsoils.

Even if this criterion is not satisfied, design of the drain is possible following in Section 5.4.2.3. When the coefficient of permeability decreases to less than that given by Equation (5.5), the installation spacing of the drain becomes quite small as indicated in Section 5.4.2.3. For the extreme case, total replacement of the natural subsoil with gravel/crushed stone becomes necessary.

It is necessary to avoid clogging of a gravel drain in order to maintain its required coefficient of permeability. The clogging of the gravel drain refers to a decrease in the permeability of the drain due to migration of the natural soil into the drain. The condition of clogging can be classified into the following three categories.

a) *Clogging during drain installation.* When the compaction method is applied after installing the gravel drain nearby, excess pore water pressure of the sandy soil adjacent to the gravel drain increases due to vibration from the compaction and pore water is discharged into the drain. This may cause clogging of the gravel drain. In implementing the compaction with drain installation simultaneously as in gravel compaction, there is also a possibility of clogging in the same manner.

b) *Clogging after the drain installation but before earthquake.* There is a possibility that clogging of the drain may gradually progress due to various factors occurring over a long period of time such as mixing of fine grained particles due to fluctuations in the ground water level, or breeding of microorganisms. This aspect of the clogging has not yet been studied.

c) *Clogging during earthquakes.* During an earthquake, pore water is discharged into the drain so that there is a possibility that the drain may be clogged along with the discharge.

Out of the categories noted above, a and c are based on a similar mechanism and there is a large amount of research related to these types of clogging [44-47]. Those research projects were undertaken based on the assumption that the natural soil directly in contact with the gravel drain will not undergo liquefaction.

Clogging conditions in a or c can be identified by a laboratory test [44]. The following proposals may also be used as a condition in which clogging does not occur [45-47].

– $d_{15}/D_{85} \leq 5$ (where d_{15}: 15% diameter of the drain material (mm), D_{85}: 85% diameter of the natural soil (mm)) [45];

– $d_{25}/D_{75} \leq 9.5$ or $d_{15}/D_{85} \leq 6.4$ (where d_{25} and D_{75} are similarly defined as noted above for 25 and 75% diameters) [46];

– $d_{15}/D_{85} \leq 9$ ($d_{15}/D_{85} \leq 5$ was proposed as a conclusion, however, by allowing an adequate factor of safety) [47].

5.4.2.2 *Soil survey/soil test*

For design of the gravel drain, it is necessary to determine suitable constants in addition to the coefficient of permeability of the drain obtained in Section 5.4.2.1, including:

– Liquefaction characteristics of the soil;
– Coefficient of permeability in the horizontal direction of the subsoil;
– Coefficient of volume compressibility of the soil.

The survey/test for determining the required constants is discussed below.

a) *Liquefaction characteristics of the soil.* Liquefaction characteristics of the soil are obtained from cyclic triaxial tests as discussed in Chapter 4. In designing the gravel drain, the test results are summarized as shown in Figures 5.19a and 5.19b.

Figure 5.19a shows the relationship between cyclic shear stress ratio R and number of load cycles N_l to achieve liquefaction. The method for obtaining this curve was shown in Chapter 4. In Figure 5.19b, the horizontal axis shows the normalized number of load cycles N_l, and the vertical axis shows the normalized excess pore water pressure u with respect to the initial confining pressure σ'_c. Curves plotted in this figure may vary according to the level of cyclic stress ratios R used in the test; an average curve is obtained for design.

The results are used for estimating the rate of excess pore water pressure

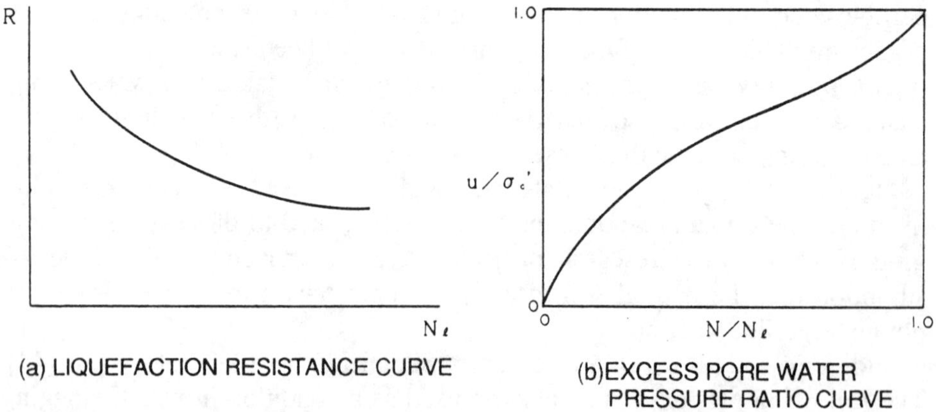

Figure 5.19. Liquefaction characteristics of soil.

increase during earthquakes in Section 5.4.2.3 to be noted later.

b) *Coefficient of permeability of the soil.* The coefficient of permeability of the soil often differs in the vertical and horizontal directions. During earthquakes, pore water mainly flows in the horizontal direction toward the gravel drain. It is necessary to obtain the coefficient of permeability of the soil in the horizontal direction when designing the gravel drain. The coefficient of permeability of the soil obtained is used in Section 5.4.2.3. When the soil is stratified horizontally, it is necessary to obtain the coefficient of permeability for each layer.

The coefficient of permeability of the soil in the horizontal direction is generally greater than that in the vertical direction. When it is difficult to obtain the coefficient of permeability in the horizontal direction, a conservative design is obtained by substituting the value of the coefficient of permeability in the vertical direction.

For a uniform soil, it is possible to obtain, with reasonable reliability, a value corresponding to the coefficient of permeability in the horizontal direction from permeability tests conducted at the site. When the soil is composed of horizontal strata, however, the permeability test performed at the site, in most cases, results in an apparent coefficient of permeability of all the soil layers rather than the coefficient of permeability of each individual layer. When the soil is stratified horizontally, careful evaluation is necessary with regard to the suitability of the permeability test method performed at the site.

For an approximate value of the coefficient of permeability of the soil, refer to Tables 5.3 and 5.4 [48]. With regard to relatively uniform soil having a coefficient of uniformity of $U_c < 5$ and an effective diameter of D_{10} = 0.1 to 0.3 mm, refer to the following equation [49].

$$k = C \times D_{10}^2 \tag{5.6}$$

where: k = coefficient of permeability (cm/sec), C = coefficient (= 100/(cm/sec)), D_{10} = effective diameter (cm).

c) *Coefficient of volume compressibility of the soil.* The coefficient of

Table 5.3. Diameter of sand particles and coefficient of permeability [49].

Soil type	Particle size (mm)	Coefficient of permeability (cm/s)
Very fine sand	0.05~0.10	0.001~0.005
Fine sand	0.10~0.25	0.005~0.01
Medium sand	0.25~0.50	0.01~0.1
Coarse sand	0.50~1.00	0.1~1.0
Small pebbles	1.00~5.00	1.0~5.0

Table 5.4. Coefficient of permeability estimated from D_{20} proposed by Craiger [48].

D_{20} (mm)	k (cm/s)	Soil type
0.005	3.00×10^{-6}	Coarse clay
0.01	1.05×10^{-5}	Fine silt
0.02	4.00×10^{-5}	Coarse silt
0.03	8.50×10^{-5}	
0.04	1.75×10^{-4}	
0.05	2.80×10^{-4}	
0.06	4.60×10^{-4}	Very fine sand
0.07	6.50×10^{-4}	
0.08	9.00×10^{-4}	
0.09	1.40×10^{-3}	
0.10	1.75×10^{-3}	
0.12	2.6×10^{-3}	Fine sand
0.14	3.8×10^{-3}	
0.16	5.1×10^{-3}	
0.18	6.85×10^{-3}	
0.20	8.90×10^{-3}	
0.25	1.40×10^{-2}	
0.3	2.20×10^{-2}	Medium sand
0.35	3.20×10^{-2}	
0.4	4.50×10^{-2}	
0.45	5.80×10^{-2}	
0.5	7.50×10^{-2}	
0.6	1.10×10^{-1}	Coarse sand
0.7	1.60×10^{-1}	
0.8	2.15×10^{-1}	
0.9	2.00×10^{-1}	
1.0	3.60×10^{-1}	
2.0	1.80	Fine gravel

volume compressibility of soil is obtained from a cyclic triaxial test. The cyclic triaxial test is performed based on the procedure outlined in Chapter 4, using undisturbed samples. The cyclic load is interrupted, however, when the excess pore water pressure ratio u/σ'_c reaches a specified level, below 0.5, and thereafter, the coefficient of volume compressibility during consolidation is obtained by opening a drain valve of the cyclic triaxial test device. If the excess pore water pressure at the beginning of consolidation is u_{max} (kgf/cm^2), the volume of pore water discharged by consolidation is ΔV(cm^3), and the volume of the specimen is ΔV(cm^3), then the coefficient of volume compressibility m_v(cm^2/kgf) is obtained from the following equation.

Table 5.5. Typical values of coefficient of volume compressibility of sand.

Type of sand	Coefficient of volume compressibility (cm²/kgf)*	References (year)
Sacramento River Sand	2×10^{-3}	Lee et al. (1974) [50]
El Monte Sand (D)	2×10^{-3}	Lee et al. (1974) [50]
El Monte Sand (E)	2×10^{-3}	Lee et al. (1974) [50]
Akita Port Sand	$3 \sim 4 \times 10^{-3}$	Zen et al. (1984) [44]
El Monte Sand (C)	4×10^{-3}	Lee et al. (1974) [50]
Monterey Sand	4×10^{-3}	Lee et al. (1974) [50]
Fuji River Sand	6×10^{-3}	Ishihara et al. (1978) [51]
El Monte Sand (B)	8×10^{-3}	Lee et al. (1974) [50]
Ogishima Sand	10×10^{-3}	Ono et al. (1983) [52]

*Measured at $u / \sigma'_c < 0.5, \sigma'_c = 1\ \text{kgf} / \text{cm}^2$.

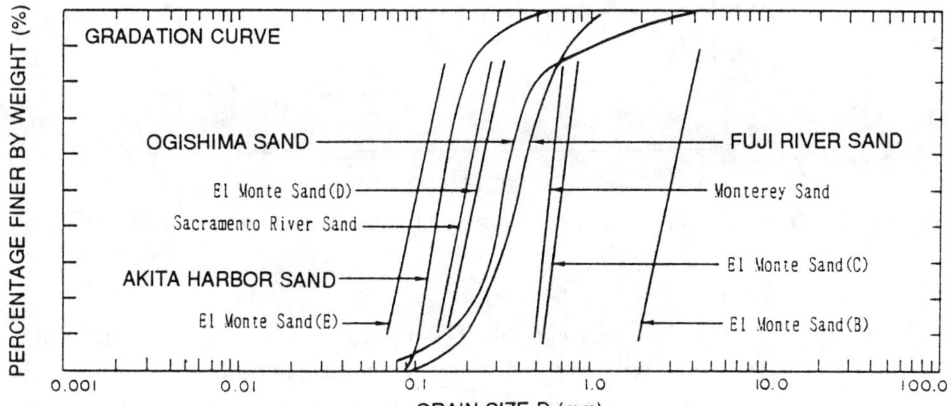

Figure 5.20. Gradation curves of the sands used for obtaining the coefficient of volume compressibility.

$$m_v = \frac{\Delta V}{V} \frac{1}{u_{max}} \tag{5.7}$$

Typical values of the coefficient of volume compressibility are shown in Table 5.5. These values were obtained at the initial confining pressure of σ'_c = 1 kgf/cm^2 and excess pore water pressure ratio $u/\sigma'_c < 0.5$. The gradation curves for the sands used in these tests are shown in Figure 5.20.

5.4.2.3 *Radius and installation spacing of the drain*

The depth of installation of the gravel drain is the bottom of the deepest soil layer that is assessed to be subject to liquefaction. The radius and installation spacing of the gravel drain are designed using the following procedure [7, 53].

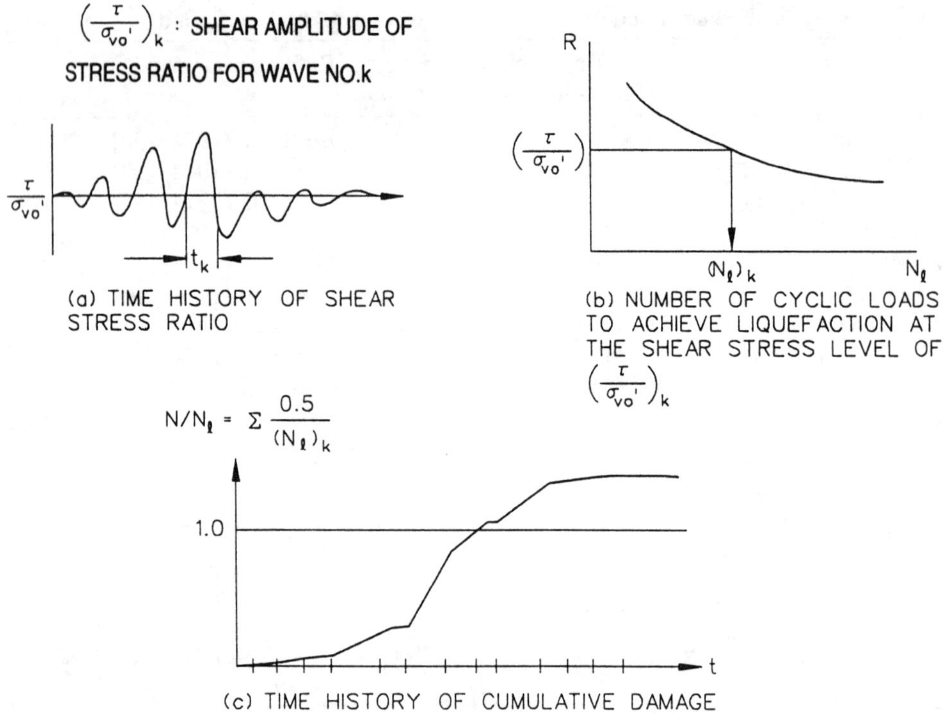

Figure 5.21. Schematic diagrams for computing the time history of cumulative damage.

First, the liquefaction process of the natural soil under an undrained condition is determined with the following procedure according to Steps 1-3.

– *Step 1*. If earthquake response analysis has not already been done for assessing liquefaction potential, it should be performed, based on the procedure outlined in Chapter 4.

– *Step 2*. Cumulative damage index N/N_l is obtained by the following procedure, based on the results of the earthquake response analysis and liquefaction strength curve (Fig. 5.19a) using the method indicated in Figure 5.21.

1. The amplitude of wave No. k in time history of the shear stress ratio is $(\tau/\sigma'_{vo})_k$;

2. N_l with respect to $R = (\tau/\sigma'_{vo})_k$ is obtained using the liquefaction strength curve and is denoted $(N_l)_k$;

3. Cumulative damage index N/N_l is obtained using the following equation.

$$\frac{N}{N_l} = \Sigma \frac{0.5}{(N_l)_k} \tag{5.8}$$

where: Σ means the sum up to the current wave in the time history.

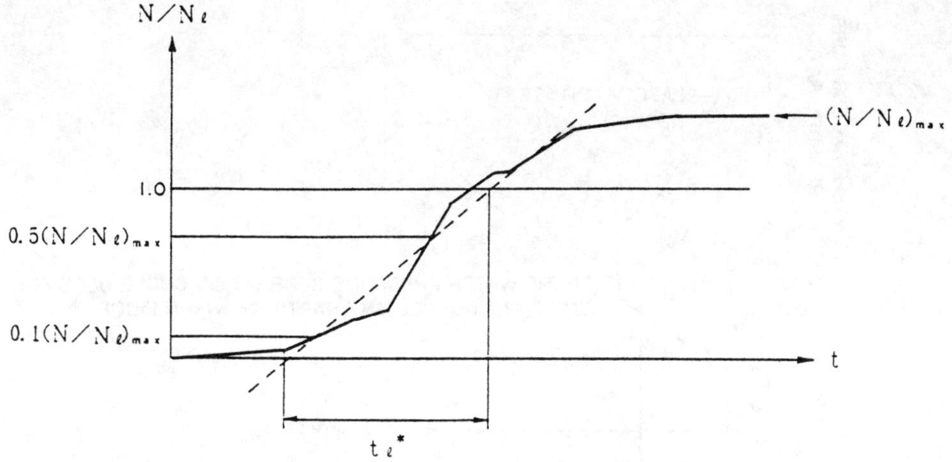

Figure 5.22. Method for obtaining equivalent time t_l^* to attain liquefaction.

– *Step 3*. Equivalent time t_l^* to attain liquefaction is obtained with the following equation from the time history of N/N_l (refer to Fig. 5.22).

$$t_l^* = \frac{t_{0.5} - t_{0.1}}{0.4\,(N/N_l)_{\max}} \tag{5.9}$$

where: t_l^* = equivalent time to attain liquefaction (sec), $t_{0.5}$ = time (sec) for achieving N/N_l = $0.5(N/N_l)_{max}$, $t_{0.1}$ = time (sec) for achieving N/N_l = $0.1(N/N_l)_{max}$, $(N/N_l)_{max}$ = maximum value of cumulative damage index N/N_l.

The equivalent time to attain liquefaction obtained by the procedure outlined above is used for determining the installation spacing of the gravel drains by Steps 4-11 shown below.

– *Step 4*. The excess pore water pressure ratio $(\bar{u}/\sigma'_{vo})_{max}$ allowable during earthquakes is established in the design. $(\bar{u}/\sigma'_{vo})_{max}$ means the maximum value of the time history of the excess pore water pressure ratio averaged in the horizontal cross section of the natural subsoil profile around the gravel drain.

– *Step 5*. Equivalent time t_l^* to attain liquefaction under undrained conditions is corrected as follows:

1. If Section 4.5.2 results in zone I shown in Figure 4.41 but $(N/N_l)_{max}$ in the present computation does not reach 1.0, then t_l^* will be multiplied by $(N/N_l)_{max}$ to give the corrected t_l. This correction is equivalent to correcting $(N/N_l)_{max}$ to 1.0 so that consistency with the liquefaction prediction/determination based on gradation and N-value can be achieved.

2. Correction with respect to the excess pore water pressure generation characteristics is determined using the following equation to obtain t_l (refer to Fig. 5.23).

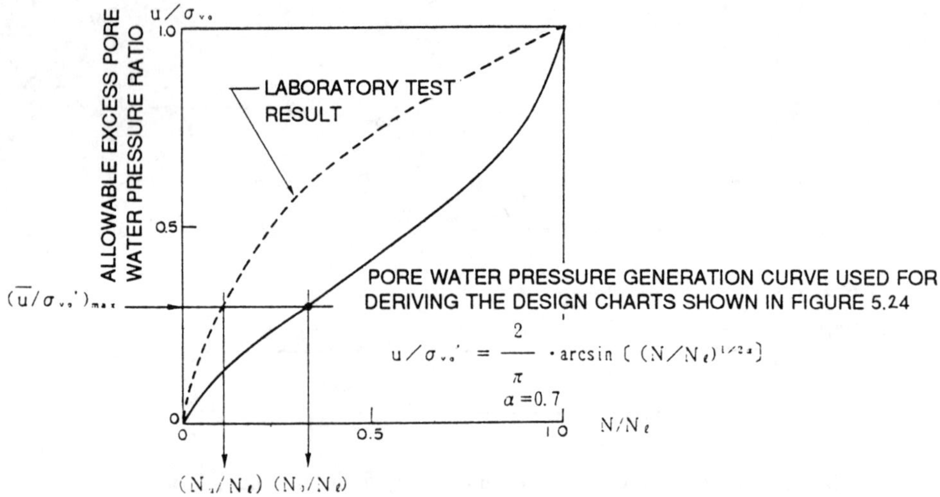

Figure 5.23. Schematic diagram for obtaining the correction factor for the time to attain liquefaction.

$$t_l = \frac{(N_u / N_l)}{(N_o / N_l)} \times t_l^* \tag{5.10}$$

where: t_l = equivalent time (sec) to attain liquefaction after the correction, N_u/N_l = Value of N/N_l corresponding to $(\bar{u} / \sigma'_{vo})_{max}$ in the excess pore water pressure generation curve obtained from the laboratory test, $(N_o/N_l) = (\sin(\pi/2\,(\bar{u} / \sigma'_{vo})_{max}))^{2\alpha}$, where $\alpha = 0.7$.

– *Step 6*. Radius a (cm) of the gravel drain is determined by considering clogging, construction equipment, etc.

– *Step 7*. Time factor T_l is obtained from the following equation.

$$T_l = \frac{k_s t_l}{m_v \gamma_w a^2} \tag{5.11}$$

where: k_s = permeability of the soil in the horizontal direction (cm/sec), m_v = coefficient of volume compressibility of the soil (cm^2/kgf), γ_w = unit weight of water (= 10^{-3} kgf/cm^3).

– *Step 8*. Well resistance factor R is obtained from the following equation.

$$R = \frac{8}{\pi^2}\left(\frac{k_s}{k_d}\right)\left(\frac{h}{a}\right)^2 \tag{5.12}$$

where: π = the circular constant (= 3.14), k_d = coefficient of permeability of the drain (cm/sec), h = length of the drain (cm).

– *Step 9*. Pile diameter ratio a/b of the drain is obtained using the design charts shown in Figure 5.24. b (cm) is the radius of the equivalent effective

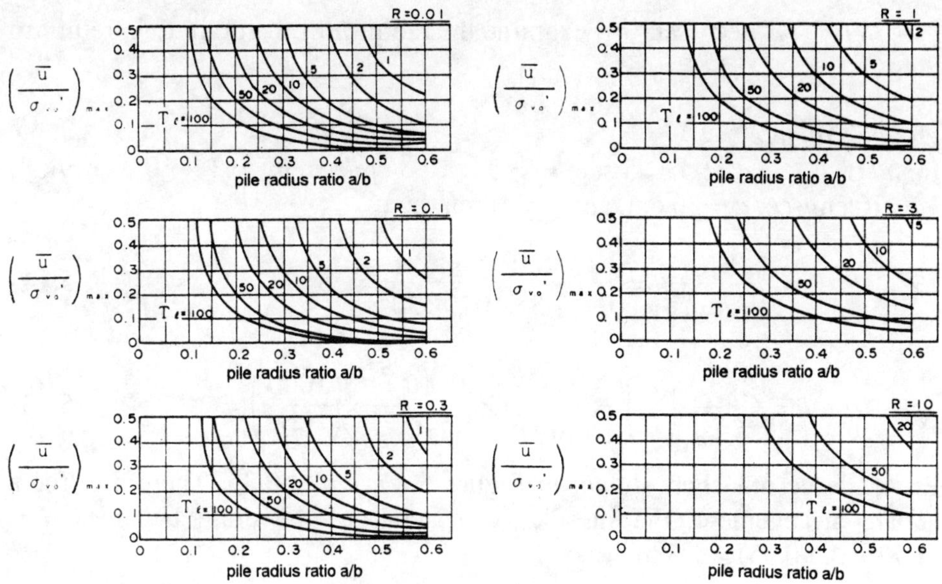

Figure 5.24. Design chart for a gravel drain spacing.

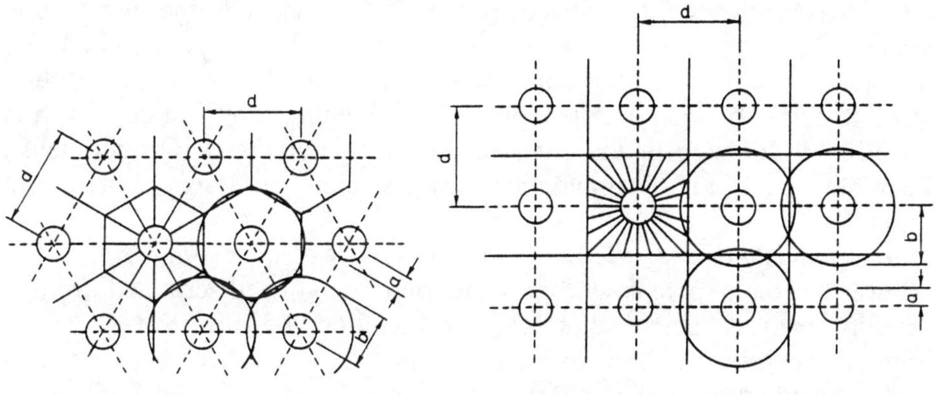

(a)TRIANGLE INSTALLATION (b)SQUARE INSTALLATION

Figure 5.25. Schematic diagram of equivalent effective circular area and installation configuration of the drain.

circular area of the drain (as shown in Figure 5.25, a circular area with the same area as that of a square or a hexagon).

– *Step 10.* Installation spacing of the drain d (cm) is obtained from the following equation.

$$\begin{aligned} d &= 1.9\,b \text{ (for equilateral triangular arrangement)} \\ d &= 1.77\,b \text{ (for square arrangement)} \end{aligned} \tag{5.13}$$

– *Step 11.* If necessary, re-examine the radius a of the drain then return to Step 2.

Design example
If $(\bar{u}/\sigma'_{vo})_{max} = 0.2$, $t_l = 5$ sec, $k_s = 3 \times 10^{-2}$ cm/sec, $m_v = 3 \times 10^{-3}$ cm²/kgf, $k_d = 10$ cm/sec, $a = 30$ cm, and $h = 10$ m, then

$$T_l = \frac{k_s t_l}{\gamma_w m_v a^2} = \frac{3\times10^{-2}\times5}{10^{-3}\times3\times10^{-3}\times30^2} = 55 \tag{5.14}$$

$$R = \frac{8}{\pi^2}\frac{k_s}{k_d}\left(\frac{h}{a}\right)^2 = \frac{8}{(3.14)^2}\times\frac{3\times10^{-2}}{10}\times\left(\frac{1000}{30}\right)^2 = 2.7 \tag{5.15}$$

From the design chart shown in Figure 5.24, $a/b = 0.35$. Therefore, for a square arrangement, the installation spacing of the drain becomes $d = (1.77 \times 0.3/0.35) = 1.5$ m.

5.4.2.4 *Remarks on design*

a) *In the procedure shown in Section 5.4.2.3, the parameter t_l plays a key role.* The parameter t_l is affected by the slightest error in the liquefaction strength curve. For example, based on the study by Tanaka et al. [54], an error of 75% is induced in t_l when there is an error of 10% in the liquefaction strength curve. In addition, the liquefaction strength curve is affected by the slightest disturbance in the sampling of the natural soil. Caution must, therefore, be taken in obtaining undisturbed samples for design of the gravel drain.

b) *Establishing the allowable excess pore water pressure ratio.* The allowable excess pore water pressure ratio may be determined from a stability analysis which takes into consideration the influence of excess pore water pressure. For example, if $(\bar{u}/\sigma'_{vo}) < 0.5$, earth pressure = seismic earth pressure* + excess pore water pressure is obtained (here, *... $\gamma' \times (1.0 - (\bar{u}/\sigma'_{vo})_{max})$ is used instead of buoyant unit weight γ', and the equivalent seismic coefficient is computed using this value). The value for excess pore water pressure is, however, strongly affected by the error in the liquefaction strength curve as indicated in Section 5.4.2.3. When considering overall precision, therefore, it may not be necessary to perform a detailed evaluation regarding the allowable excess pore water pressure ratio. The influence of the excess pore water pressure can be considered noticeable when it is $(\bar{u}/\sigma'_{vo})_{max} > 0.5$, and the influence of excess pore water pressure can be considered minimal in the case of $(\bar{u}/\sigma'_{vo})_{max} < 0.1$. It is, therefore, preferable to use a suitable value in the range of 0.1 to 0.5 as the allowable value of $(\bar{u}/\sigma'_{vo})_{max}$ for design. In order to maintain stability with respect to any error in parameter t_l, 0.25 is recommended as a standard value for $(\bar{u}/\sigma'_{vo})_{max}$ in design.

c) *Dependence of coefficient of volume compressibility on pressure.* The coefficient of volume compressibility depends on the effective vertical pressure σ'_{vo} and generally, m_v becomes larger with decreasing σ'_{vo}. When the effective vertical pressure is particularly small, for example, for a soil near the seabed, caution is necessary with regard to how effectively the drain will function.

d) *Sensitivity analysis with respect to the constants used in design.* For the design method noted in Section 5.4.2.3, the equivalent time necessary for liquefaction (corrected value) t_l, the coefficient of permeability in the horizontal direction of the soil k_s, and the coefficient of volume compressibility of the soil m_v are used as basic constants. The accuracy of these constants may not always be satisfactory in current practice. In design, therefore, it is preferable to perform a sensitivity analyses with respect to these constants, allowing for an adequate margin of safety factor in the design.

e) *When the permeability of the soil in the vertical direction is nominal.* The design method noted in Section 5.4.2.3 is based on the results of analysis for the condition where the coefficient of permeability in the vertical direction of the soil is ignored. This is because the permeability in the vertical direction is often noticeably reduced by the existence of a thin layer such as silt, etc. often included within a sand layer. For a sand layer interposed between gravel layers, however, the influence on the permeability in the vertical direction is nominal. In this case, a more economical design may be made by numerical analysis of the gravel drain rather than following the design method noted in Section 5.4.2.3.

f) *Details in the top part of the gravel drain.* For the gravel drain method, it is necessary not to prevent drainage from the top part of the drain. Therefore, a drainage mat (gravel/crushed stone layer) using the same gravel/crushed stone as the material used for the drain is often provided on top of the drain. For the pavement above the gravel/pebble mat, one example is where a drain hole is provided at the location of each drain installation, another example using a small number of drain holes with each one collecting water flowing from several drains, and another having water storage connected to the drain hole.

g) *Earthquake induced residual settlement of natural soil improved by the gravel drain method.* Settlement of natural soil improved by the gravel drain method is induced by earthquakes. Generally, the amount of settlement within the gravel drain and the natural subsoil profile is not always identical, resulting in differential settlement after earthquakes. The tests conducted at the Ports and Harbours Research Laboratory indicate that protrusions of a gravel drain were created on the ground surface due to differential settlement after excitation [53].

The amount of residual settlement in the portion of the natural soil improved by the gravel drain is, in theory, computed using the following equation.

$$S = \sum h_i \varepsilon_i \tag{5.16}$$

where: S = settlement (cm), h_i = thickness of layer i (cm) ($i = 1, ..., l$) (where l = number of layers), ε_i = volumetric strain of layer i ($i = 1, ..., l$).

Volumetric strain ε_i of each layer in Equation (5.16) is obtained, in theory, using the following equation.

$$\varepsilon_i = m_{vi}\sigma'_{vo\,i}\left(\frac{N}{N_l}\right)_{\max i}\frac{(\bar{u}/\sigma'_{vo})_{\max i}}{(N_u/N_l)_i} \tag{5.17}$$

where: m_{vi} = coefficient of volume compressibility of layer i (cm^2/kgf), σ'_{vo} = effective vertical pressure of layer i (kgf/cm^2), $(N/N_l)_{\max i}$ = maximum cumulative damage of layer i, $(N_u/N_l)_i$ = value of N/N_l corresponding to $(\bar{u}/\sigma'_{vo})_{\max i}$ in the excess pore water pressure generation curve for layer i, $(\bar{u}/\sigma'_{vo})_{\max i}$ = maximum value of the time history of excess pore water pressure ratio averaged over horizontal cross section for layer i. If $(\bar{u}/\sigma'_{vo})_{\max i} < 0.5$, the allowable excess pore water pressure ratio $(\bar{u}/\sigma'_{vo})_{\max i}$ established in Section 5.4.2.3 may be used as a substitute for $(\bar{u}/\sigma'_{vo})_{\max i}$. With regard to $(N/N_l)_{\max i}$ and $(N_u/N_l)_i$, refer to Section 5.4.2.3.

According to laboratory tests, the amount of residual settlement obtained with Equation (5.16) ranges from approximately the same to about half of that measured so that it is necessary to include a factor of about 2.0 as a safety factor [53, 54].

When the excess pore water pressure ratio generated during earthquakes exceeds 0.5, the value of the coefficient of volume compressibility changes considerably. For example, the value of the coefficient of volume compressibility when liquefaction is generated becomes more than 10 times as large [7, 50].

5.4.3 *Design of plastic drain*

5.4.3.1 *Material for the plastic drain*

The material used for the plastic drain should satisfy the following conditions:

1. Have a sufficiently high permeability for liquefaction remediation;
2. Have a sufficient strength to resist anticipated earth pressures on the drain; and
3. Have a suitable device to prevent clogging.

Some of the recently developed methods classified in this category are shown in Appendix 4.

For liquefaction remediation based on plastic drains, it is important to measure the coefficient of permeability of the drain and use the measured value in design even if the drain is made of a hollow tube or pipe. The coef-

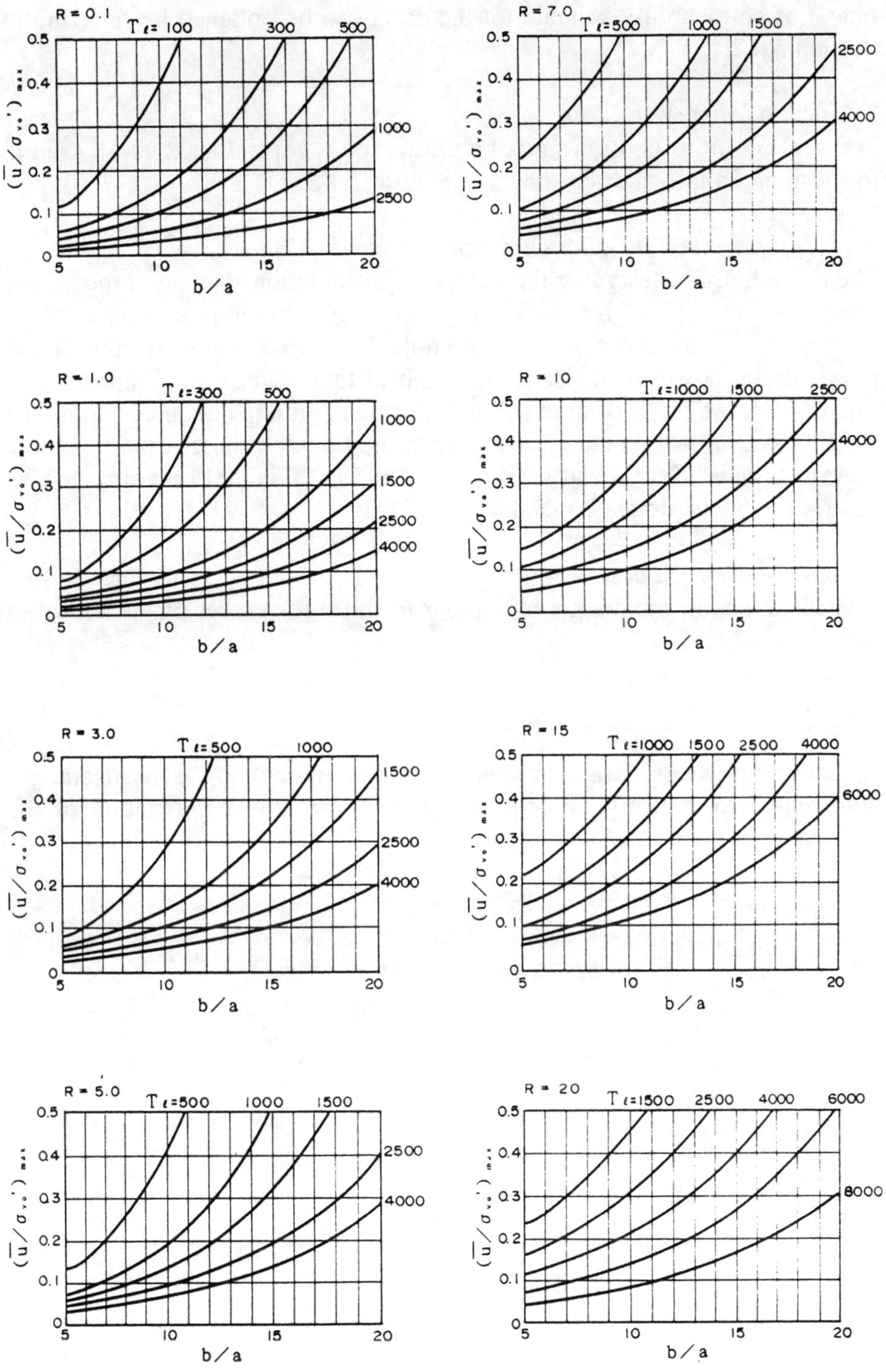

Figure 5.26. Design charts for plastic drains.

ficient of permeability suitable for the drain can be obtained by referring to Section 5.4.2.1.

5.4.3.2 *Soil survey/soil test*

Soil survey/soil tests for design of a plastic drain are similar to those used in the design of a gravel drain shown in Section 5.4.2.2.

5.4.3.3 *Radius and installation spacing of plastic drain*

The procedure for designing the radius and installation spacing of the plastic drain is similar to that for the gravel drain shown in the procedure in Steps 1-11 noted in Section 5.4.2.3. The equivalent cross sectional area of the plastic drain, however, is about one tenth of that of the gravel drain and in Figure 4.24, the range of time factor T_l normally anticipated in designing the plastic drain is not included in Figure 5.24. For designing a plastic drain, refer to Figure 5.26 instead of Figure 5.24 [55]. It should be noted that, in this figure, *b/a* is plotted on the horizontal axis.

5.4.3.4 *Remarks on design*

Considerations in the design are similar to those for gravel drains shown in Section 5.4.2.4.

5.4.4 *Installation examples*

As noted at the beginning of this section, the drain method has disadvantages with regard to ductility. In order to overcome this, a combination of the

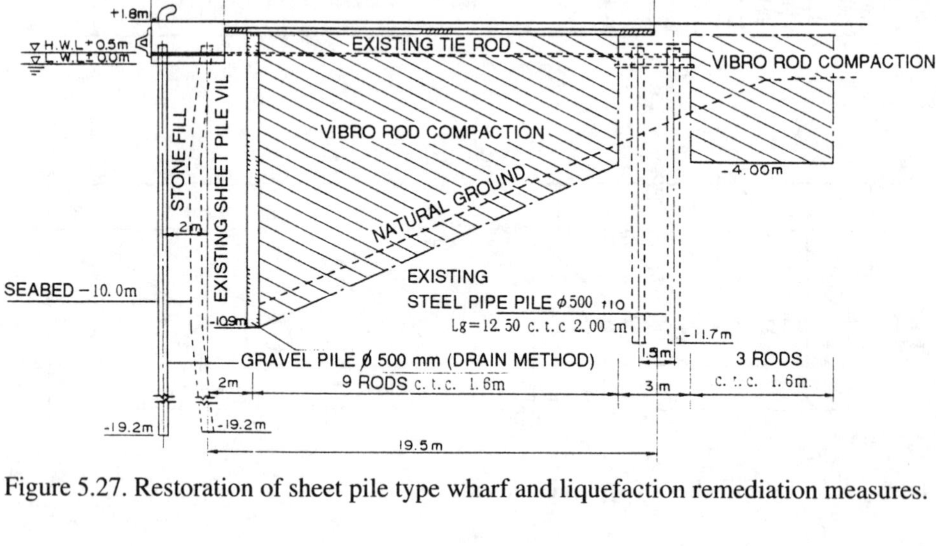

Figure 5.27. Restoration of sheet pile type wharf and liquefaction remediation measures.

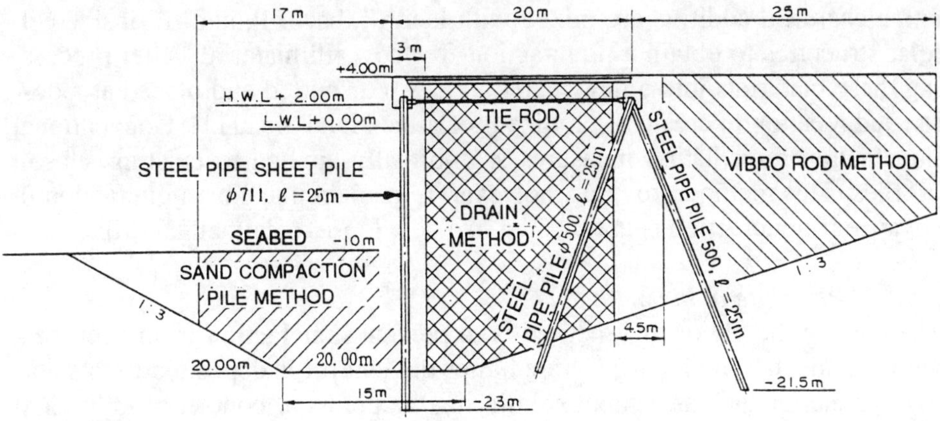

Figure 5.28. Liquefaction remediation measures for a sheet pile type wharf.

drain and compaction methods is preferable to adopting only the drain method. In the example shown in Figure 5.27, a liquefaction remediation was installed as part of the restoration of a wharf which was damaged by an earthquake [56]. As shown in the figure, ductility in the soil was achieved by using the compaction method whereas the drain method was used near the sheet pile wall to avoid any influence from installation work. In the test installation accomplished before the actual installation, the influence of the installation procedure was monitored in the gravel drain and vibro rod compaction methods 2 m from a sheet pile wall [56]. The results of the monitoring indicate an increase in bending stress in the steel sheet pile of about 250 kgf/cm^2 per gravel drain installation and an increase of 300 kgf/cm^2 per rod for the vibro rod compaction method. When compaction was done using the vibro rod compaction method after installation of the gravel drain, a noticeable accumulation of stress due to the vibro rod compaction method was not found to occur, and hence combining the gravel drain and compaction methods was considered suitable for soil improvement with respect to the existing structures.

In the example shown in Figure 5.28, the area of compaction included deep soil layers, so that the influence on the sheet pile from compaction was much greater. The area improved by the drain method was therefore much wider than the example shown in Figure 5.27 and was considered preferable in order to avoid the influence of compaction during installation [57].

5.5 DESIGN/INSTALLATION OF PREMIX METHOD

5.5.1 *Outline*

The premix method is a method where a small amount of stabilizing material and a chemical additive are mixed with landfill, backfill, and fill inside cellular structures to obtain a liquefaction-free new fill material. After processing these materials into a new material, it is transported and placed at a designated location to create a stabilized reclaimed land area [11]. Conventional liquefaction remediation measures are normally applied to existing subsoil profiles. This method provides, however, remediation against liquefaction in advance if liquefaction is anticipated in a new loose reclaimed landfill.

5.5.1.1 *Principle of the method*

The basic principle of this method is to prevent liquefaction from occurring with a cementation effect generated after the landfill is deposited by a stabilizing material such as cement solidifying the points of contact of individual soil particles.

Generally, the shear strength τ_f of soil, based on Mohr-Coulomb's failure criterion, is expressed by

$$\tau_f = c' + (\sigma - u)\tan\phi' \qquad (5.18)$$

where: c' = cohesion, ϕ' = angle of internal friction, σ = total vertical stress on shear plane, u = excess pore water pressure.

For a sand, $c' = 0$, and when excess pore water pressure is generated by an earthquake so that $\sigma = u$, liquefaction occurs and the shear strength becomes 0. When c' is added with cementation, however, the shear strength does not decrease to zero and liquefaction is not generated even when $\sigma = u$. In addition, the volumetric compression (negative dilatancy) of the soil skeleton is so restricted by the cementation effect that an increase in the excess pore water pressure itself is also suppressed.

5.5.1.2 *Application of the method*

For this method, sandy material used conventionally in landfill installation is deposited for reclamation after processing it in advance into a new material which does not undergo liquefaction. The method can, therefore, be applied to most conventional construction sites where sandy material was normally used without being processed. Examples of this application are shown in Figure 5.29. It should be mentioned that Figure 5.29 includes hypothetical construction cases. Since the installation method and characteristics of the treated sand required by the respective application differ, the method should be used only after performing laboratory/field tests and giving sufficient consideration to difference in the conditions.

5.5.1.3 *Characteristics of the method*

The following factors make this method advantageous: 1) remedial measures against liquefaction after the landfill is completed become unnecessary, the work period can be shortened and the labor force can be reduced, 2) landfill

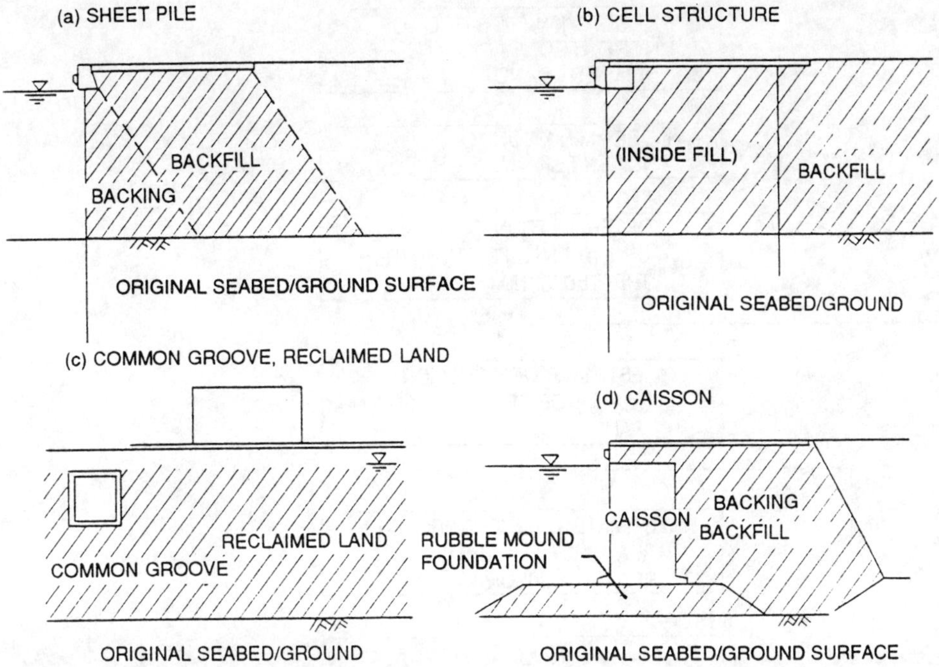

Figure 5.29. Application of premix method.

with a conveyor, chute or direct placement with bottom dump barge is possible for large, deep-water installations, 3) the strength of the treated soil can be optimized, 4) existing facilities/vessels can conveniently be used without requiring special installation equipment, 5) in addition to liquefaction remediation, the effect of a lateral earth pressure reduction can be expected. Consequently, the reduction of dimensions of a structure subject to earth pressures such as a sheet pile quay wall can be achieved. Increased economic benefit can therefore be realized, 6) there is no negative influence from noise and vibration compared to the conventional vibro-rod and sand compaction methods, 7) there is no effect on the sheet pile due to compaction of the wharf backfill, and 8) effective use of dredged sand can be achieved.

There are, however, four disadvantages: 1) variation in the strength of treated subsoil profiles is high depending on the type of sand and placement procedure, 2) some cost is associated with the stabilization material, 3) there is a need to consider environmental problems such as influence on the water quality, and 4) use is limited to excavated or replaced subsoil and newly reclaimed land.

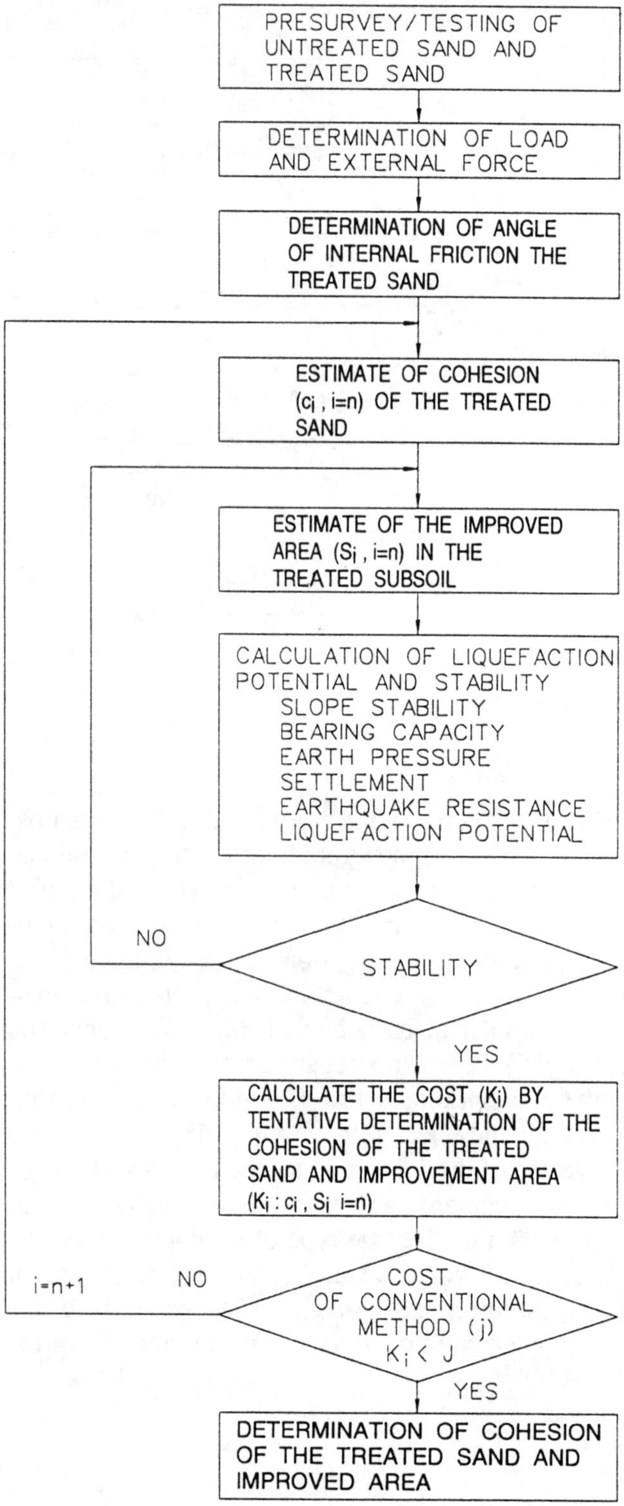

Figure 5.30. Design procedure.

5.5.2 *Design method*

5.5.2.1 *Design procedure*

This method differs from the conventional remedial measures against liquefaction which are usually applied to the existing subsoil after the completion of reclamation. It is a method which advances reclamation work while performing quality control of the treated landfill material so that reclaimed land with the necessary strength is obtained just after the completion of reclamation. Consequently, at the time of design, the area to be reclaimed does not exist. In the design, therefore, the strength of the treated sand and treated area necessary for resisting earth pressures, and providing adequate bearing capacity and preventing liquefaction of the newly reclaimed land must be determined considering the type of structures and facilities. The percentage of cement additive which satisfies the design conditions is assumed and the installation cost is approximately estimated. Soil strength and area of improvement are related to the percent of cement additive to be used and a number of different values of cement additive are assumed until the final percent cement additive is determined by considering the economic and installation factors. The design procedure is shown in Figure 5.30.

5.5.2.2 *Soil survey/test*

The soil survey and test are executed at each stage for planning, design, construction (construction work management), and post construction (quality control of reclaimed land). The survey and test at the time of planning are generally performed according to Section 4.2. With regard to newly reclaimed land constructed using the premix method, however, it is necessary to perform testing of the landfill material before and after treatment. Estimation of the SPT *N*-value for the reclaimed land is also necessary. For design, evaluating the relationship between cohesion, density, and the percent of cement additive, and understanding of the characteristics of both the treated and untreated landfill become the main objectives. The items to be tested and content of the treated and original landfill materials are shown in Table 5.6. Based on the items indicated in Table 5.6, the survey, test items, and frequency are determined according to the objectives for using this method. The unconfined compressive strength of the treated sand is one index related to the strength characteristics, as noted in Section 5.5.2.3b, which is evaluated in order to determine the approximate value of cohesion. Treated sand is brittle and therefore caution must be paid to undisturbed sampling to ensure that cracks are not created in the sample. Evaluation of the treated sand after the landfill is deposited is made by making an overall determination from the site investigation results such as elastic wave velocity, SPT *N*-value, and unconfined compression test results of undisturbed samples.

Table 5.6. Test items.

Landfill material		Test items	Test results
Untreated sand		Specific density test for soil particles	Specific gravity
		Moisture content test	Water content
		Grain size analysis	Grain size accumulation curve
		Maximum density/minimum density test	Maximum density/minimum density
		Triaxial compression test (CD or $\overline{C}\overline{U}$ test)	Angle of internal friction (ϕ', ϕ_d)
		Cyclic triaxial test	Liquefaction strength, dynamic deformation/damping characteristics (when untreated sand remains)
Treated sand	Laboratory trial for mix proportions	Specific density test for soil particles	Wet density, dry density Percent of cement additive, density, and strength
		Unconfined compression test	Unconfined compressive strength Percent of cement additive, density, and strength
		Triaxial compression test (CD or $\overline{C}\overline{U}$ test)	Angle of internal friction (ϕ', ϕ_d) Cohesion (c', c_d)
	During mixing	Calcium analysis (EDTA titration method)	Percent cement additive
		Unconfined compression test	Unconfined compressive strength
	After landfill	Standard penetration test	SPT N-value
		Elastic wave exploration	P-wave velocity, S-wave velocity
		Density test	Wet density, dry density
		Unconfined compression test	Unconfined compressive strength
		Triaxial compression test (CD or $\overline{C}\overline{U}$ test)	Angle of internal friction (ϕ', ϕ_d) Cohesion (c', c_d)
		Calcium analysis (EDTA titration method)	Percent of cement additive
		Plate loading test	Yield and ultimate bearing capacity, deformation coefficient
		Borehole logging	Yield and ultimate bearing capacity, deformation coefficient
		In-situ permeability test	Coefficient of permeability

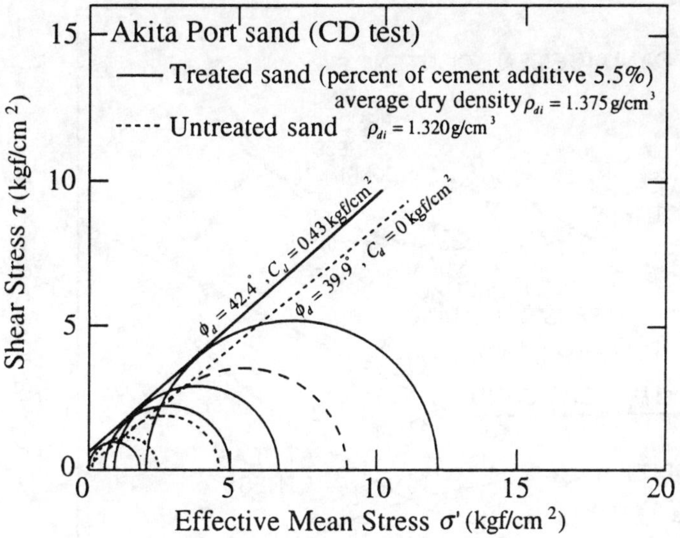

Figure 5.31. Rupture envelope (age 28 days).

5.5.2.3 *Shear strength characteristics of the treated sand [58]*

a) *Shear strength.* For sand treated with low cement content, the angle of internal friction becomes about the same or slightly greater than that of untreated sand. Figure 5.31 indicates the rupture envelope in the CD test for treated and untreated sands. The rupture envelope for treated sand is shown in parallel to that for untreated sand. A similar result is obtained for the $\overline{CU}$ test. In design, therefore, it is possible to use the same angle of internal friction for treated and untreated sand. A comparison of strength constants for the CD test of the treated and untreated sands is shown in Figure 5.32. According to this figure, it is apparent that the angle of internal friction after treatment is slightly greater or more or less equal when compared to that before treatment, and that the additional cohesion increases in proportion to the dry density of the specimen. Figure 5.33 shows the strength constants for peak and residual stress conditions (axial strain is 15%) in the CD test. The angle of internal friction is maintained at more or less the same value as that at the peak even in the residual state whereas the cohesion becomes noticeably smaller than that at the peak. The angle of internal friction is about the same as that for untreated sand even when the treated sand is crushed by mechanical grinding to provide the same grain size distribution as untreated sand. The treated sand has a strength at least equal to untreated sand, which is a unique characteristic for treated sand.

b) *Unconfined compressive strength.* The drainage condition for an unconfined compression test of treated sand differs from that of clay. The treated sand specimen is sheared under a drained condition. The result obtained, therefore, should be considered as an index which gives the cohesion

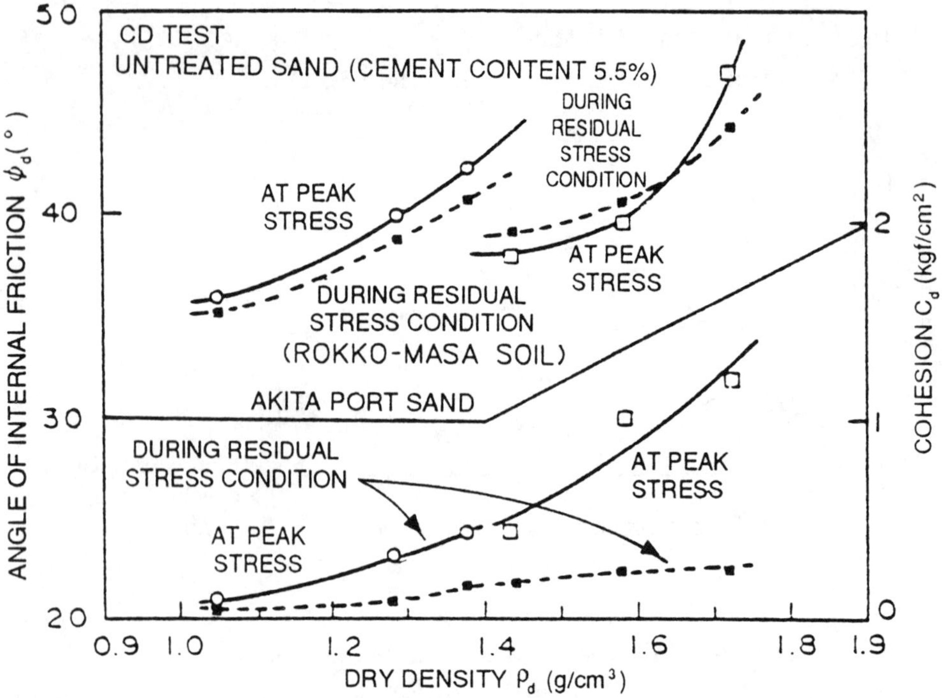

Figure 5.32. Strength constants for treated and untreated sands.

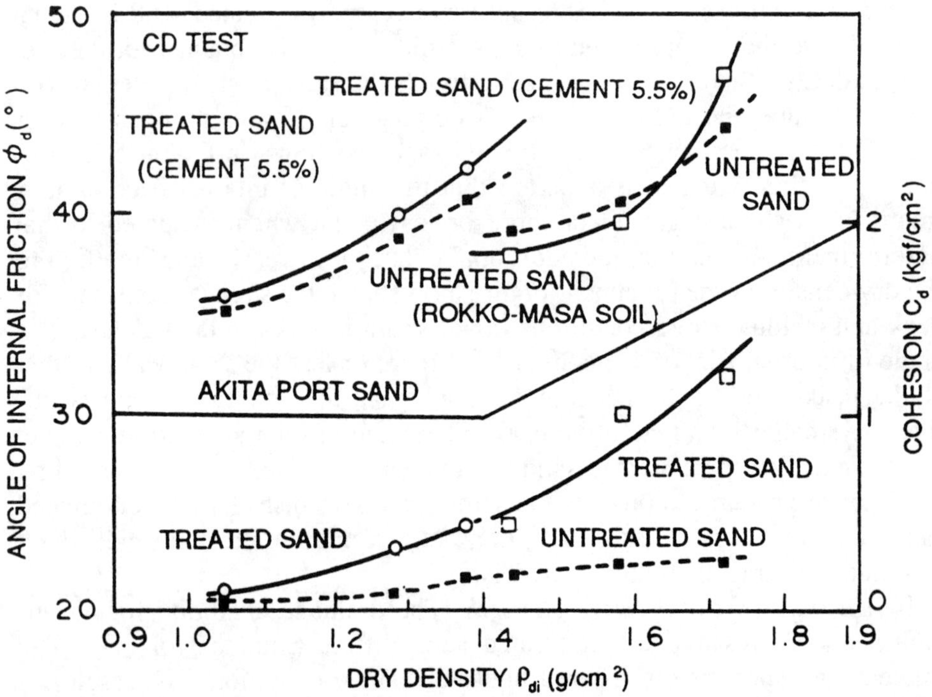

Figure 5.33. Peak strength constants during residual stress conditions (age 28 days).

component. If the angle of internal friction of treated sand and untreated sand is equal, the relationship for unconfined compressive strength q_u and cohesion c_d of treated sand is expressed by the following equation with the angle of internal friction for untreated sand as ϕ_{do}.

$$c_d \fallingdotseq \frac{q_u}{2} \frac{(1 - \sin \phi_{do})}{\cos \phi_{do}} \tag{5.19}$$

The cohesion of treated sand is estimated using Equation (5.19) from the angle of internal friction for untreated sand and unconfined compressive strength.

5.5.2.4 *Strength constants for design*

a) *Shear strength.* In designing the stability of a treated subsoil profile, the shear strength is basically calculated by using both the angle of internal friction and cohesion, using the following equation.

$$\tau_f = c_d + \sigma' \tan \phi_d \tag{5.20}$$

where: τ_f = shear strength, c_d = cohesion from CD test, ϕ_d = angle of internal friction from CD test, σ' = effective vertical stress on shear surface.

In treated sand, just as for untreated sand, $c_d \fallingdotseq c'$, $\phi_d \fallingdotseq \phi'$, where, c' and ϕ' are respectively cohesion and angle of internal friction in terms of effective stress.

b) *Drainage condition.* To calculate the stability of a treated subsoil profile against static external force, it is possible to use the strength constants c_d and ϕ_d measured under drained conditions because the coefficient of permeability and the coefficient of volume compressibility for the treated sand both decrease and therefore the consolidation coefficient, which controls the drainage characteristics of the treated sand, decreases only slightly compared to untreated sand. For the undrained state such as that during earthquakes, if the density of the treated sand is not very low, the axial differential stress in the $\overline{\text{CU}}$ test is larger than the axial differential stress in the CD test. Using the shear strength obtained in the CD test leads to a conservative design.

c) *Strength constants.* Although the brittle state increases in treated sand, the strength is defined with Mohr-Coulomb's failure criterion as in other untreated soils. Except for the problem of strength scatter in treated subsoil, the same evaluation of strength constants as other untreated soils can be performed even in treated sand in the design. Treated sand has both the angle of internal friction and cohesion as noted above. There is some concern about too conservative design using the existing design method with either of the strength constants c_d or ϕ_d when calculating the bearing capacity and earth pressure for treated subsoils. The stability analysis of treated subsoil, therefore, is basically made by using both the angle of internal friction and cohesion [59].

d) *Evaluation of scatter in the strength of treated subsoil profile.* Scatter in the strength of the treated subsoil profile, which is mainly caused by the scatter of cohesion, is greatly affected by the reclamation method. With regard to evaluation of the scatter, there are a number of concepts, but when the premix method is limited to use as a liquefaction remediation measure, it is possible to use only the angle of internal friction for the untreated sand and design without taking the additional cohesion into consideration, although this is considerably conservative. When the cohesion is taken into consideration in the calculation of earth pressure, circular slip failure, bearing capacity, etc., the following approaches are considered: 1) design using the cohesion and angle of internal friction with the lower bound value for the scatter of the cohesion, 2) design using the cohesion and angle of internal friction by using the average value as in the clayey soils, and 3) design by using the cohesion and the angle of internal friction by giving consideration to the standard deviation in the cohesion.

5.5.2.5 *Earthquake resistant design, treatment area, and strength of the treated subsoil profile*

Treated sand has relatively high rigidity when compared with both sand and clay. It behaves like a rigid body during earthquakes. In design, therefore, not only the internal stability against the earthquake-induced stresses in the treated subsoil profile but also the overall stability of the treated sand during earthquakes must be evaluated. In the evaluation of the overall stability of the treated sand, the treated subsoil profile is considered as a rigid mass and the static design method based on the seismic coefficient method is used. When evaluating the overall stability, the design seismic coefficient, dead weight of mass, earth pressure, hydraulic pressure, residual water level, and the coefficient of friction between the mass and original subsoil profile are considered. The sliding, overturning, and bearing capacity are evaluated with respect to the entire mass of the treated subsoil profile. On the basis of these results, the area of the treated subsoil profile necessary to provide earthquake resistance is determined. For evaluating internal stability, the dynamic design method based on earthquake response analysis is used. By calculating the acceleration, strain, and stress within the treated subsoil profile, the strength of the treated sand necessary for earthquake resistance is obtained. For seismic response analysis, the computer program FLUSH for example can be used [60]. The response analysis shows that the strain in a treated subsoil profile is less than the strain in an untreated subsoil profile and a negative influence on earthquake resistance is not observed. With regard to the determination of the actual treatment area and strength of the treated sand, consideration is given to the stability and installation not only during earthquakes but also for static load conditions. An overall decision must be made.

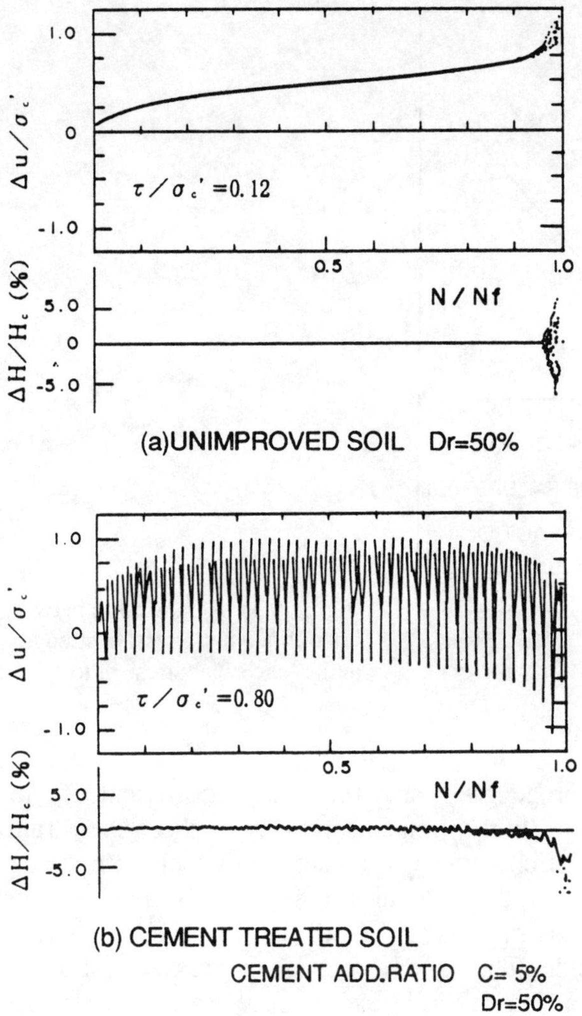

Figure 5.34. Excess pore water pressure ratio ($\Delta u/\sigma_c'$) and axial strain ($\Delta H/H_c$).

5.5.2.6 *Trial for mix proportion*

a) *Liquefaction-free fill material.* The percent of cement additive required for treatment to produce a material which will not undergo liquefaction differs according to the soil type. It can be determined by performing a mixing test. According to existing test results, however, no liquefaction occurs if the cement content is about 5% [61]. Also if the unconfined compressive strength is 0.5 to 1.0 kgf/cm^2, it can not undergo liquefaction [62]. This is apparent from the fact that the liquefied state of a specimen of treated sand is not observed in a cyclic triaxial test to which a small percentage of cement is added. The specimen remains in the initial solid state unless extensive failure was created by necking. Figure 5.34 shows, as one example, the axial strain $\Delta H/H_c$ and excess pore water pressure ratio $\Delta u/\sigma_c'$ in a cyclic triaxial

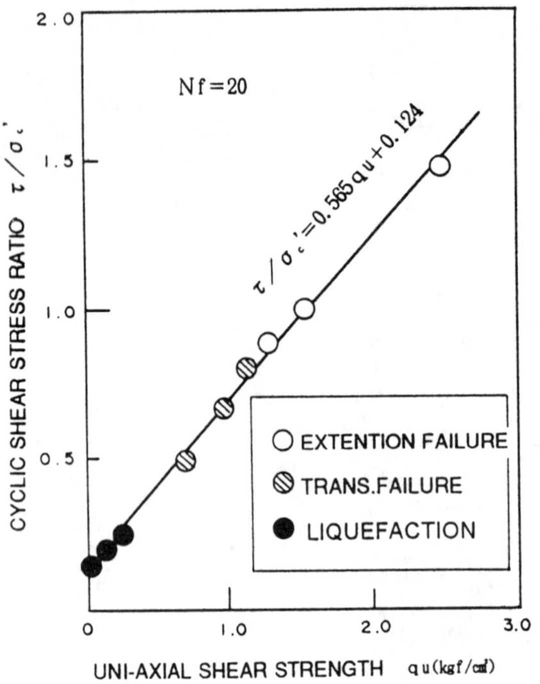

Figure 5.35. Relationship between unconfined compressive strength and cyclic shear stress ratio.

test on Otaru Port sand (average grain size 0.26 mm, coefficient of uniformity 2.23, fines content 5%). For untreated sand (cyclic shear stress ratio $\tau/\sigma'_c = 0.12$), a typical liquefaction pattern is indicated in which both the excess pore water pressure ratio and the axial strain suddenly increase. For treated sand (cyclic shear stress ratio $\tau/\sigma'_c = 0.80$), the residual axial strain increases on the extension side. The excess pore water pressure ratio also differs from untreated sand and a negative excess pore water pressure results. This specimen maintains a solid state even after the test and no liquefied state in which the sand particles are dispersed is observed. Figure 5.35 shows the relationship between the unconfined compressive strength and cyclic shear stress ratio by giving consideration to this type of failure pattern. For Otaru Port sand, liquefaction was clearly generated at an unconfined compressive strength below 0.5 kgf/cm^2, while liquefaction was not generated at an unconfined compressive strength of 1.0 kgf/cm^2 or more. When the cement content creating the specimen with unconfined compressive strength of 1.0 kgf/cm^2 is estimated from the relationship between the percent of cement additive and the unconfined compressive strength obtained separately, it corresponds to a cement content of 3 to 4 %.

Figure 5.36 shows the relationship between the number of load cycles and the cyclic shear stress ratio when the amplitude of the axial strain exceeds 5% (for Awaji-masa soil (decomposed granite), the brittle state is high.

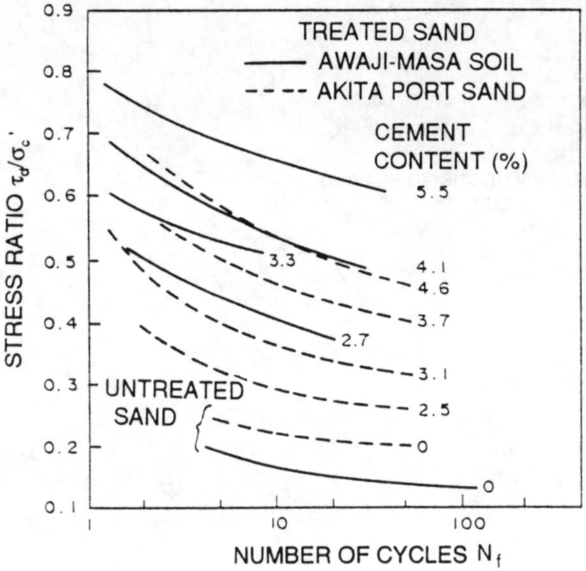

Figure 5.36. Cyclic shear stress ratio and number of load cycles (age 28 days).

When it exceeds 1%, the axial strain amplitude increases instantaneously to failure. The axial strain amplitude of 1% was therefore adopted). In accordance with Figure 5.36, the cyclic stress ratio increases noticeably even when the percent of cement additive is only a few percent. The curves in Figure 5.36 do not always express the liquefaction strength, but do include the extensive failure. When the failure mode in extension noted above is considered, it is on the conservative side. By mixing 5% cement, it is possible to process a fill material which will not undergo liquefaction with a factor of safety.

Figure 5.37 shows results based on the shaking table test (container inner dimension: 3 m × 3 m × 1.2 m) for untreated sand and treated sand in Niigata Eastern Port [63]. The percent of cement additive of the treated sand is 1%, curing time is 7 days, and the average unconfined compressive strength is 0.3 kgf/cm^2. Both layers are formed similarly by placing underwater. Compaction is not performed. The acceleration is a maximum of 250 Gal. In the treated layer, hardly any generation of excess pore water pressure is found and liquefaction as in untreated sand is not caused. Even with a low percent of cement additive, it is evident that effective liquefaction remediation is achieved.

The relationship between the cyclic shear stress ratio and unconfined compressive strength is shown in Figure 5.38. According to Figure 5.38, a corrclation is determined between the cyclic shear stress ratio and unconfined compressive strength. When the unconfined compressive strength increases, the cyclic stress ratio also increases.

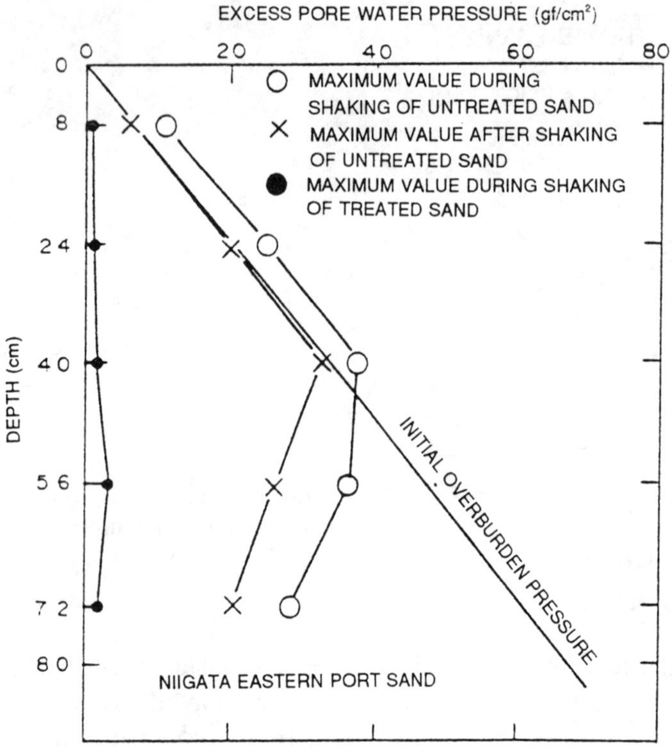

Figure 5.37. Distribution of maximum excess pore water pressure.

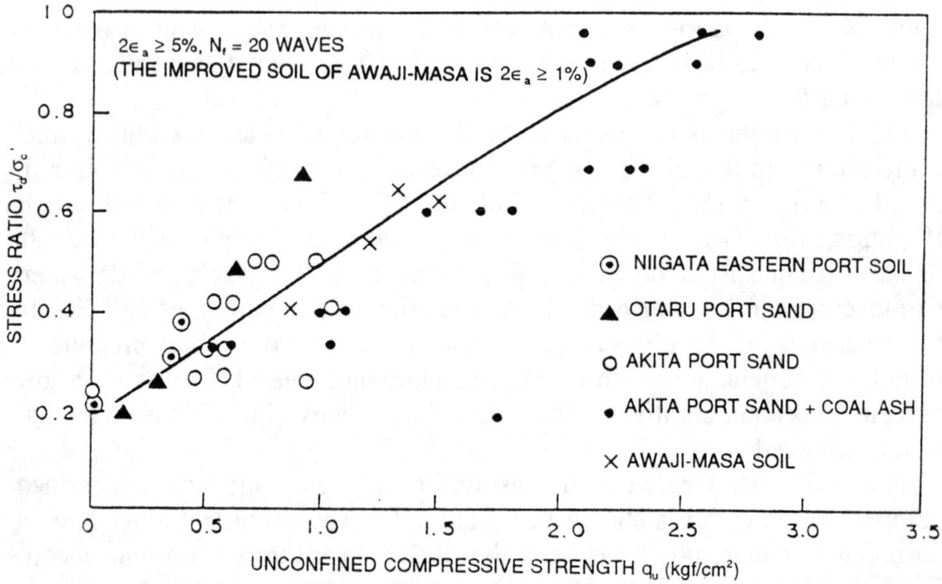

Figure 5.38. Correlation of cyclic shear stress ratio and unconfined compressive strength.

b) *Cement content.* The relationship between the percent of cement additive and strength of the treated sand is obtained from a laboratory test. At such a time, the number of days for curing the treated sand is based on 28 days. It is preferable that the strength of the treated sand be obtained from a triaxial compression test but it is possible to calculate the cohesion added conveniently using the unconfined compression test. In this case, Equation (5.19) is used for calculating the cohesion. The necessary cement content corresponding to the designated cohesion is obtained from the laboratory test.

For an actual installation, differences in the degree of mixing the cement, separation and dispersion of the cement caused by underwater placement, difference in curing environment and density of the deposit cause loss of cement. When determining the cement content for actual construction, therefore, the cement content obtained from the laboratory test must be increased. The percent of cement additive is determined from the relationship between the unconfined compressive strength at the site and the unconfined compressive strength using the laboratory test as an index. The ratio of unconfined compressive strength at the site and in the laboratory test is defined as a coefficient expressed by the following equation.

$$\alpha = \frac{\bar{q}_{ul}}{\bar{q}_{uf}} \tag{5.21}$$

where: $\bar{q}_{ul}$ = average unconfined compressive strength from the laboratory test (kgf/cm^2), $\bar{q}_{uf}$ = average compression strength at the site obtained with the same mix proportion as the laboratory test (kgf/cm^2).

Up to the present time, the unconfined compressive strength of the treated subsoil profile created at the site has been found to be between 1/1.1 and 1/2.2 of the laboratory unconfined compressive strength with respect to the same percent of cement additive [64, 65]. In order to maintain strength at the site for the actual installation, the percent of cement additive with a margin of about 1.1 to 2.2 times is determined by converting unconfined compressive strengths. There is considerable margin in the percent of cement additive and actually it depends greatly on the condition at the site such as the mixing method used in the installation and the reclamation method. It is preferable to determine the percent of cement additive by performing a field test before installation.

5.5.3 *Installation method*

5.5.3.1 *Installation work and facility*

The procedure for the installation is: 1) transporting landfill material, 2) mixing the cement, 3) adding the separation inhibitor, 4) transporting the

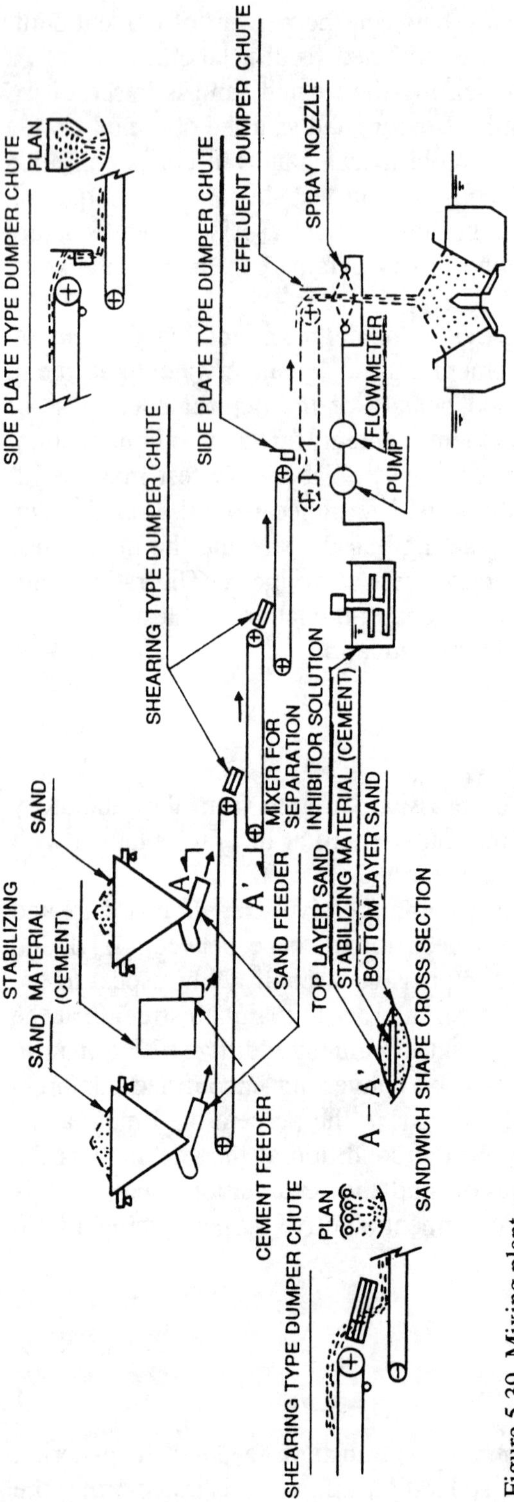

Figure 5.39. Mixing plant.

Table 5.7 Installation work.

Method of installation	Installation content
Mixing method	Mixing with belt conveyor
	Mixing with mechanical mixer
	Mixing with large equipment
Mixing location	Mixing on land
	Mixing in open water
Landfill method	Method using open bottom type barges
	Placement using bulldozers
	Method using a chute

treated sand, and 5) filling land. With regard to the mixing and filling land, there are a number of methods indicated in Table 5.7. The selection of a suitable combination of the methods is made based on the conditions at the construction site.

a) *Mixing work.* There are three mixing methods as follows:

1. Belt conveyor mixing method with dumper chutes installed on the connection with each belt conveyor (Fig. 5.39);
2. Mechanical mixing using a screw type mixer or rotary drum type mixer;
3. Heavy machine method using excavation equipment such as a backhoe.

In this section, the newly developed belt conveyor method [66] is introduced. This mixing method is economically and efficiently applied when the landfill material is dry and the water content is less than the natural water content. As shown in Figure 5.39, the facility is composed of a series from the sand hopper to the belt conveyor and separation inhibitor spray nozzle. The characteristics of the facility are as follows: 1) the sand is thinly spread from two sand hoppers on the belt conveyor, 2) the cement is sandwiched between the thin layers of sand on the belt conveyor, and 3) the two types of dumper chutes, called shearing type dumper chute and plate type dumper chute, are provided at the connection to each belt conveyor. There are advantages of this system. First, effective mixing is possible with minimal mixing energy because sand is spread thinly during the initial stage of mixing. Dispersion of the cement is prevented by sandwiching. A large volume can be mixed by running the belt conveyor continuously. Because of these advantages, efficient mixing equal to conventional mechanical mixing is attained.

Mixing may be done on land or sea according to the site conditions. It is best to install the mixing plant as close as possible to the landfill site because of the time required for solidification of the treated sand [67]. When a landfill site is close to the loading location for the fill material, it is possible to incorporate the mixing plant in the loading facility. In mixing at sea, a plant may be installed on the barge. A scheme for the mixing plant vessel is shown in Figure 5.40.

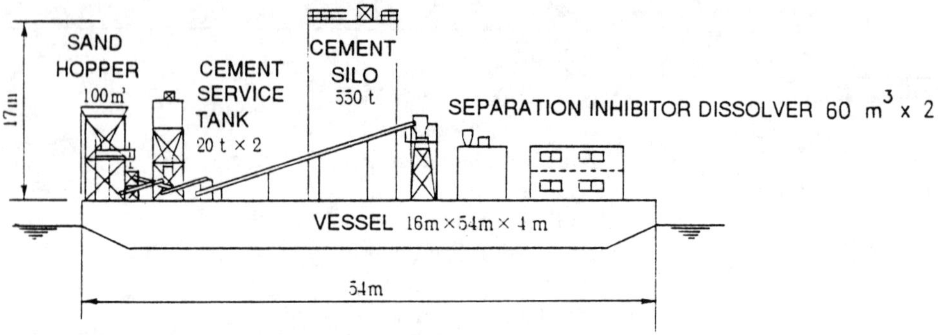

Figure 5.40. Schematic diagram of mixing plant.

The separation inhibitor prevents the added cement from separating from the sand particles underwater; an aqueous solution of nontoxic strong anionic polyacryl amide is used. It is sprayed onto the treated sand as it is discharged from the belt conveyor. If the amount of separation inhibitor is excessive, the density of the reclaimed land may decrease. Normally, a separation inhibitor of 50 to 100 gf is used per a dry weight of 1.0 tf for the treated sand.

b) *Landfill placement.* Several methods can be applied for landfill using treated sand;

– Method 1: Spreading method;
– Method 2: Chute method;
– Method 3: Direct placement method with bottom dump type barges.

The spreading Method 1 is suitable for reclamation from land with equipment such as a bulldozer. The direct placement Method 3 is suitable for the reclamation within an open water area enclosed by a bulkhead. These two methods are the same as the conventional method presently used. The chute Method 2 is rather effective for preventing water pollution. Figure 5.41 shows a chute for underwater installation. In this chute, the outside is accordion-like, made of cloth. The inside pipe is the telescopic type which can be expanded or shortened in accordance with a change in the height of the created sand mound. The bottom end of the chute during the installation is always maintained just above the sand surface. This method has the characteristics of being able to reduce the dispersion of turbidity with the outside cloth.

5.5.3.2 *Inspection in construction work*

a) *Inspection during mixing.* The strength of the reclaimed land depends not only on the percent of cement additive but also on the density. For land reclamation, however, density control is not usually done and hence the percent of cement additive must be strictly controlled in order to acquire the desig-

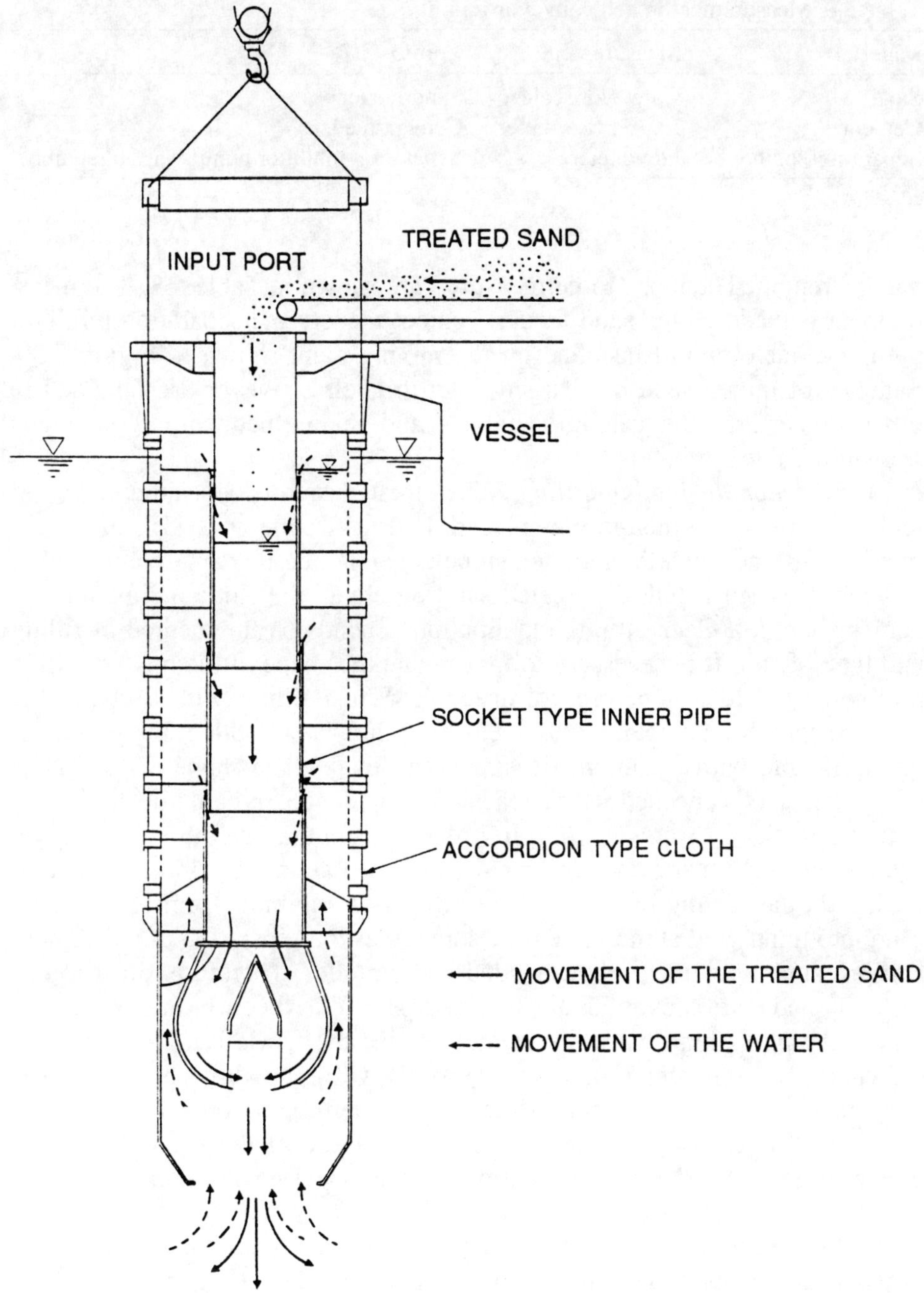

Figure 5.41. Chute for underwater fill placement.

Table 5.8. Measurement in belt conveyor mixing.

Material	Measurement	Control
Sand	Conveyor scale	Sand feeder
Cement	Impact flowmeter	Cement feeder
Separation inhibitor	Flowmeter	Separation inhibitor pump (variable pump)

nated strength. The items to be measured are shown in Table 5.8. The measurement is made with a sand feeder, cement feeder, and separation inhibitor pump (variable pump) installed on the mixing plant shown in Figure 5.39. Samples of mixed sand discharged from the belt conveyor are obtained in order to perform the calcium analysis and unconfined compression test shown in Table 5.6.

b) *Inspection during landfilling*. When treated sand is deposited underwater, the slope of the mound thus placed depends on the grain size of the fill material. In fine sand, the inclined slope is gentle but in coarse grain sand, it is steeper. When filling the treated sand by chute, the slope inclination becomes about l:4. Generally, the inclination depends on the method of filling and type of soil. It is necessary to investigate the shape of the mound with a preliminary field test before construction. When filling with a chute, the mound shape can be obtained by measuring the water depth at the perimeter of the mound with an automatic sounding device. The shape of the entire mound created is checked with a lead sounding device or bathometer.

c) *Inspection of water quality*. In order to maintain favorable water quality, the installation area is surrounded with a contamination-prevention cloth during the landfilling operation. The separation inhibitor forms the treated sand into lumps and at the same time suppresses the separation of the cement and promotes settlement of suspended soil particles. Therefore, diffusion of turbidity and elution of alkali are suppressed. Within the turbidity prevention fabric, the pH is high but the seawater outside the fabric has a neutralizing effect so that by reducing the increase in pH within an allowable level, it is possible to eliminate the worst influence on water quality outside the turbidity prevention cloth. Figure 5.42 shows a decrease of the pH in the alkali eluate by seawater dilution. It is apparent that seawater dilution of about 7 times achieves a decrease below pH 8.6 [68]. Water quality inspection during installation is executed with a focus on the turbidity and pH at representative locations inside and outside the installation area, inside and outside the turbidity prevention cloth, and near the fill placement.

d) *Inspection of quality of treated sand/fill*. For quality inspection of the treated sand and fill, the survey and test indicated in Table 5.6 are performed. Boring and sampling of the treated subsoil profile during reclamation are advised. Inspection of the treated sand discharged from the mixing plant is also necessary.

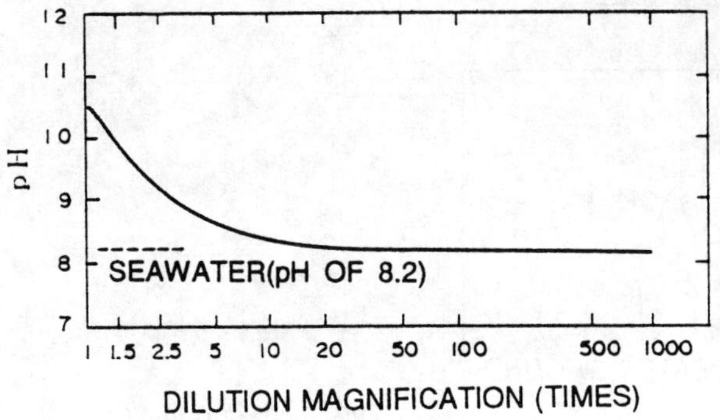

Figure 5.42. pH decrease in alkali elution due to seawater dilution.

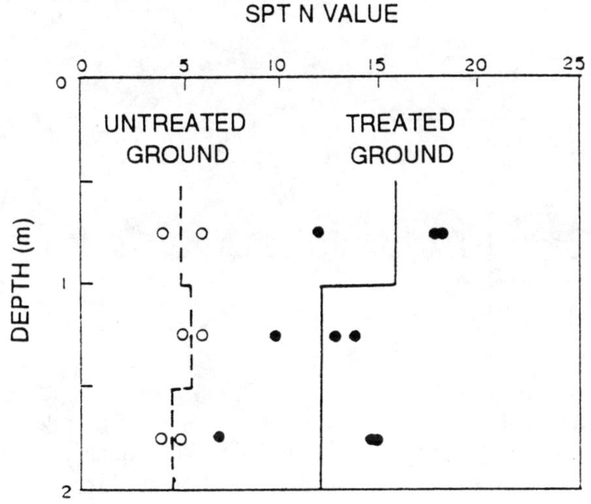

Figure 5.43. Standard penetration test results (Tokushima gravelly sand).

5.5.3.3 *Preliminary field test before construction*
A preliminary field test is recommended if no construction experience exists at the site. The test is performed with a prearranged mixing plant, equipment, and reclamation procedure, using the determined mix proportion for the treated sand. Verification is made that the designated percent of cement additive is included in the treated subsoil profile and that the required strength is obtained. In addition, the field test is designed to confirm the appropriate mix proportion, construction, and inspection methods.

5.5.3.4 *Examples of reclamation test*
a) *Filling by backhoe*. Treated gravelly sand (grain size below 300 mm) processed by mixing 4% of slag cement on land using a backhoe was placed in open water using a front-end loader (for preventing influence of compaction by the equipment). The SPT *N*-value of the treated landfill is shown in Figure 5.43. As shown in Figure 5.43, a large SPT *N*-value was obtained in

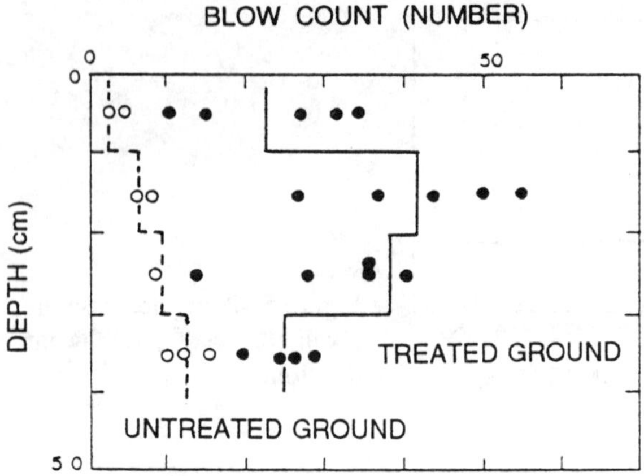

Figure 5.44. Portable dynamic cone penetration test results.

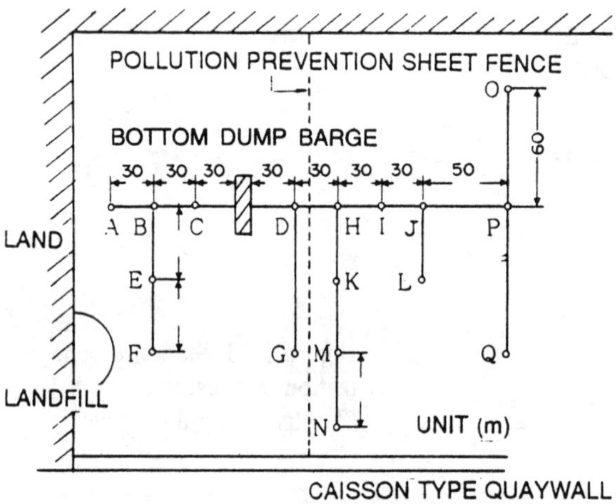

Figure 5.45. Water quality measurement point (direct placement of treated soil in open water).

the treated landfill sand compared to untreated material due to the cementation effect even when the material was not compacted.

b) *Disposal with bottom dump barge*. The treated sand was prepared by mixing slag cement 5.5% with Tokushima gravelly sand (grain size below 50 mm). The mixing was done in this case under dry conditions by a concrete mixing plant vessel. The treated sand was then placed in open water at a depth of 10 m. Treated sand deposited in a container which was set in advance on the seabed was raised by crane in order to evaluate the strength of the deposit. The results of the portable cone penetration test are shown in Figure 5.44.

Figures 5.45 and 5.46 show respectively the points of measurement of water quality in the test area and the measured pH, turbidity, SS, and COD. In the direct placement test, these values increased along the perimeter of the

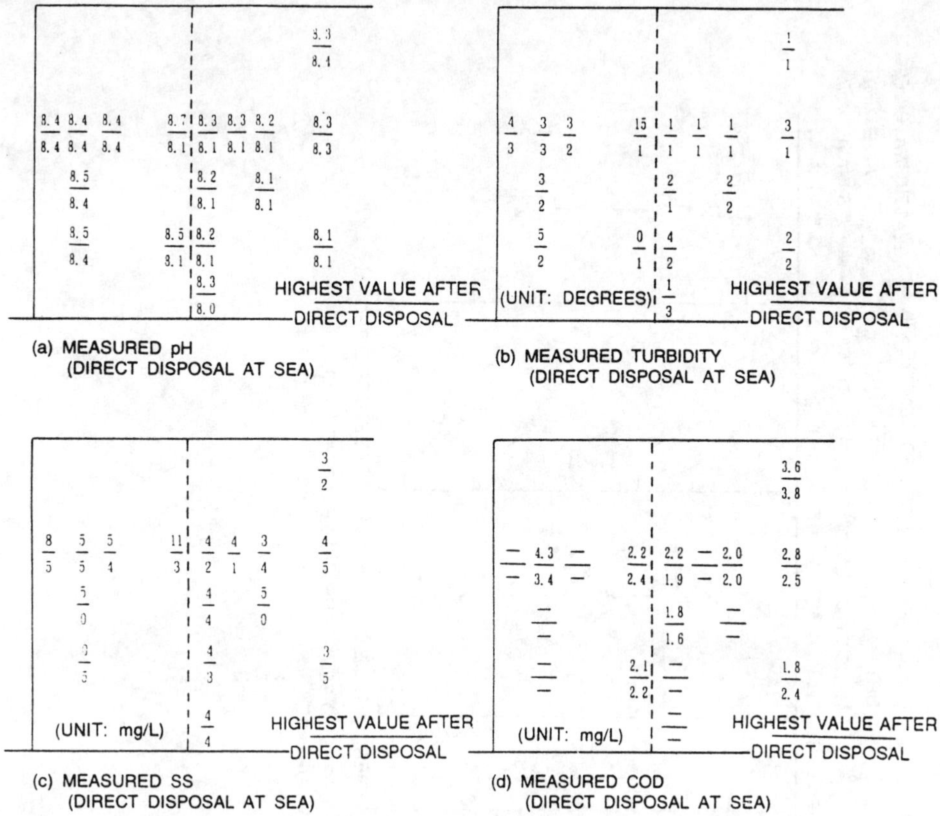

Figure 5.46. Measured water quality.

placement area within the turbidity prevention cloth, but a significant water quality change was not observed outside the turbidity prevention cloth. Generally, the water quality outside the turbidity prevention cloth is believed to depend also on the amount of treated sand placed. Depending on the situation, therefore, it will be necessary to use a neutralizing agent in the underwater placement.

Figure 5.47 indicates the result of a direct disposal test in a large water tank with a water depth of 3 m. The slag cement of 5.5% was mixed with Awaji-masa sand by belt conveyor. The total amount of fill material was 500 m^3. Average dry density γ_d was 1.56 gf/cm^3. The SPT N-value, however, showed an increase to about 20 after 28 days. After five months, an unconfined compressive strength of about 4.6 kgf/cm^2 was obtained.

c) *Filling by chute.* A fill test using a chute was performed in a water tank with a depth of 10 m using Chiba clean sand. The percent of cement additive was about 7.5% (100 kgf/m^3) and the belt conveyor method was used for mixing. The mixing and installation speed was 80 to 200 m^3/h. The total

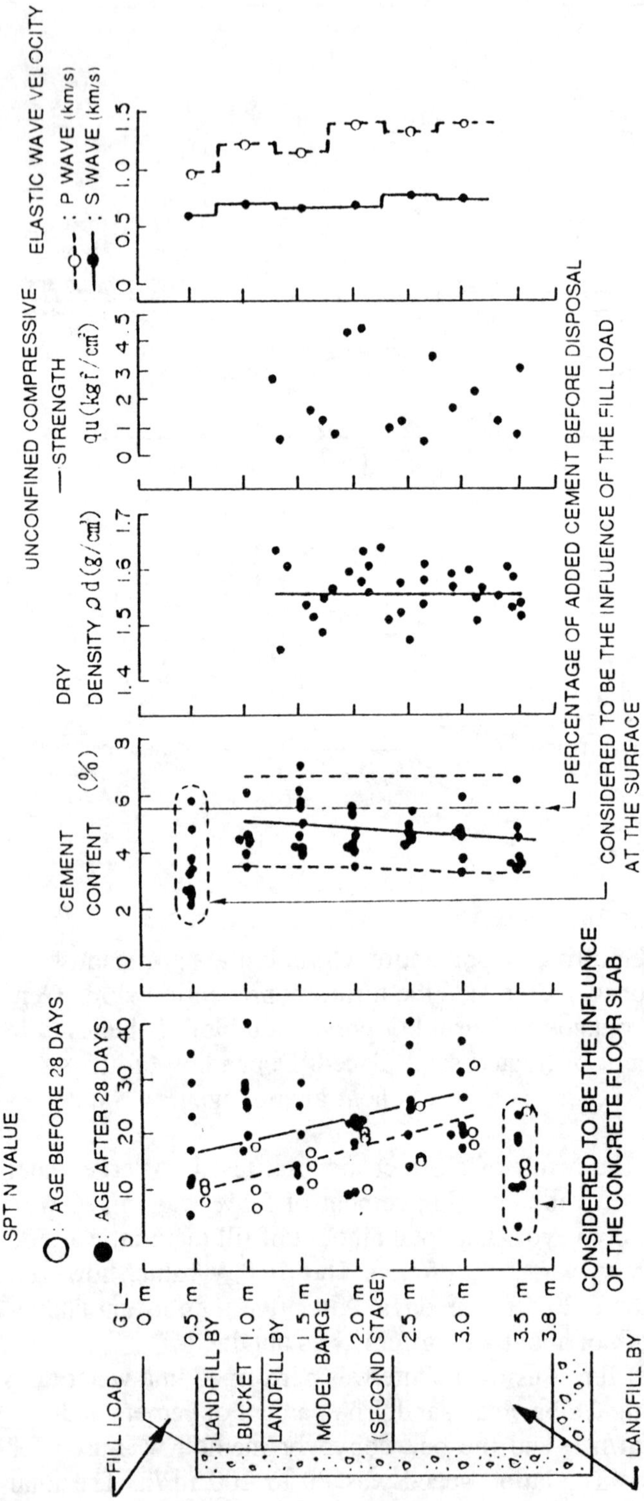

Figure 5.47. Characteristics of treated subsoil profile.

Figure 5.48. Excavation for treated subsoils.

Figure 5.49. Block sample of treated sand.

amount of landfill soil was 500 m^3. The excavated cross section of the treated sand is shown in Figure 5.48 and the block sample of the treated sand is shown in Figure 5.49. Unconfined compressive strength of 3 to 4 kgf/cm^2 was obtained after 28 days and 6 to 8 kgf/cm^2 was obtained after 91 days.

5.6 DESIGN/INSTALLATION OF PRELOAD METHOD

5.6.1 *Outline*

The preload method for liquefaction remediation is based on laboratory findings that overconsolidation of sand increases liquefaction resistance [15, 69]. The overconsolidation can be made either by placing a surcharge or by lowering the ground water level by pumping.

The preload method is adopted for liquefaction remediation in the four cases: 1) when the subsoil profile consists of sandy soil with a high fines content and it is difficult to apply compaction and/or drain methods, 2) when the subsoil profile consists of combined layers of sand and clay and it is necessary to improve the clay layer, 3) when it is necessary to lower the ground water level for excavation, and 4) when liquefaction remediation under an existing structure is necessary.

The design/installation for the preload method has not been sufficiently established for liquefaction remediation. It is preferable, therefore, to verify in advance with an in-situ test installation.

5.6.2 *Liquefaction characteristics of overconsolidated soil*

5.6.2.1 *Increase in liquefaction resistance by overconsolidation [69]*

The liquefaction resistance of sandy soil increases by overconsolidation. Figure 5.50 shows cyclic triaxial test results for Niigata East Port sand. Overconsolidation ratio (OCR) is defined by a ratio of σ'_p over σ'_c, where σ'_p is the maximum consolidation pressure that the sample has been subjected to and σ'_c is the current consolidation pressure as shown in Equation (5.22).

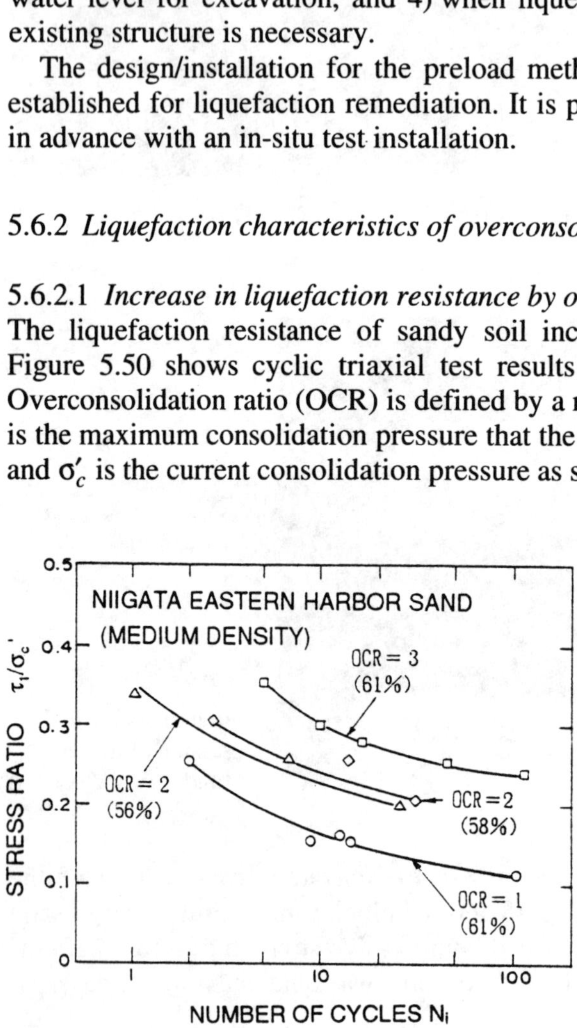

Figure 5.50. Cyclic shear stress ratio and number of load cycles (influence of overconsolidation ratio).

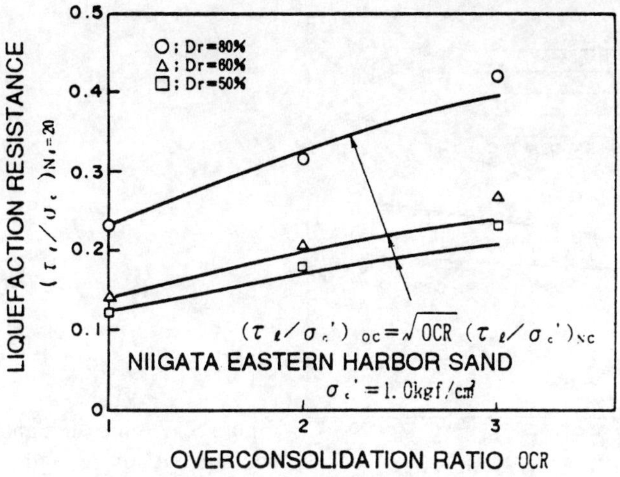

Figure 5.51. Overconsolidation ratio and liquefaction strength.

$$\mathrm{OCR} = \frac{\sigma'_p}{\sigma'_c} \tag{5.22}$$

By determining the cyclic shear stress ratio at 20 load cycles using Figure 5.50 (hereafter the stress ratio is referred to as liquefaction strength), the relationship between the overconsolidation ratio and liquefaction strength is determined as shown in Figure 5.51. As shown in this figure, the liquefaction strength by overconsolidation is in proportion to the square root of the overconsolidation ratio as follows [70].

$$\left(\frac{\tau_l}{\sigma'_c}\right)_{OC} = \left(\frac{\tau_l}{\sigma'_c}\right)_{NC} \times \sqrt{\mathrm{OCR}} \tag{5.23}$$

where: $(\tau_l/\sigma'_c)_{NC}$ = liquefaction strength for normally consolidated state, $(\tau_l/\sigma'_c)_{OC}$ = liquefaction strength for overconsolidated state.

The solid line in Figure 5.51 was obtained using Equation (5.23). If the overconsolidation ratio is OCR = 2.0 in Equation (5.23), for example, it is apparent that the liquefaction strength becomes 1.4 times higher; it becomes 2 times higher if OCR = 4.0.

5.6.2.2 *Result of repetition of pre-consolidation*

The influence of consolidation history (repetition of consolidation) with respect to the overconsolidation effect noted above is indicated by the ◇ mark in Figure 5.50. The ◇ mark is the result of a case when the consolidation history (σ'_c = 1.0 kgf/cm^2 ⟷ 2.0 kgf/cm^2, OCR = 2) was repeated 5 times. Since the consolidation history of OCR = 2 does not differ greatly from the mark for a single consolidation, it is possible to consider that there is hardly any influence due to repetition of pre-consolidation.

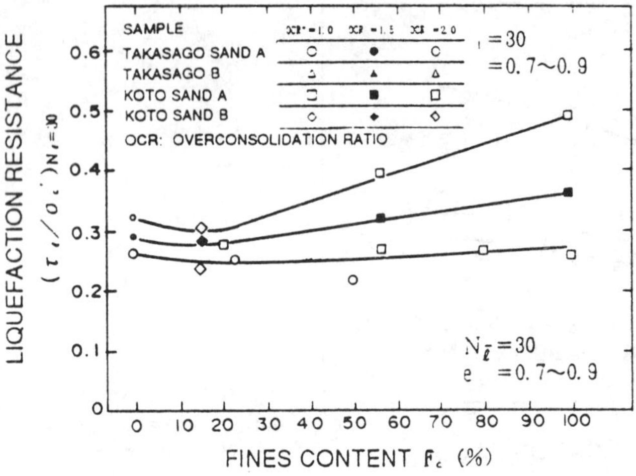

Figure 5.52. Fines content and liquefaction strength.

5.6.2.3 *Influence of fines content*

The effect of consolidation history on the mechanical behavior of soil is generally greater in clay than sand. If sandy soil contains clay and silt, the overconsolidation effect may be higher. This fact can be studied in terms of the effect of fines content F_c on the overconsolidation effect. Figure 5.52 shows the result of Ishihara et al. [71] on the relationship between the fines content and the influence of the overconsolidation ratio on the liquefaction strength (defined in terms of cyclic shear stress ratio at a number of load cycles of 30 times). It is apparent in this figure that the effect of overconsolidation on liquefaction strength increases with increasing fines content.

5.6.3 *Design/installation procedure*

The design/installation flow chart for the preload method is shown in Figure 5.53. The design/installation procedure can be roughly classified into the following four categories (Sections 5.6.3.1-5.6.3.4).

5.6.3.1 *Establishing the overconsolidation ratio*

The overconsolidation ratio applied to the natural deposit is determined by the procedures in 1-5.

1. An undisturbed sample is obtained from the field before improvement. Cyclic triaxial test is performed with respect to the normally consolidated specimen to obtain the liquefaction strength before improvement;

2. Using the cyclic triaxial test results, the liquefaction safety factor F_L is obtained based on the procedure shown in Section 4.5.3;

3. The soil undergoes liquefaction during earthquakes when the liquefaction safety factor F_L is below 1. If the liquefaction strength of this soil is in-

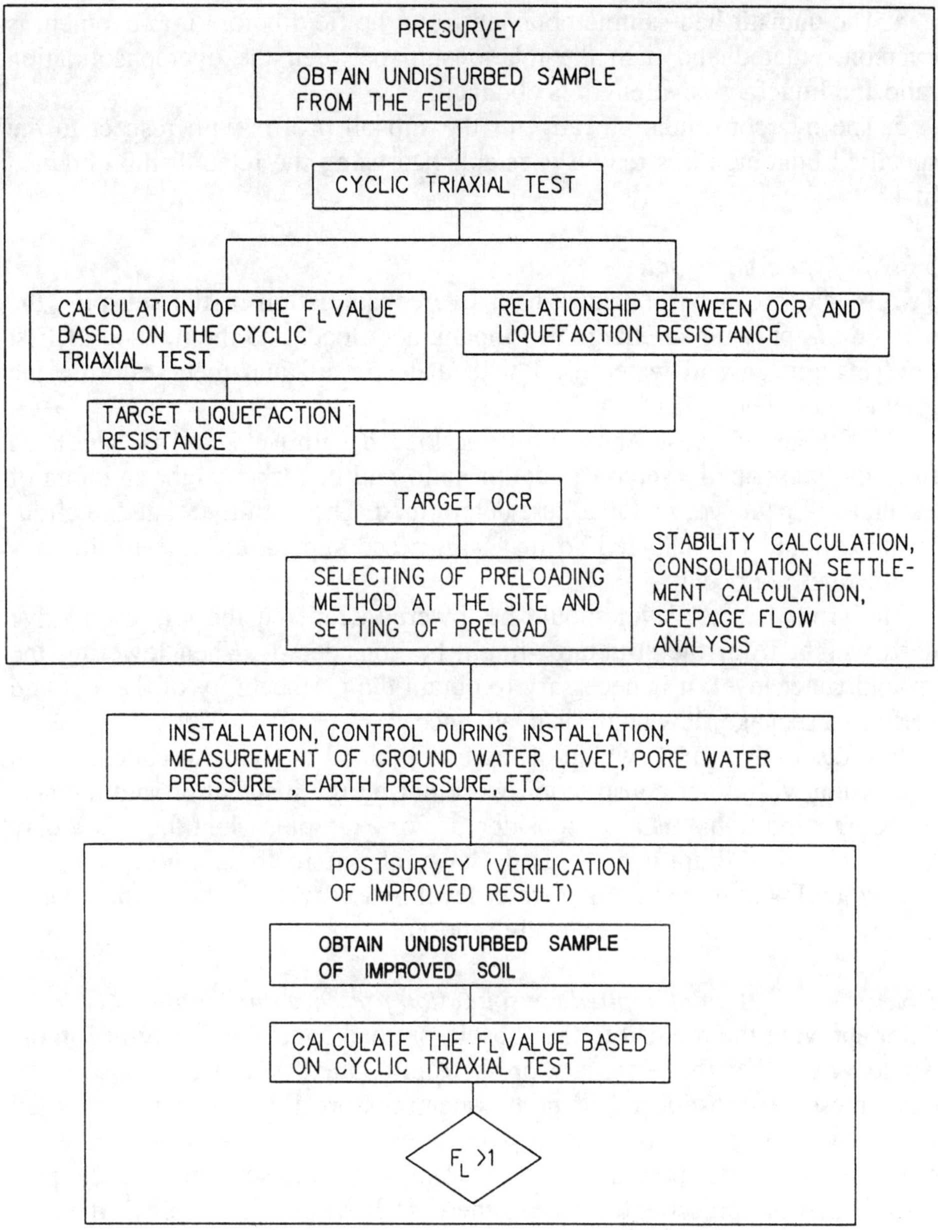

Figure 5.53. Design/installation flow chart.

creased by a factor greater than the inverse of F_L (i.e. by the factor of $1/F_L$), the new safety factor for liquefaction of this soil becomes $F_L > 1$ and liquefaction will not occur. The liquefaction strength to be achieved by the overconsolidation, therefore, should be larger than the factor of $1/F_L$ before improvement;

4. The undisturbed sample obtained from the field before improvement is overconsolidated and then the relationship between the overconsolidation ratio and liquefaction strength is obtained;

5. The overconsolidation ratio for the subsoil profile with respect to the specified liquefaction strength is established using the relationship obtained in 4.

5.6.3.2 *Types of preload*

Two methods can be considered for overconsolidation of the subsoils; the method of applying surcharge by embankment load, and the method of first lowering the ground water level with a deep well and then restoring the ground water level [72].

For the surcharge, the pressure to be loaded within the soil is calculated from the prescribed overconsolidation ratio and then the height and area of the embankment (vertical load) are determined. The stability related to circular slip should be analyzed so that subsurface slip failure due to the surcharge does not occur.

The ground water level should be lowered deeper if the larger effective dead weight from the structure should be considered. When lowering the ground water level, it is necessary to obtain the permeability of the soil and perform a seepage flow analysis in advance.

In order to determine which of the two preload methods to choose from, interaction with other construction work during the installation and the surrounding conditions must be considered. For example, when there is a clay layer and a sand drain is to be used, the surcharge method is normally chosen. When lowering the ground water level by excavation, the method based on lowering the ground water level is chosen.

5.6.3.3 *Verification of applied consolidation pressure during preload*

After applying the preload to the subsoil, preload control is performed in order to verify that the necessary pressure is applied within the subsoil. An earth pressure transducer and pore water pressure transducer are installed within the soil, or an observation well is installed for monitoring the earth pressure, pore water pressure, ground water level, and settlement in the preload. When the prescribed pressure increase has not been applied, the surcharge should be increased.

5.6.3.4 *Verification of overconsolidation effect*

It is necessary to verify that liquefaction strength has exceeded the specified value by overconsolidation. The verification is made by obtaining undisturbed samples from the construction site and performing cyclic triaxial tests. By combining the standard penetration test, the effect of liquefaction remediation by overconsolidation is verified using the method noted in Section 4.5.

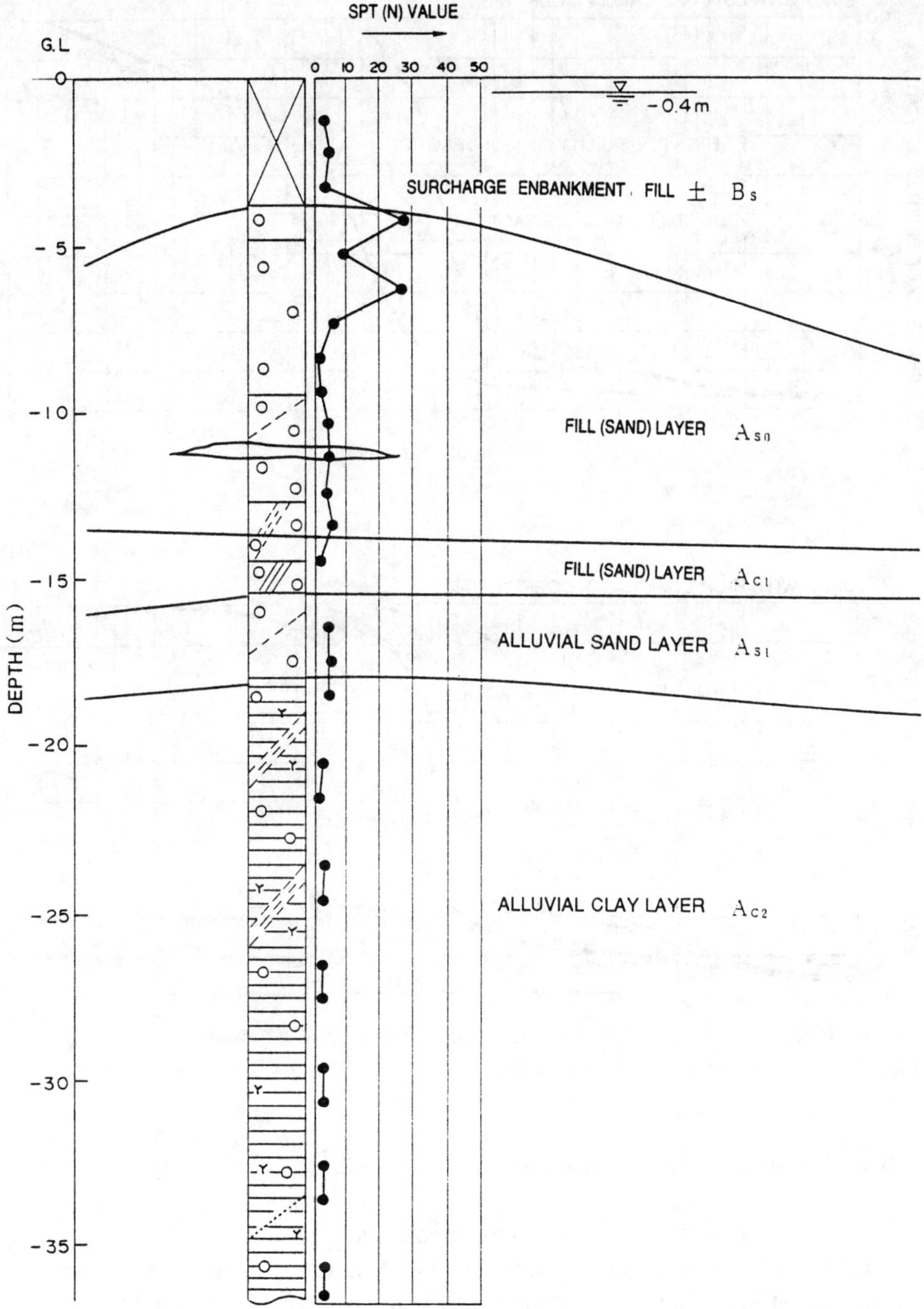

Figure 5.54. Soil boring log and SPT *N*-value.

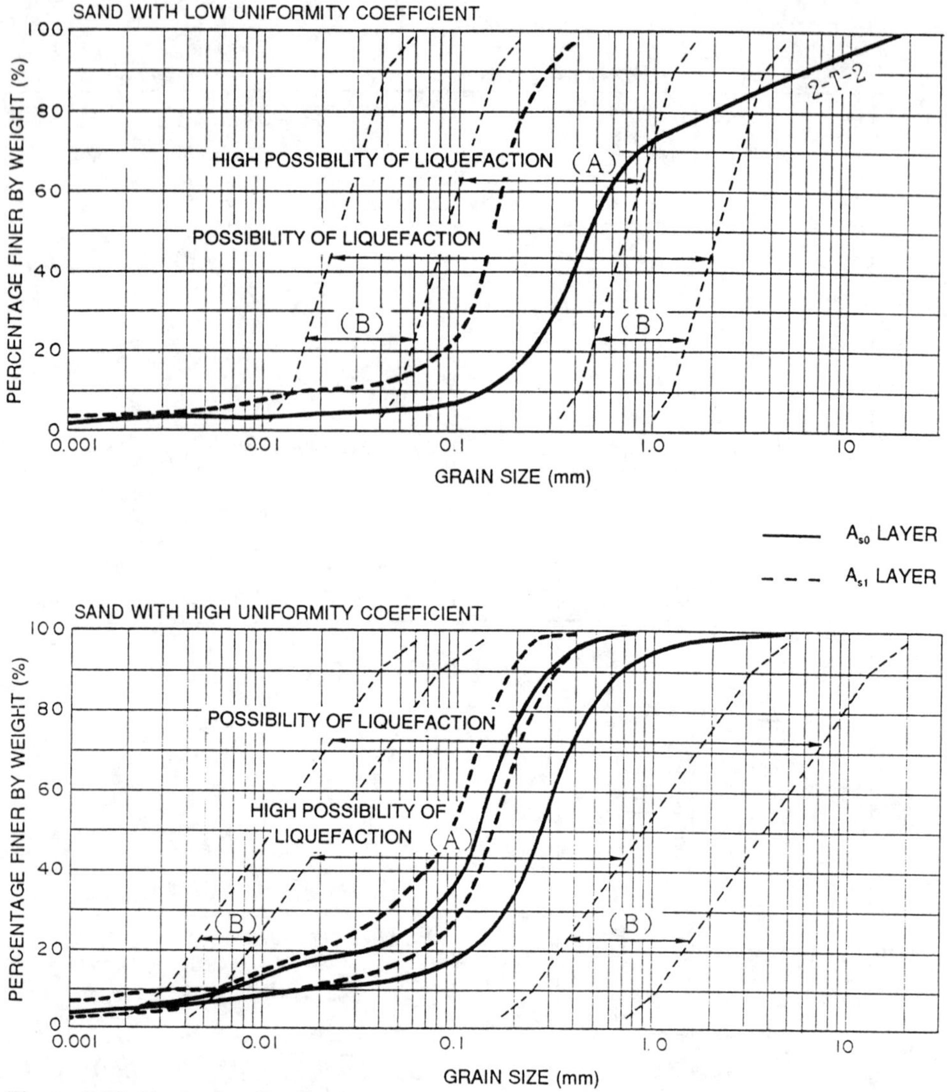

Figure 5.55. Grain size distribution curve.

5.6.4 *Design/installation example*

5.6.4.1 *Site conditions and liquefaction potential*

Figure 5.54 shows a case requiring liquefaction remediation measures (liquefaction remediation is required for the A_{S1} layer and A_{S0} layer) in a subsoil profile when constructing an underground structure at a depth of –10 m. The subsoil profile consisted of a diluvium sand layer (D_S layer: deeper than GL –37 m), alluvial clay layer (A_{C2} layer: GL –37 to –18 m), alluvial sand layer (A_{S1} layer: GL –18 to –16 m), landfill clay layer (A_{C1} layer:

GL –16 to –14 m), fill sand layer (A_{S0} layer: GL –14.0 to –4.0 m), and embankment for surcharge (B_S layer: GL = –4.0 to 0 m). The ground water level was located at a depth of –0.4 m.

The SPT *N*-values before improvement were 1 to 5 in the A_{S0} layer below GL –7 m and about 7 in the A_{S1} layer as shown in Figure 5.54. The gradation shown in Figure 5.55 indicated that the A_{S1} layer contained a large fines content of 30 to 35%. When liquefaction evaluation was performed based on the grain size/SPT *N*-value method, the A_{S1} layer was considered not likely to liquefy, but the A_{S0} layer was evaluated to undergo liquefaction. Thus, with respect to the A_{S0} layer, a liquefaction remediation measure was necessary.

5.6.4.2 *Prediction/determination based on the cyclic triaxial test results*

The F_L values of the A_{S0} layer and A_{S1} layer were calculated based on Section 4.5.3. The cyclic triaxial test results (○mark) on an undisturbed sample obtained from the construction site are shown in Figure 5.56. It was apparent that the liquefaction strength for the A_{S0} layer at a number of load cycles $N_l = 20$ was as low as 0.13, and that for the A_{S1} layer was 0.24 and was slightly higher. A seismic response analysis of the subsoil profile was performed using an input earthquake motion recorded at Hachinohe Port during the 1968 Tokachi-oki earthquake with a peak acceleration of $\alpha = 300$ Gal. The analysis using the SHAKE program resulted in a waveform of the shear stress of the impact type. The maximum shear stress ratio within the subsoil profile was about 0.18. In this study, the earthquake wave attenuated considerably within the thick deposit A_{C2} layer and hence the shear stress generated in the A_{S0} layer was smaller than that in the ordinary sandy soil deposit.

The result of calculation for the F_L value at each depth is indicated with an ○ mark in Figure 5.57. The A_{S0} layer had an F_L value ranging from 0.78 to 0.99 with an average of about 0.89 and hence it was determined that it would undergo liquefaction. The A_{S1} layer had an F_L value of 1.55 and hence it was determined that it would not undergo liquefaction. These results were consistent with those obtained by the liquefaction evaluation based on the gradation/SPT *N*-values mentioned earlier.

According to the results of the liquefaction evaluation, a liquefaction remediation was necessary in the A_{S0} layer as mentioned earlier. In the selection of the liquefaction remediation measure, the preload method by lowering the ground water level was considered suitable for three reasons: 1) the soil improvement method based on compaction was difficult since the specified soil layer contained a relatively large fines content, 2) a large vertical load was not applied from the structure, and 3) pumping was necessary for installation of this underground structure.

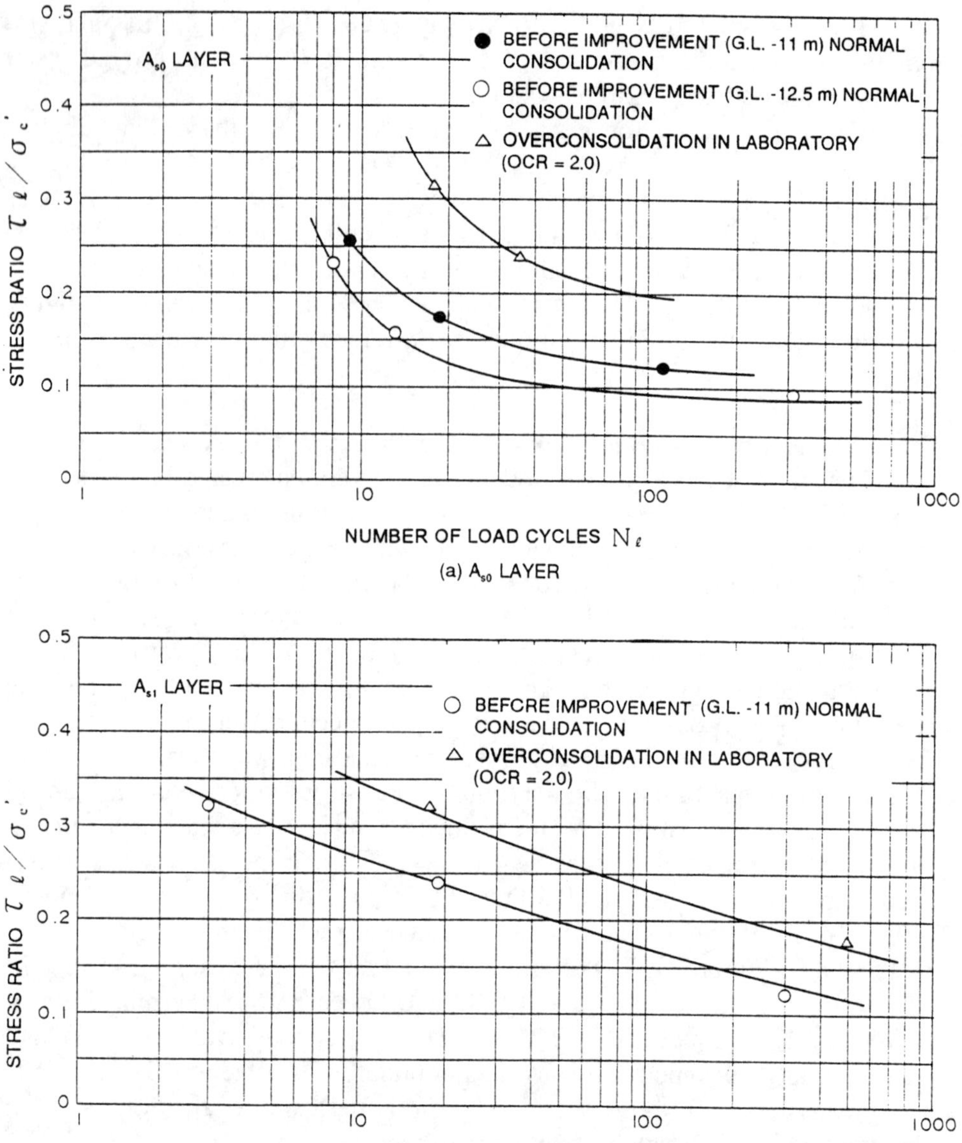

Figure 5.56. Cyclic shear stress ratio and number of load cycles.

5.6.4.3 *Determination of overconsolidation ratio*

Cyclic triaxial tests were performed in order to obtain the overconsolidation ratio required for the A_{S0} layer. The cyclic triaxial tests were performed on the specimen for the overconsolidated state of OCR = 2. The specimen was initially isotropically consolidated under a confining pressure of twice the

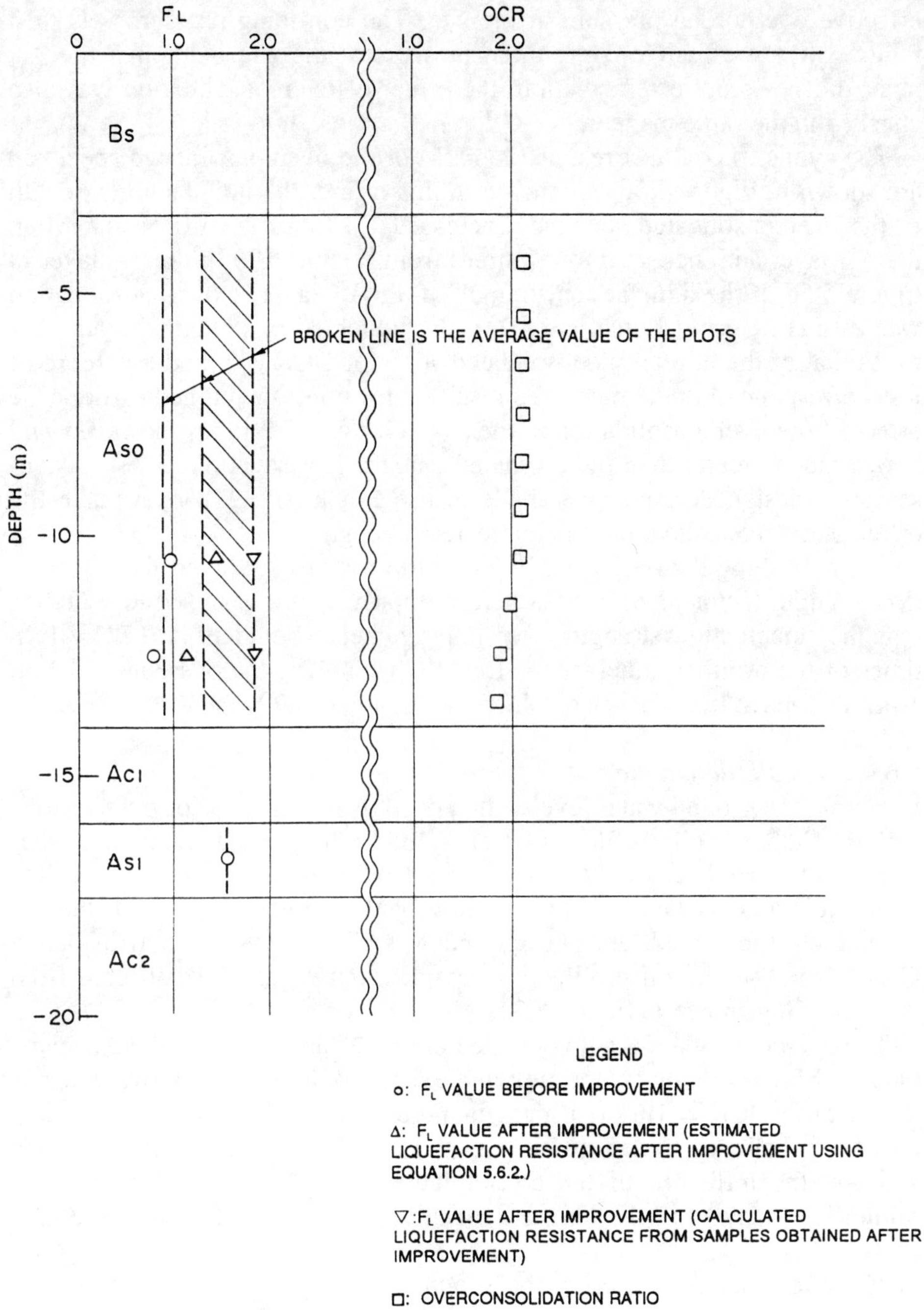

Figure 5.57. Overconsolidation ratio and F_L-value.

effective overburden pressure at the site. The confining pressure was then reduced to the effective overburden pressure at the site. Although the A_{S1} layer did not require remediation, the effect of overconsolidation was also checked in the same manner.

The cyclic triaxial test results (Δ mark) of the overconsolidated specimen are shown in Figure 5.56. As shown in this figure, the liquefaction strength of the overconsolidated specimen increased 1.3 times from 0.24 to 0.31 in the A_{S1} layer and increased to 2.2 times from 0.13 to 0.29 in the A_{S0} layer in Figure 5.56. If the liquefaction strength of the A_{S1} layer in overconsolidation was estimated using Equation (5.23), the liquefaction strength became 1.4 times that of the normally consolidated state for OCR = 2, and this more or less corresponded to the measured result of 1.3 times mentioned earlier. The effect of the overconsolidation in the A_{S0} layer was 2.2 times, however, and it was much greater than the estimated value using Equation (5.23). A conservative design decision was made to use Equation(5.23) to evaluate the effect of overconsolidation on liquefaction strength.

In establishing the overconsolidation ratio for the A_{S0} layer, the liquefaction strength factor of F_L = 0.78 before improvement was used to establish that the liquefaction strength after improvement should be 1/0.78 = 1.28 times of the original liquefaction strength. Therefore, the overconsolidation ratio required at the site was obtained by Equation (5.23) as OCR = 1.64.

5.6.4.4 *Installation at the site*

Lowering the ground water level with a deep well was used for overconsolidation of the subsoil profile. In order to lower the ground water level efficiently, the specified area was surrounded with sheet piles driven into the bottom of the clay layer, thereby inflow of the ground water from the perimeter was prevented, and the ground water level was lowered to below GL –10 m. The subsoil profile was complex and Figure 5.58 differed from the boring log shown in Figure 5.54.

The overconsolidation ratio obtained in the A_{S0} layer after preload became OCR = 1.82 at the lowest location and exceeded the necessary overconsolidation ratio OCR = 1.64 (□ mark in Figure 5.57). The overconsolidation effect was believed to depend only on the pre-consolidation pressure of the soil and the influence of the consolidation period was considered to be minimal. 10 days was the longest consolidation period allowed in the process.

Measurements for lowering the ground water level were performed by installing a pore water pressure transducer in the soil. The results of the pore water pressure measurements during lowering of the ground water level are shown in Figure 5.58. The left side of the sheet pile in Figure 5.58 was the remediation area, where there was a reduction in the pore water pressure of about 10 tf/m^2 from the initial value due to lowering the ground water level by about 10 m.

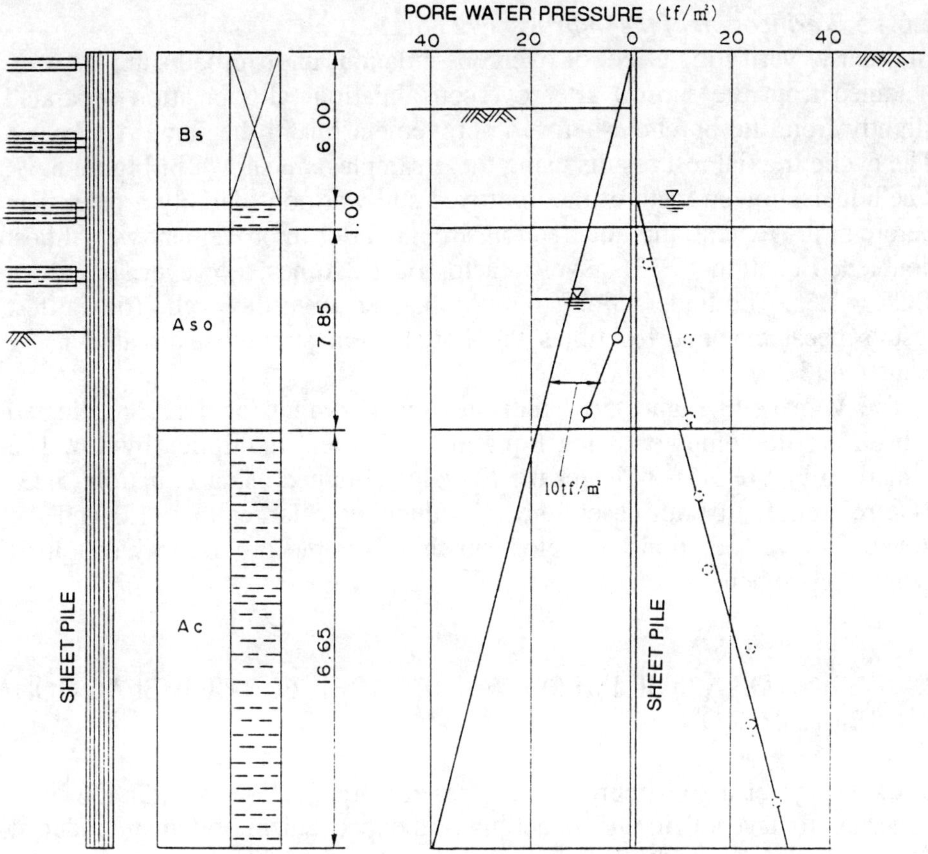

Figure 5.58. Pore water pressure in subsoil profile.

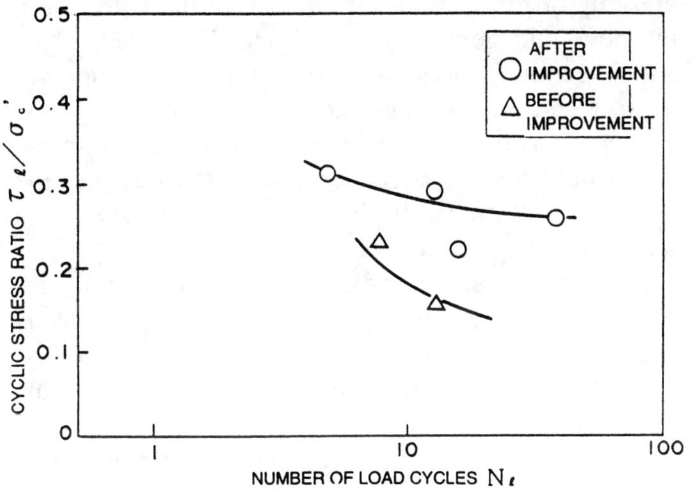

Figure 5.59. Cyclic shear stress ratio and number of load cycles.

5.6.4.5 *Verification of the improvement effect*

In order to verify the effect of overconsolidation, undisturbed samples were obtained from the subsoil after overconsolidation (at a position separated slightly from the borehole before improvement but in the same A_{S0} layer). The cyclic triaxial test results using these samples are shown in Figure 5.59. The liquefaction strength of the improved soil was considerably greater than before improvement; the liquefaction strength after improvement was almost doubled. Equation (5.23) gave a factor of 1.2 times, however, based on OCR = 1.55. The liquefaction strength obtained from the cyclic triaxial test results became about 1.6 times the liquefaction strength estimated using Equation (5.23).

The ∇ mark in Figure 5.57 indicates the F_L value for overconsolidated subsoil profile computed using Equation (5.23) and multiplied by 1.6. The Δ mark in Figure 5.57 indicates the F_L value obtained using Equation (5.23) as a reference. In both cases, the F_L values were larger than 1.0 and the liquefaction remediation by preload in this example proved to have sufficient improvement effect.

5.7 DESIGN OF A SOIL IMPROVEMENT AREA FOR LIQUEFACTION REMEDIATION

In the design of a soil improvement area for liquefaction remediation, it is important to have a firm understanding of the mechanism of damage induced by liquefaction; in particular, the understanding of whether the damage will be caused by reduction in shear resistance of the liquefied soil or by a new external force due to excessively large displacement of liquefied subsoil as noted in Section 2.3. The remediation measures against the former cause of damage have a relatively long history whereas those against the latter cause of damage are relatively new, still being researched and developed. Thus, the discussion given in this section focuses on the design of a soil improvement area against the former cause of damage (i.e. decrease in shear resistance of soil). There are many cases, however, in which remediation measures with respect to the former cause in effect serve simultaneously as the remediation for the latter cause (excessive displacement of the soil) such as when compaction is adopted as a remediation measure.

It should be noted that, for buried structures such as lifeline facilities including gas, water and telecommunication pipes, special consideration is necessary with respect to the latter cause of damage (excessively large displacement of the soil). Appendix 2 presents examples for coping with this type of damage in lifeline facilities.

Generally, even when the soil undergoes liquefaction over a wide area, the area requiring soil improvement is limited to the area which controls the stability of the structure. For example, that portion of the subsoil which contributes to the stability of spread foundation structures is the part directly below and on the perimeter of the structure; that portion of the subsoil far away from the structure does not contribute to the stability of shallow foundations. The question in the design to be specifically answered is 'exactly how wide the soil improvement area should be?'.

It is necessary to determine the region for soil improvement in both the vertical and horizontal directions. With regard to depth, standard practice is to improve down to the deepest part of the liquefiable soil layer. If soil improvement is performed only at the top part of the soil which will undergo liquefaction, upward seepage flow of pore water is generated if the unimproved layer below it liquefies, and stability at the top layer may be lost, resulting in liquefaction. With regard to the area in the horizontal direction, it can be determined according to the procedure noted below [73, 74].

The design procedure for a soil improvement area indicated in this section of the handbook is for the compaction method. The procedure may, however, be adopted for other methods with appropriate modification by considering the characteristics of improved soil such as permeability and cyclic strength/deformation.

Hereafter, 'liquefaction' means that state in which the excess pore water pressure ratio of loose and medium sand reaches 1.0 as noted in Table 2.1.

a) *Propagation of excess pore water pressure.* At the boundary of the soil to be improved by compaction adjacent to the unimproved soil, excess pore water pressure is applied with liquefaction of the adjacent unimproved soil, resulting in propagation into the improved soil. The influence of propagation of excess pore water pressure is very complex because the deformation characteristics of dense saturated sand are complex. As a simplified design procedure, this phenomenon will be treated as follows.

First, for excess pore water pressure ratio $u/\sigma'_v < 0.5$, the effect of excess pore water pressure increase may be ignored in the design because laboratory test data indicate a very small strain generation at this level of pore water pressure increase. For $u/\sigma'_v > 0.5$, it is necessary to take into account the effect of excess pore water pressure increase. According to the shaking table tests and seepage flow analysis, the square ABCD in Figure 5.60 is where the pore water pressure ratio becomes $u/\sigma'_v > 0.5$. In this area, a reduction in shear resistance has to be considered in design. In particular, the triangle ACD in the figure exhibited unstable characteristics in shaking table tests. Thus, in evaluating the stability of the soil/structure, it is appropriate to consider this part as liquefied.

An exception is when a drain or impermeable sheet or zone has been installed at the perimeter of the improved area in order to shut out the inflow of pore water from the liquefied perimeter subsoils into the improved soil. In these conditions, the area corresponding to the square ABCD need not be included in the improvement area.

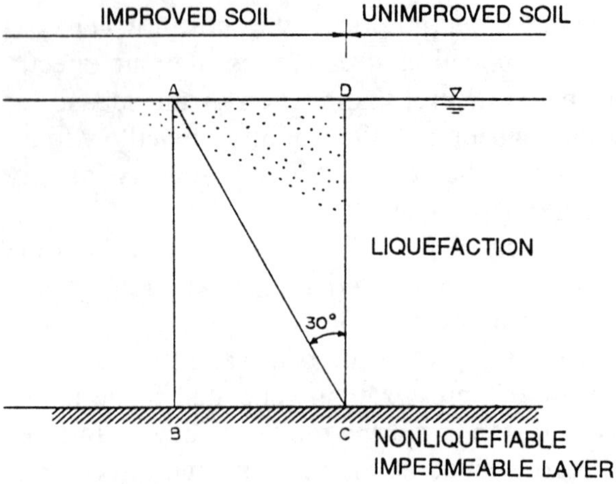

Figure 5.60. Area of softening due to seepage flow.

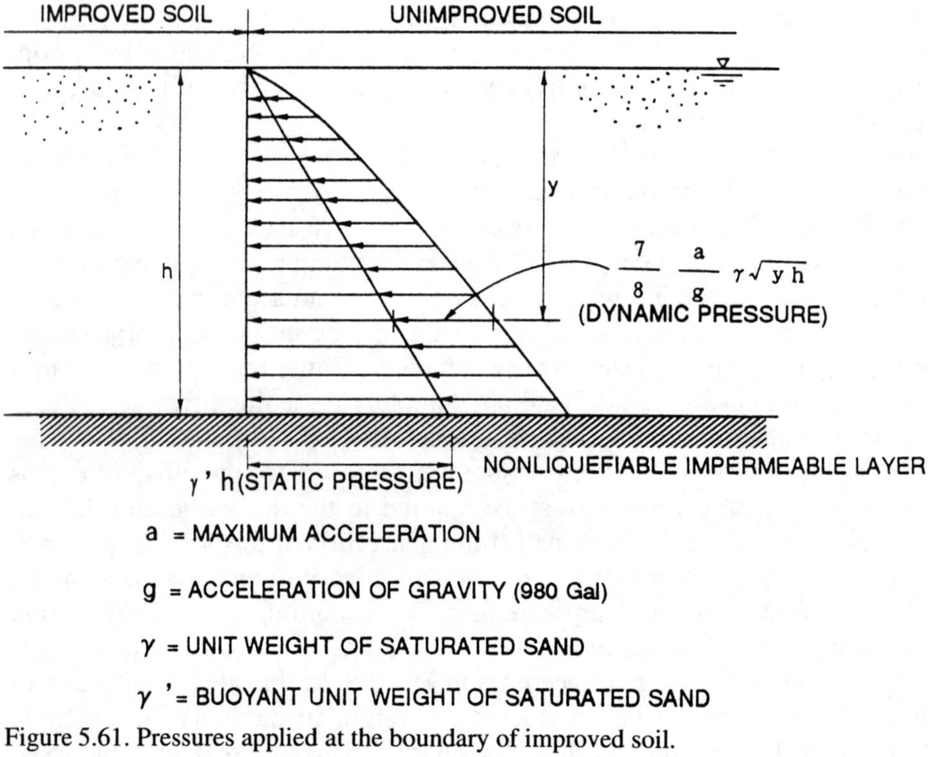

Figure 5.61. Pressures applied at the boundary of improved soil.

b) *Pressure applied by the liquefied sand layer.* At the boundary, a dynamic pressure indicated in Figure 5.61 and a static pressure corresponding to an earth pressure coefficient $K = 1.0$ are applied due to liquefaction of unimproved soil as shown in the figure. For a retaining structure that resists

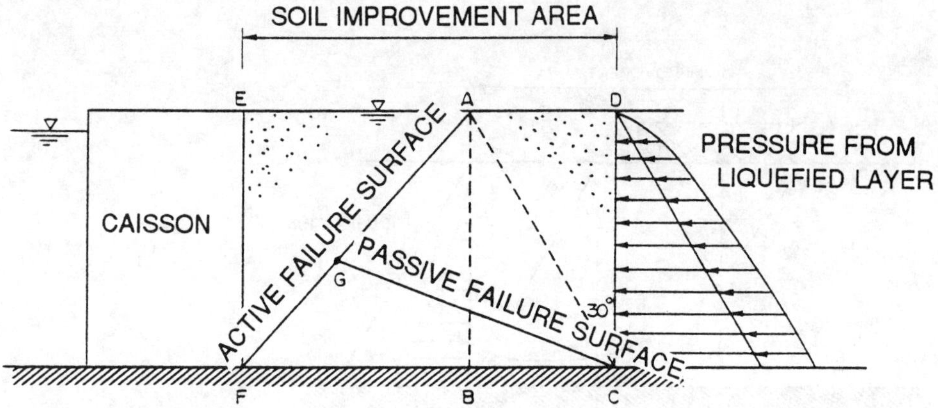

Figure 5.62. Schematic diagram for investigation of stability with respect to pressures applied from the liquefied sand layer.

the active earth pressure at the back of the structure EF as shown in Figure 5.62, it is necessary to design an improvement area so that there is no influence of liquefaction in the active failure zone. In particular, earth pressure at the boundary EF is active earth pressure whereas that at the boundary CD is much greater than this. Thus, it is preferable to evaluate whether or not sufficient shear resistance is mobilized along the passive failure surface GC.

c) *Loss of shear strength in liquefied sand layer*. In the simplified design procedure, the shear strength of liquefied unimproved soil should be considered completely lost (namely, to consider it as a liquid) if the liquefied soil density is loose or medium. Since the shear strength in the triangular shaped ACD portion noted in a is not anticipated (namely, this part is also determined to have liquefied), the improvement area should be wide enough to obtain sufficient bearing capacity from the shear resistance along the solid lines EFG and HI in Figure 5.63. Pressure from the liquefied sand layer may contribute to the stability of the structure so that the static pressure corresponding to an earth pressure coefficient $K = 1.0$ with subtraction of dynamic pressure can be included in the design for stability analysis of the structure as shown in Figure 5.63. The surface on which the pressure from the liquefied sand layer is applied can be considered the GG' surface or II' surface for convenience. Although this figure shows only five cases of foundations, the variation can easily be made for other foundations by referring to these examples.

d) *Change in dynamic response in the compacted portion*. In the compacted portion of the subsoil profile, its dynamic response will be changed by the influence of liquefaction in the adjacent unimproved soil and excess pore water pressure propagation, resulting in a change in the inertia force applied to the compacted portion of the subsoil profile. In order to evaluate

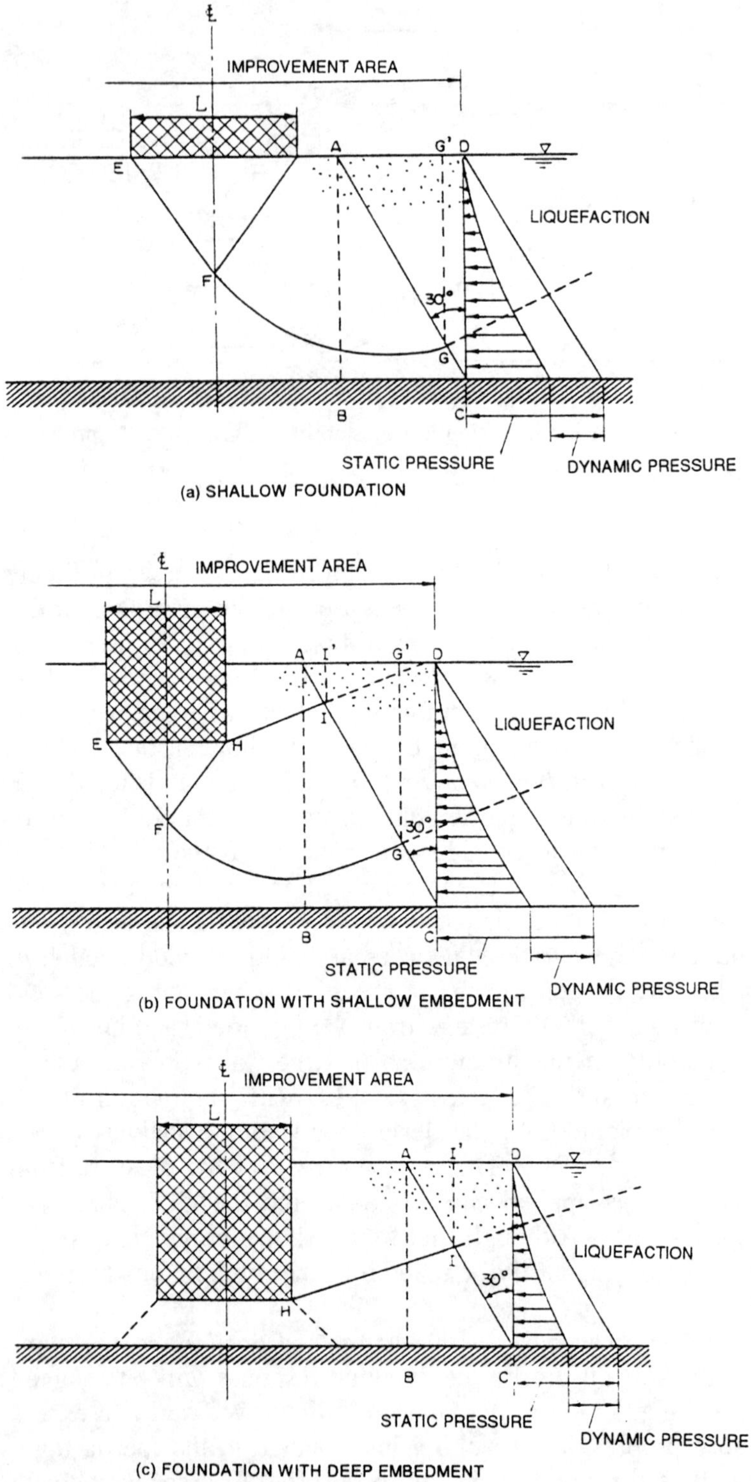

Figure 5.63. Schematic diagrams for investigation of stability for determining the soil improvement area.

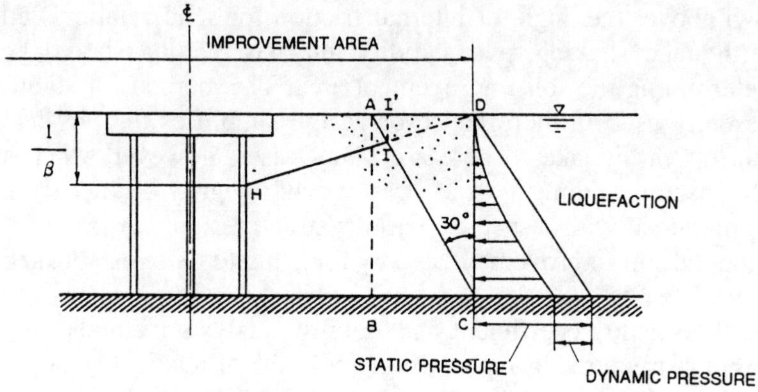

(d) PILE FOUNDATION

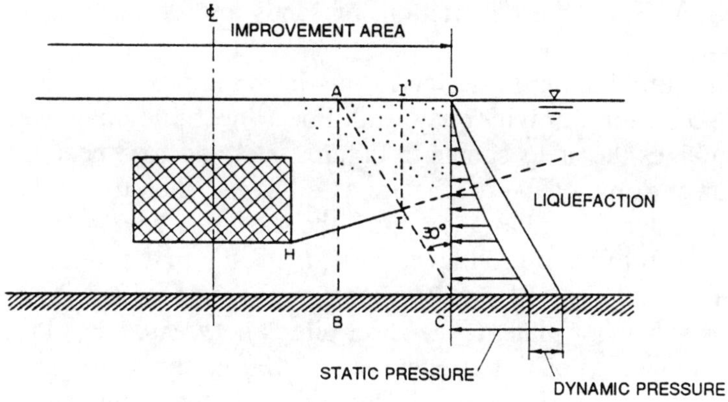

(e) UNDERGROUND STRUCTURE (WHEN PASSIVE RESISTANCE, IS NECESSARY IN ORDER TO MAINTAIN STABILITY OF THE STRUCTURE)

Figure 5.63. Continued.

the influence of this change on the stability of the structure and the compacted subsoil profile, a separate detailed evaluation may be necessary. For example, evaluation based on two-dimensional earthquake response analysis (refer to Section 4.4) may be necessary. If the area in which the excess pore water pressure ratio becomes more than 0.5 is small compared to the entire improved area, and if the cross-section of the entire improved area is relatively flat, then a change in the dynamic response of the compacted portion of the subsoil profile is considered to be small. For example, if the flatness of the improved area is the extent shown in Figure 5.62, it is not necessary to consider the change in the dynamic response of the compacted portion of the subsoil profile.

e) *Design of soil improvement area for port and harbour structures*. In the procedure shown above, the angle of internal friction for sands, anticipated intensity of earthquake shaking, and stability analysis method should be specified for determining the soil improvement area. The method for stability analysis presently used in the practice of design simplifies the physical phenomenon during earthquakes. There are many cases, however, when a combination of constants and methods assures overall integrity of the design based on past experience with respect to various structures so that it is possible to determine the soil improvement area by referring to these established design procedures.

For example, the seismic coefficient and stability analysis methods used for port and harbour structures in Japan are given in the 'Technical standard for port and harbour facilities and commentaries' [75]. In this procedure, earth pressures during earthquakes are computed by the Mononobe and Okabe theory and stability analysis is performed by the Bishop's method using the specified angle of internal friction for sands and design seismic coefficients.

For example, this combination in design results in the improvement area indicated in Figure 5.62 for a gravity quay wall. For a sheet pile quay wall, the area which combines the areas shown in Figures 5.64 and 5.65 becomes the improvement area. Some of the details are as follows. When the anchor is made of a sheet pile and treated as a long pile, the sheet pile below a position deeper than $l_{m1}/3$ from the tie rod level (here, l_{m1} is the first contraflexure point) is ignored, and the soil improvement area is determined in the same manner as for an anchor plate. If the sheet pile is not treated as a long pile, the sheet pile below the position deeper than $l_{m1}/2$ from the tie rod level is ignored, and the soil improvement area is determined in the same manner as for an anchor plate.

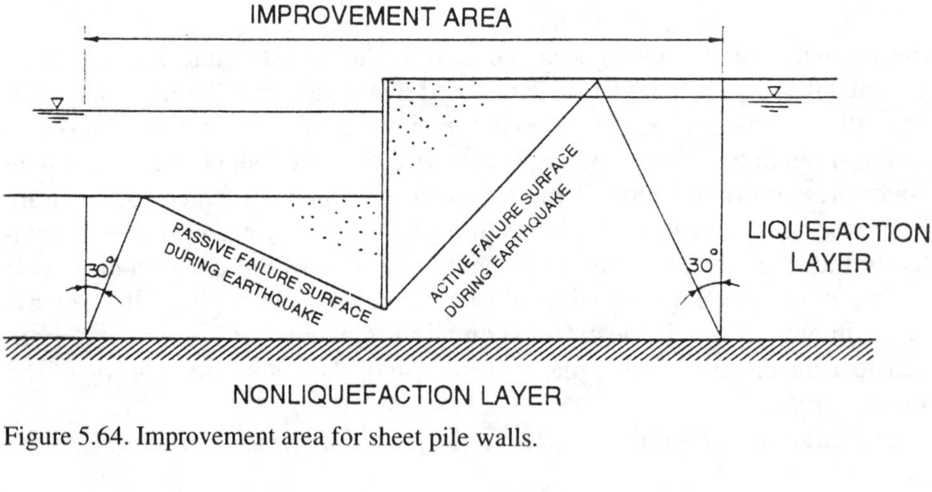

Figure 5.64. Improvement area for sheet pile walls.

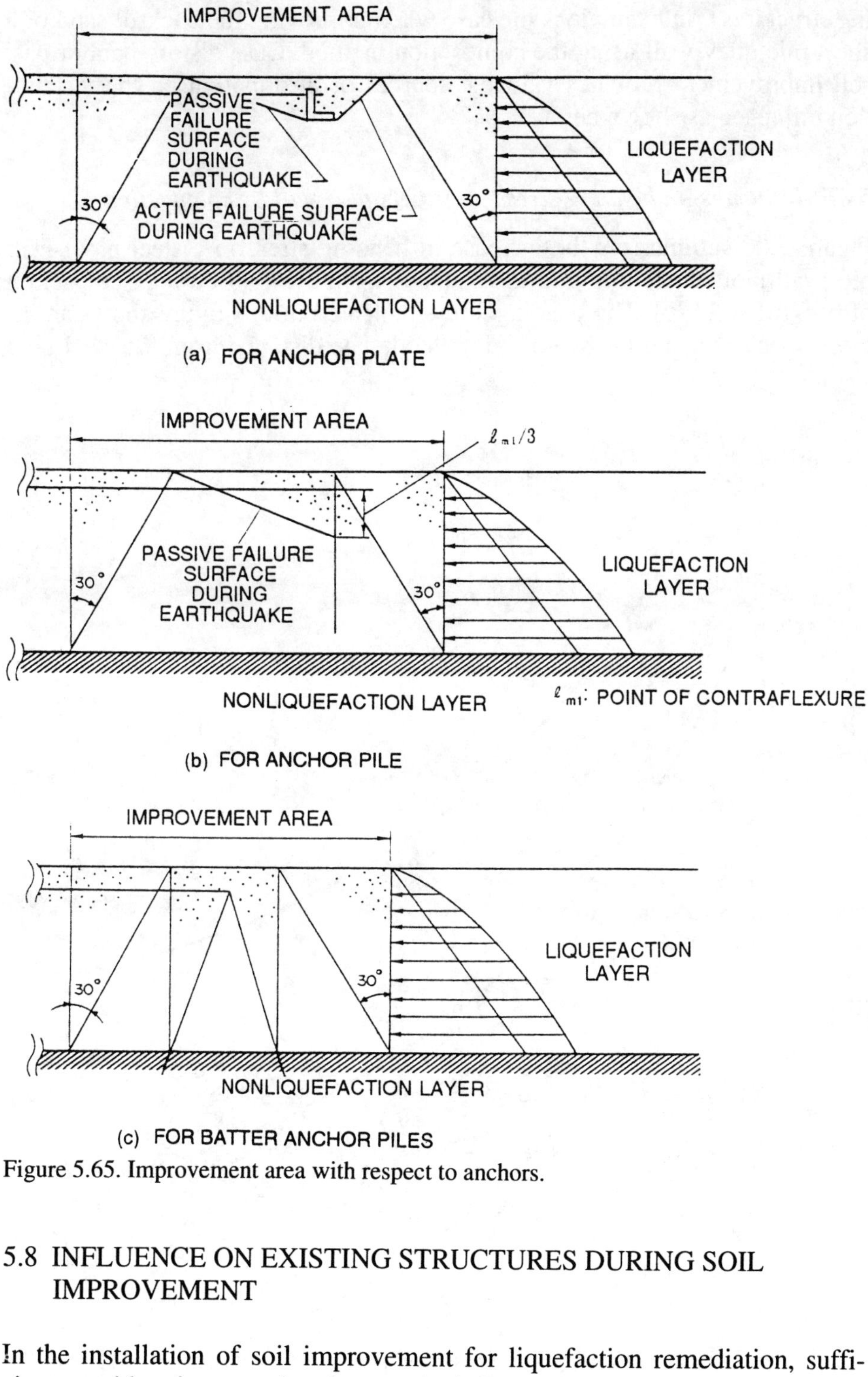

Figure 5.65. Improvement area with respect to anchors.

5.8 INFLUENCE ON EXISTING STRUCTURES DURING SOIL IMPROVEMENT

In the installation of soil improvement for liquefaction remediation, sufficient consideration must be given to the influence of the installation on existing structures. An example is the case when improving the backfill sand of a sheet pile quay wall using the compaction method. Case history data of past soil improvement provides valuable sources of information on the installation influence as shown below.

5.8.1 *Influence on bending stress in a sheet pile wall by compaction*

Figure 5.66 summarizes the increase in bending stress in a steel pipe sheet pile wall plotted according to the distance from the wall during compaction of backfill soil [76]. The bending stress indicated in this figure shows an increase per rod using the vibro rod method. The distance from the steel pipe

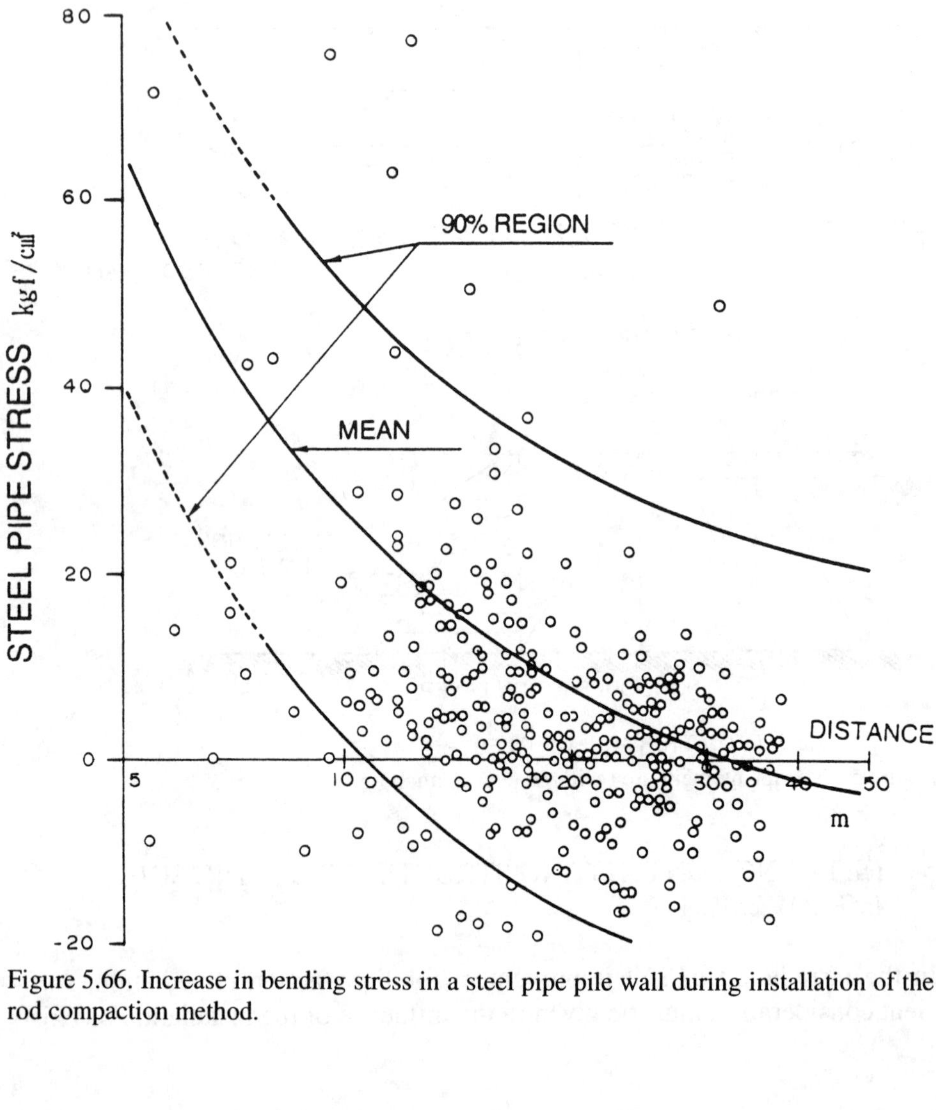

Figure 5.66. Increase in bending stress in a steel pipe pile wall during installation of the rod compaction method.

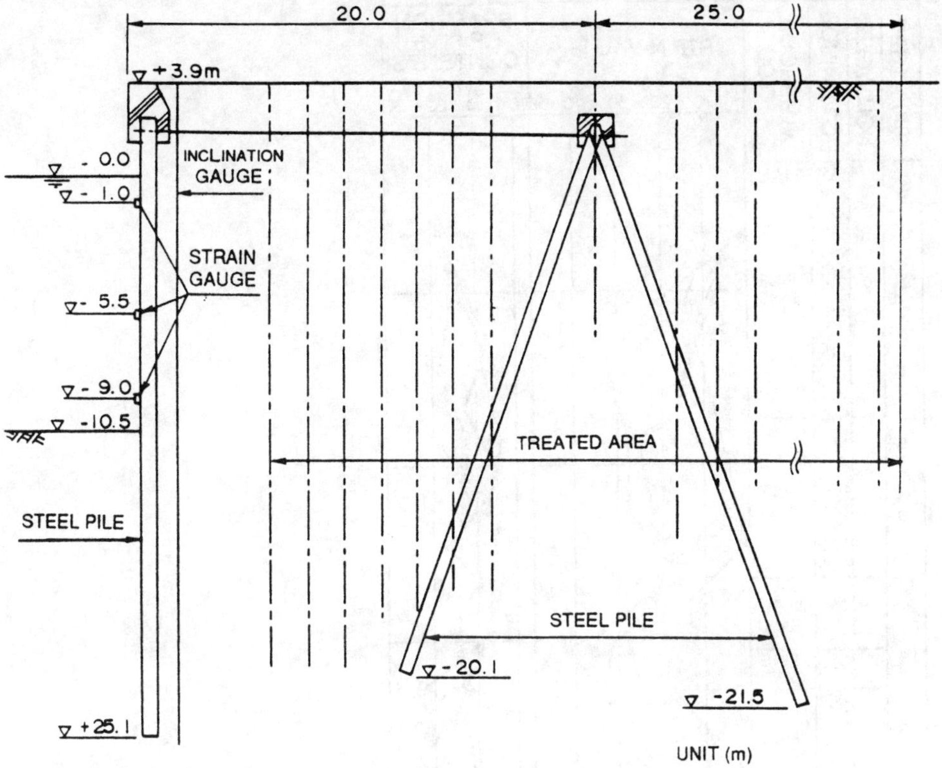

Figure 5.67. Location of instruments for monitoring the effect of compaction installation for a quay wall.

sheet pile indicated in the figure is the distance measured from the vibro rod driven position. The bending stress in the steel pipe pile wall increases gradually in accordance with the accumulation of the number of installation rods in the vibro rod method. It should be noted that the figure includes the effect measured when the direction of the position of the vibro rod relative to the point of measurement on the sheet pile wall is not normal to the face line of the wharf.

The results shown in this figure are obtained under the following conditions for subsoil and structures. First, the structural cross section was a steel pipe sheet pile type quay wall (–10 m) as shown in Figure 5.67. Soil improvement was accomplished by the vibro rod method at the position and depth (maximum depth –20 m) indicated by the broken line in the figure. The soil condition was as shown in Figure 5.68 [4, 76].

In Figure 5.66, the influence closer than 5 m is not shown. Other case records are available at 2 m from the sheet pile for the influence of the gravel drain and vibro rod installation. The case history records at 2 m distance indicated an increase of 300 kgf/cm^2 in the bending stress of the steel sheet

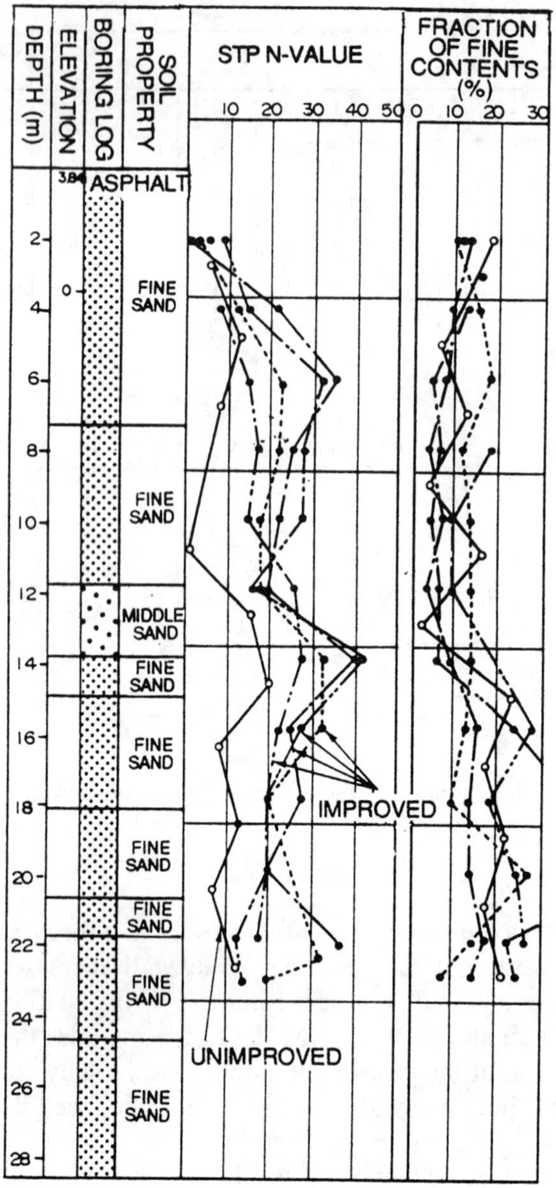

Figure 5.68. Soil conditions at monitoring site.

pile per rod using the vibro rod method and an increase of 250 kgf/cm^2 per gravel pile in the gravel drain [56]. When compaction was implemented by the vibro rod method behind the wall after the gravel drain installation, a noticeable accumulation of stress by the vibro rod method was not found, indicating that combined use of the gravel drain and the compaction method may be suitable for the installation of soil improvement with respect to existing structures.

These case history records may be used as a rough guideline for liquefaction remediation on existing structures. The actual stress generated, however, differs according to soil conditions, conditions of structure, installation conditions, etc., so that it is preferable to perform a test installation for monitoring the effects of the installation.

5.8.2 *Excess pore water pressure and vibration resulting from compaction*

Figure 5.69 indicates the measured acceleration during compaction of subsoils at a number of ports and harbours in Japan. Though the acceleration shown in the figure is quite large, vibration generated by compaction has a high frequency of about 10 Hz; direct comparison of these figures with earthquake intensity normally used in earthquake resistant design of structures is inappropriate. Based on this figure, a sudden increase in acceleration is found at a distance closer than 5 to 10 m so that installation at these distances requires sufficient caution with regard to the influence of the installation on the surrounding area.

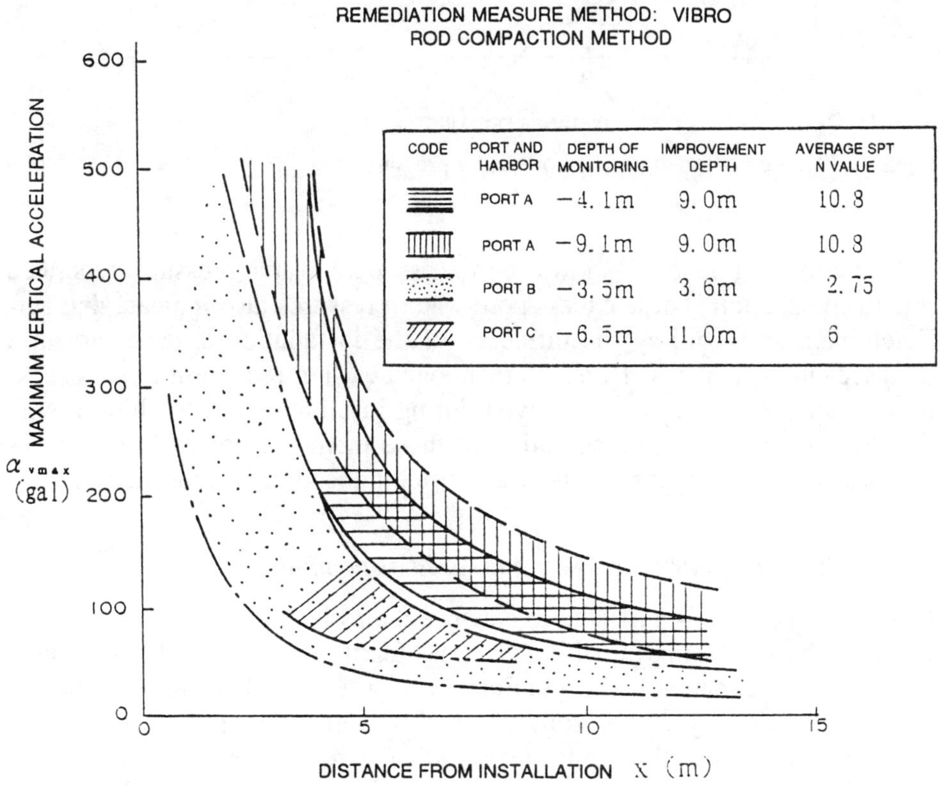

Figure 5.69 Attenuation of acceleration.

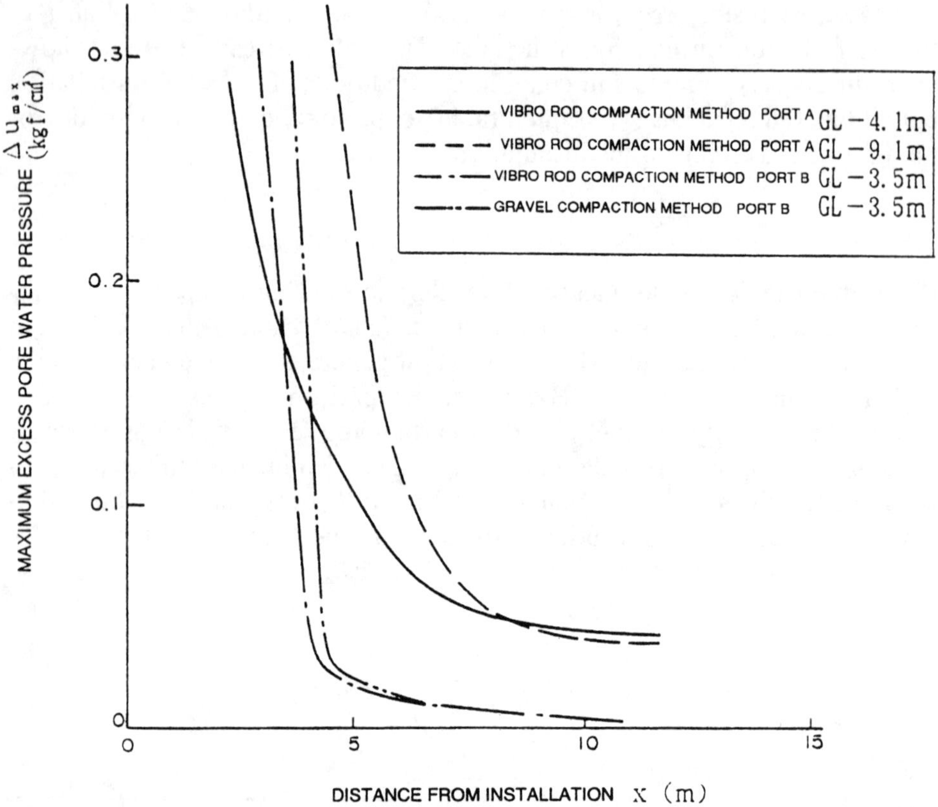

Figure 5.70. Attenuation of excess pore water pressure.

Similarly, Figure 5.70 shows the excess pore water pressure generated due to compaction. These excess pore water pressures also indicate that sufficient caution regarding the influence of the installation is necessary at a distance closer than 5 to 10 m. As mentioned earlier, reduction in the excess pore water pressure can be achieved during installation by combining with the drain method. In a combination of these methods, however, caution is necessary so that clogging of the drain is not induced during installation.

5.8.3 *Influence of earth pressure increased by compaction*

Compaction of soil is expected to enhance the resistance with respect to liquefaction by increasing the effective stress (earth pressure) in the horizontal direction and enhancing the confining pressure simultaneously while increasing the density of the soil. When this method is applied to existing structures such as a sheet pile quay wall, the resistance of the sheet pile during earthquakes may decrease due to an increase in the bending stress

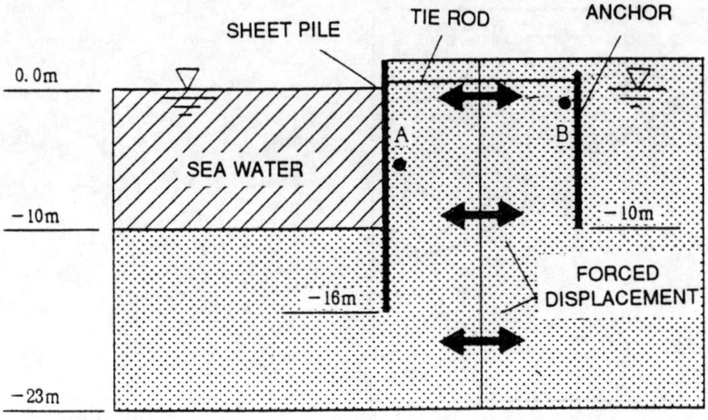

Figure 5.71. Cross section of quay walls for the analyses.

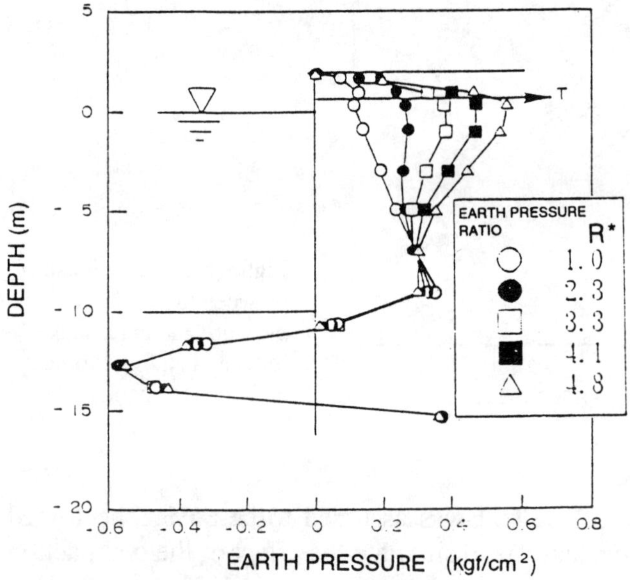

* DEFINITION OF EARTH PRESSURE RATIO: RATIO OF THE INDUCED EARTH PRESSURE TO THAT AT NORMAL CONSOILIDATION STATE FOR +3.5m ELEVATION.

Figure 5.72. Earth pressure distribution after induced displacement.

generated during installation. Determining whether a favorable effect due to the increase in the effective stress of the soil or an unfavorable effect due to an increase in the bending stress predominates is an important factor in the evaluation of the design for liquefaction remediation.

A study was conducted with respect to this point by using the effective stress model explained in Technical note 1 [77]. In this study, forced displacement corresponding to the deformation of the soil accompanying the compaction was induced to the portion indicated with an arrow in Figure

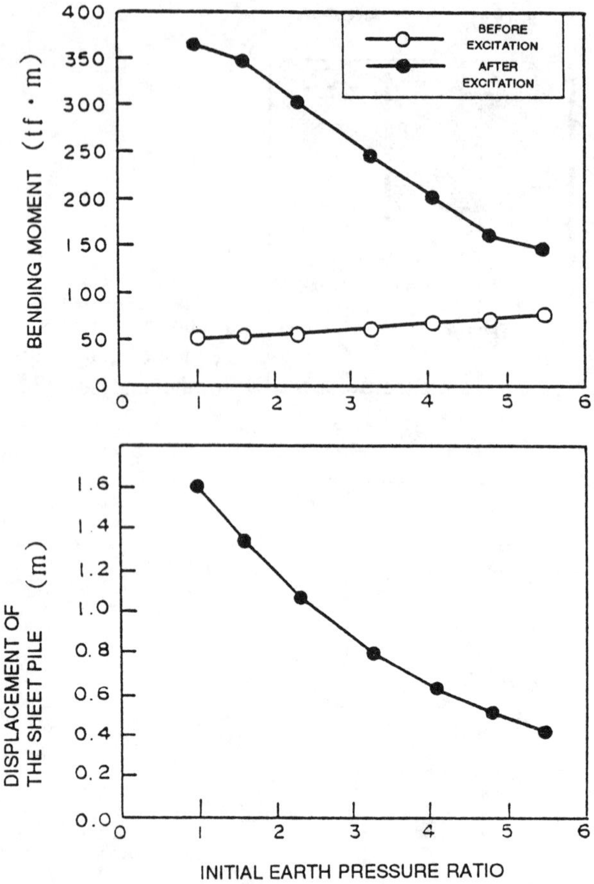

Figure 5.73. Relationship between the initial soil pressure and earthquake induced bending moment and displacement.

5.71. The relative density of the soil was assumed to be 35%. The forced displacement induced in the analysis to simulate the effect of the compaction resulted in increased earth pressures on the sheet pile wall as shown in Figure 5.72, in which the open circles in the figure indicate the earth pressure without compaction. In this figure, the earth pressure ratio indicates the ratio of the earth pressure induced by compaction over the earth pressure without compaction. In this evaluation, only the influence of the initial earth pressure was evaluated and the influence of the increase in density from compaction was ignored.

When quay walls with various initial earth pressures were shaken analytically, both the displacement and bending moment in the sheet pile wall after the earthquake were computed as shown in Figure 5.73. As shown in this figure, with an increase of initial earth pressure, the initial bending moment indicated with the open circle increased. The earth pressure after the earthquake, however, decreased. A similar tendency was noted in the displacement of the sheet pile.

These results indicate that the increase in the bending stress and earth pressure from compaction results in an increase in the overall resistance of sheet pile walls to earthquakes.

TECHNICAL NOTE 1

Recent advances and future trends in liquefaction/damage evaluation

The procedures necessary in practice for liquefaction prediction/determination have all been presented in Chapter 4 of the main text of this handbook. Studies are still continuing for enhancing the reliability of liquefaction prediction/determination and developing new techniques for predicting the extent of damage caused by liquefaction. The objective of this technical note is to present these new developments in prediction/determination of liquefaction and evaluation of extent of damage induced by liquefaction. Understanding of these new developments will lead to more suitable engineering judgement with enhanced understanding of the procedures for liquefaction prediction/determination.

T1.1 RECENT DEVELOPMENTS AND FUTURE TRENDS IN LIQUEFACTION PREDICTION/DETERMINATION

In the liquefaction prediction methods noted in this handbook, predicting liquefaction in a horizontal subsoil profile is the main objective. The techniques necessary for liquefaction prediction of horizontal subsoil profiles have been fairly well established in recent years. Those of the subsoils affected by additional static shear stresses, however, are still under study. In the soil near a structure, for example, static shear stress is constantly applied due to the dead weight of the structure. Current and future research on liquefaction and deformation characteristics of soils under these stress conditions will lead to enhanced reliability and accuracy in liquefaction prediction in practice.

Inclined surfaces of soil structures or slightly inclined subsoil profiles also undergo static shear stresses. When they are made of a loose sand (i.e. with low SPT N-value), there is a possibility of flow failure triggered by slight external disturbances. This type of flow failure often occurs in loose silty sand having a low plasticity index. Examples in Japan are the flow failure of railway embankments during the Niigata earthquake of 1964 as mentioned in Section 2.2 and the flow failure of tailing dams at Mochikoshi Mines during the 1974 Izu Oshima Kinkai earthquake.

In the prediction method based on gradation and SPT *N*-value discussed in the main text, the effect of static shear stress is taken into account by allowing a large safety factor with respect to soils with low SPT *N*-values. This is based on engineering judgement. In future, suitable confirmation must be made by undertaking detailed research on this point.

In the method based on the cyclic triaxial test, the effect of static shear stress is often not considered. Further enhancement of precision in the liquefaction potential evaluation, therefore, is needed in the cyclic triaxial test on the influence of static shear stress.

T1.2 RECENT ADVANCES AND FUTURE TRENDS IN EVALUATION OF LIQUEFACTION INDUCED DAMAGE

T1.2.1 *Outline of new techniques for evaluation of liquefaction induced damage*

Active research is ongoing for developing techniques for evaluating the extent of damage induced by liquefaction. This field, however, is still developing. Some techniques work well but others do not. With suitable guidance by specialists in these research fields as well as sound engineering judgement based on past case histories, reasonable predictions can be made on the extent of damage induced by liquefaction. Engineers should be aware, however, that some of the techniques proposed in this field are not ready for application in practice; it is important to evaluate in advance the applicability of the proposed methods. The comparative studies done by the Japanese Society of Soil Mechanics and Foundation Engineering may be a good reference for this [1, 2].

Damage evaluation techniques based on the effective stress analysis using the finite element method are also being developed by the Port and Harbour Research Institute, Ministry of Transport, Japan. A general purpose computer program using the finite element method has been developed and is widely used for static geotechnical problems for stress/deformation and stability analysis of the subsoil and soil structures [3]. A similar analysis program for earthquake and liquefaction is presently being developed. Detailed explanation of the theory which forms the basis for the effective stress analysis is presented in [4, 5]. Some of the analysis results using this method have been presented in the joint study of the Japanese Society of Soil Mechanics and Foundation Engineering [1, 2].

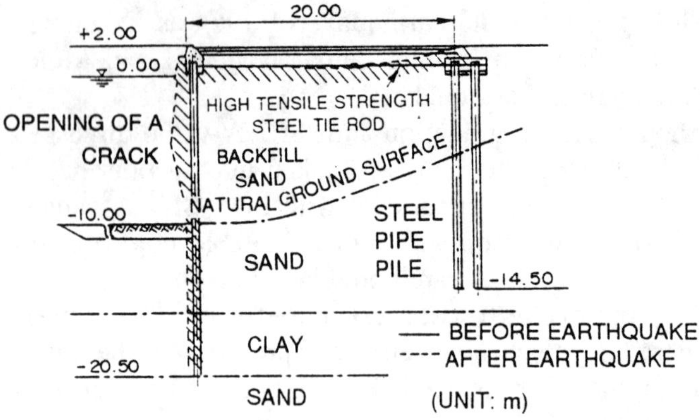

Figure T1.1. Cross section of a damaged quay wall.

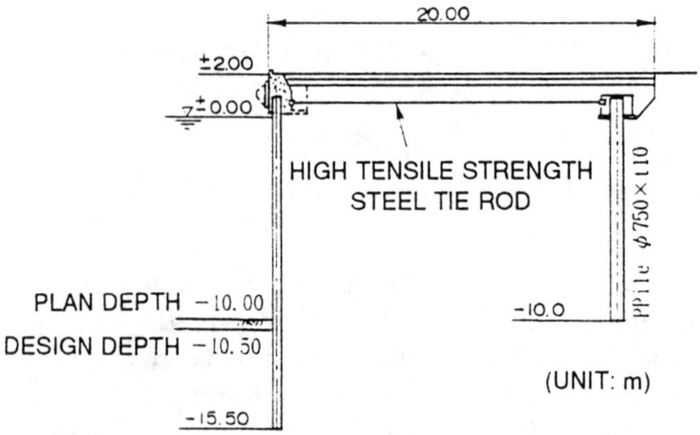

Figure T1.2. Cross section of an undamaged quay wall.

T1.2.2 *Applicability of the damage evaluation method to case histories*

As an example to indicate the applicability of the effective stress method mentioned earlier, the results of the analysis on sheet pile quay walls in the 1983 Nihonkai Chubu earthquake is shown below [6]. The quay walls chosen for the analysis are one damaged by liquefaction and another undamaged adjacent to it. As shown in Figures T1.1 and T1.2, these quay walls were identical with respect to the types of sheet piles, water depths, and positions of the anchors. The undamaged quay wall was constructed by excavating the natural subsoil in front of the sheet pile wall, whereas the damaged quay wall was constructed by backfilling sand behind the wall. From the boring logs obtained at the centers of the aprons in both of the quay walls shown in Figure T1.3, the SPT N-values of the undamaged quay wall were found to be much larger than those of the damaged quay wall from the ground surface to a depth of 10 m. The parameters of the effective stress model were determined based on these boring logs and the laboratory test results of sands sampled from the site. An analysis was performed based on the earthquake

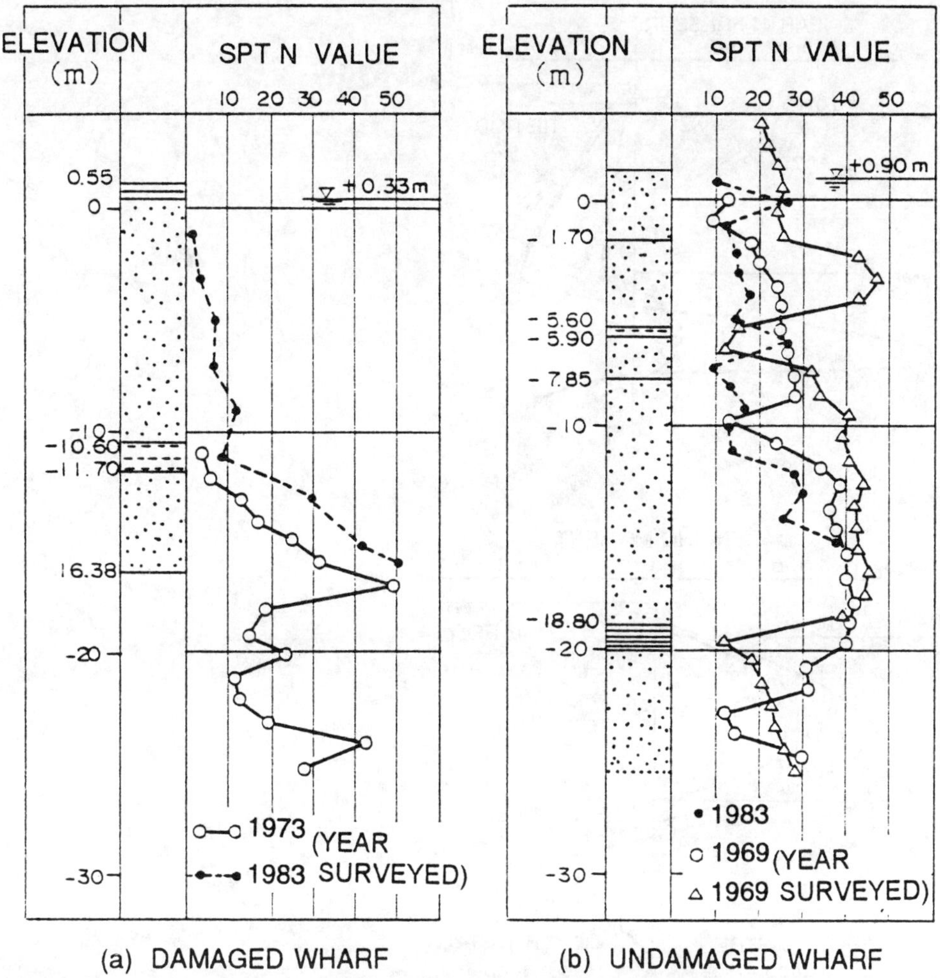

Figure T1.3. Boring logs for damaged and undamaged quay walls.

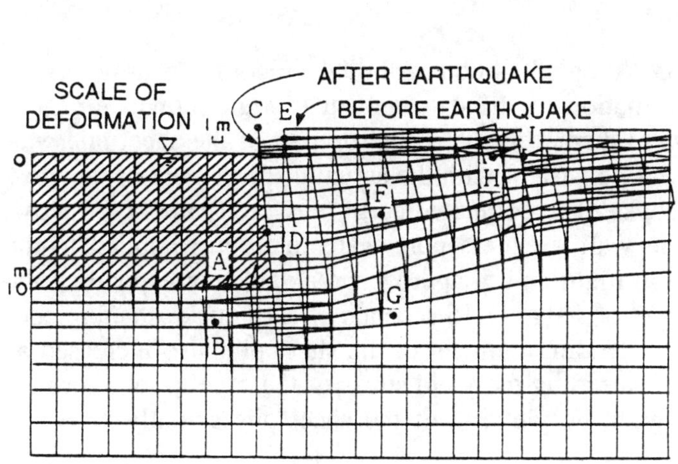

Figure T1.4. Computed deformation in a damaged quay wall.

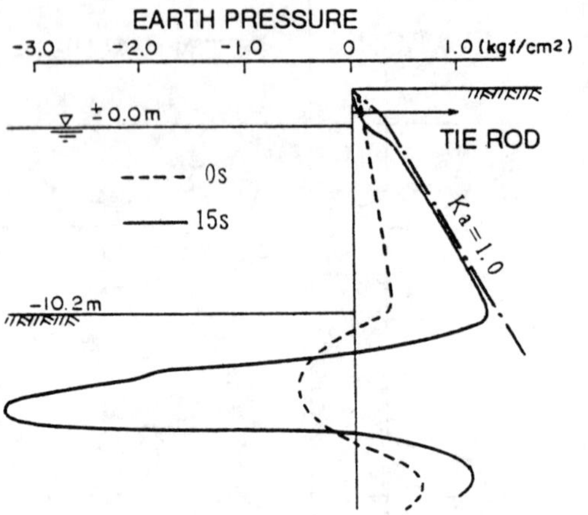

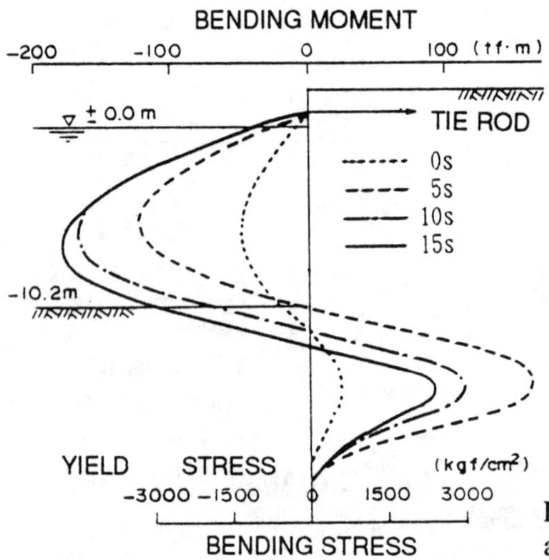

Figure T1.5. Earth pressure computed for a damaged quay wall.

motion during the 1983 Nihonkai-Chubu earthquake recorded nearby.

The computed deformation of the damaged quay wall shown in Figure T1.4 predicted a seaward displacement of the top part of the sheet pile wall of about 1.1 m. This was consistent with the actual displacement of the quay wall during the earthquake. The computed earth pressure (namely, effective normal pressure on the wall + excess pore water pressure) applied to the sheet pile wall, shown in Figure T1.5, approached a value corresponding the earth pressure coefficient of 1.0 after 15 seconds of earthquake shaking. Accordingly the computed bending moment of the sheet pile also increased as shown in Figure T1.6, exceeding the yield stress of the sheet pile. These results were consistent with the yielding of the sheet pile actually observed during the earthquake.

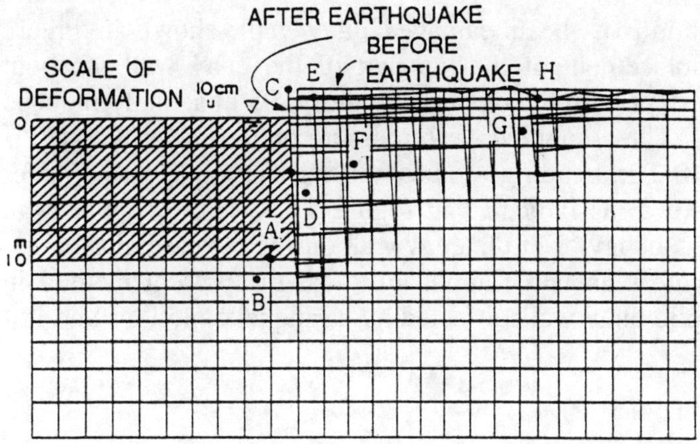

Figure T1.7. Computed deformation for an undamaged quay wall.

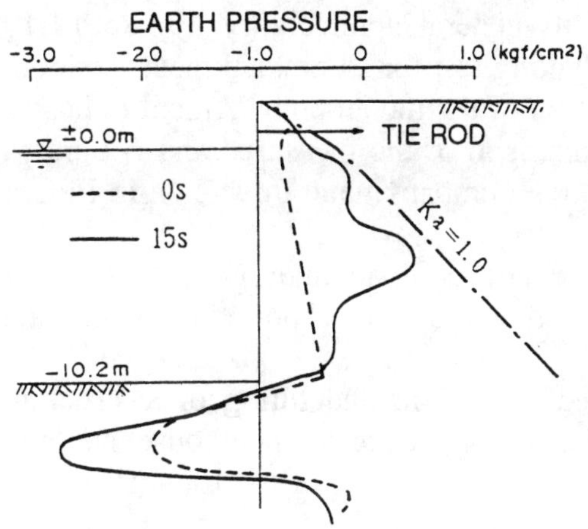

Figure T1.8. Earth pressure computed in an undamaged quay wall.

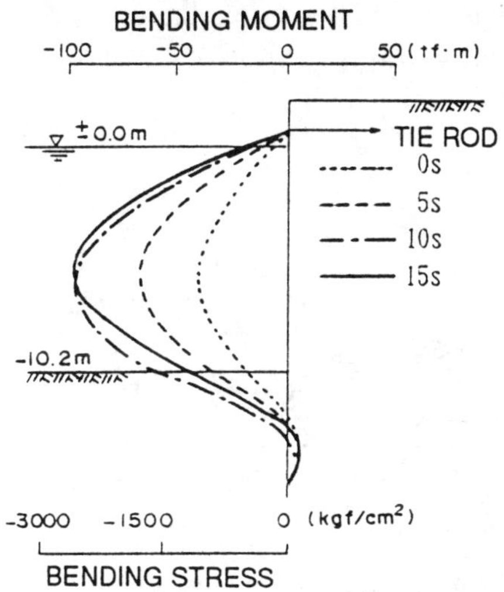

Figure T1.9. Bending moment computed in an undamaged quay wall.

Computed deformation of the undamaged quay wall, shown in Figure T1.7, predicted a displacement of the top part of the quay wall of about 0.1 m. This was consistent with the displacement of 0.05 m actually induced during the earthquake. The computed earth pressure on the sheet pile wall is shown in Figure T1.8. The bending moment of the sheet pile was smaller than the yield stress level as indicated in Figure T1.9. These results were also consistent with the behavior of the quay wall during the earthquake.

These results indicate reasonable applicability of the present method in the analysis of sheet pile quay walls affected by earthquake motions and liquefaction.

T1.2.3 *Other applications*

Other applications of the present method are presented in Figures T1.10 through T1.13 in order to indicate its general applicability. Figure T1.10 shows an example of analysis of a shaking test for an embankment model in a centrifuge. The computed deformation of the model embankment indicates lateral spread and settlement. A comparison of the measured and computed excess pore water pressures in the model embankment are shown in Figure T1.11 [1].

Figure T1.12 shows the computed heave of an underground structure caused by liquefaction. As shown in the figure, it is possible to simulate heaving phenomena [7].

Figure T1.13 shows the computed settlement and tilting of a structure resting on liquefiable ground [2]. The tilting toward the penthouse side and overall settlement of the structure due to liquefaction of the foundation soil is reasonably simulated by the analysis.

The various modes of damage in the structures shown here all conform to the modes of damage in the actual damage case histories explained in Section 2.2, giving confidence in the applicability of this method discussed here.

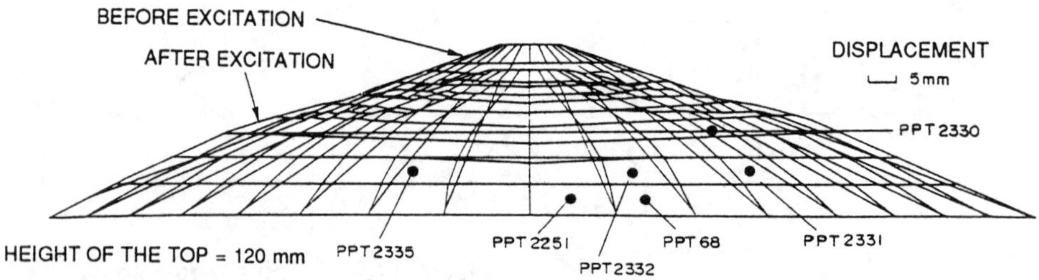

Figure T1.10. Computed deformation of an embankment (centrifuge model test).

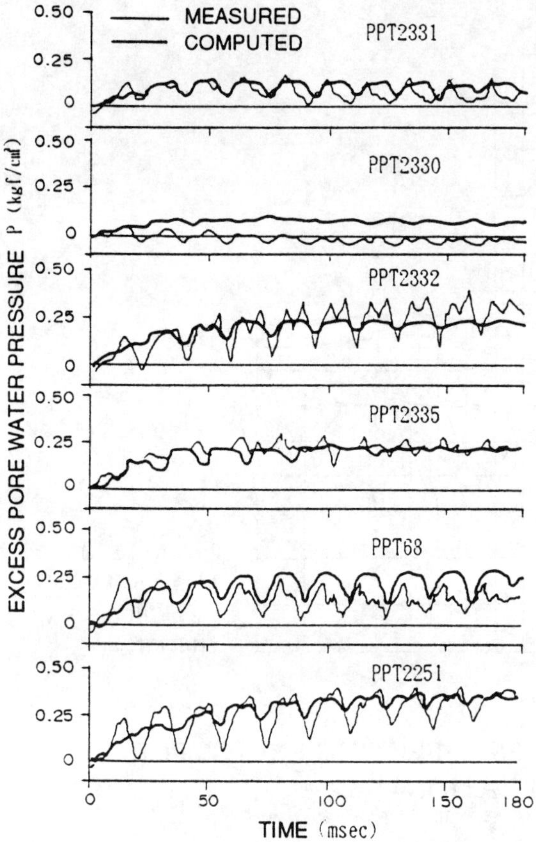

Figure T1.11. Excess pore water pressure in the embankment model.

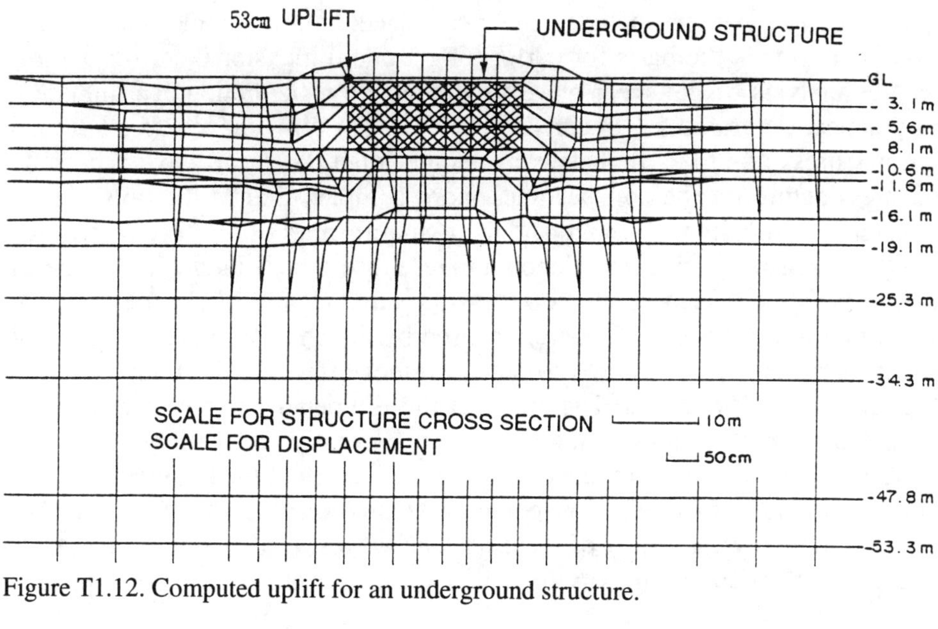

Figure T1.12. Computed uplift for an underground structure.

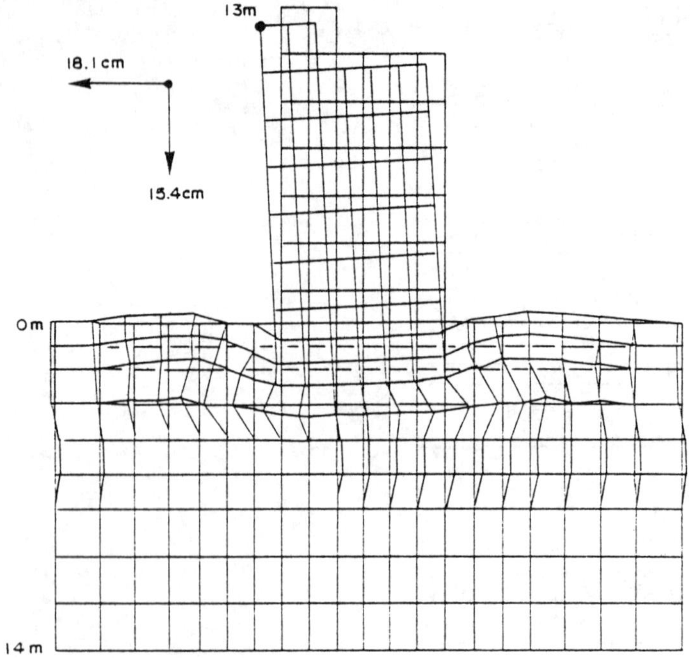

Figure T1.13. Computed settlement/tilting of a structure.

T1.3 TYPES OF SEISMIC WAVES AFFECTING EXCESS PORE WATER PRESSURE GENERATION

Seismic waves are classified into body waves (*P*- and *S*-waves) and surface waves (Love and Rayleigh waves). Generally the primary part of the earthquake motion, which includes the peak acceleration, is the shear wave (*S*-wave). This is the basis for performing a one-dimensional earthquake response analysis of the shear wave for liquefaction prediction/determination in practice. Saturated soil receives, however, not only shear stress but also normal stress due to earthquake waves other than the shear wave. The state of stress acting on the soil becomes more complex due to the presence of structures resting on the ground and the nonuniformity of the subsoil profile.

Of the seismic waves mentioned above, those which induce a change in the normal stress such as the compression wave (*P*-wave) change the pore water pressure. This change merely caused by the change in the total normal stress, is generally considered to have a minor effect on the liquefaction if the sand is fully saturated. The effect of these types of stresses for partially saturated or unsaturated sands, however, are currently under study.

The surface waves also induce shear stress and affect the liquefaction. The current practice of liquefaction potential evaluation does not take into account completely the effect of these surface waves as a routine practice because of the added complexity in analysis. The current practice of liquefaction potential evaluation, however, includes most of the contributions from the surface waves because it is established and calibrated based on the case history data during past earthquakes, which actually include the effects of surface waves generated during earthquakes.

T1.4 EFFECT OF PERMEABILITY OF SOIL ON LIQUEFACTION

T1.4.1 *Drainage of pore water from the soil under shaking*

Drainage of pore water from the soil under shaking affects the engineering properties of liquefied soil. Drainage conditions are basically determined by the relative relationship between the loading rate and the dissipation rate of pore water governed by the coefficient of permeability. For embankment construction on clayey subsoil, for example, the pore water pressure rises according to the load increase and remains at that level for some time. This process may be approximated by soil behavior under an undrained condition. In the long term, however, the pore water pressure gradually dissipates and the soil accordingly consolidates. This process should be treated as soil behavior under a partially drained condition.

In the earthquake response analysis of saturated subsoil, pore water was conventionally considered as not moving relative to the soil particles during earthquakes; the undrained condition was assumed. This assumption has the advantage of simplicity in the analysis because it is not necessary to consider additional degrees of freedom for the motion of pore water. This assumption allows a two-stage analysis, where the first stage is performed under an undrained condition for earthquake response analysis, and the second stage includes drainage effects in the analysis of the subsoil after earthquake shaking.

More accurate analysis may be performed by taking into consideration the effect of the drainage even during earthquake shaking. This may be particularly effective for the analysis of liquefaction remediation measures based on the drain method because it totally relies on the effect of drainage during earthquake shaking.

The effect of drainage may be classified into the following two categories. First is the direct effect of the seepage flow of pore water. The other is the indirect effect of the dissipation of excess pore water pressure upon the soil properties, resulting in a change in earthquake response characteristics of the subsoil profile.

Figure T1.14 shows the dynamic stiffness of saturated soil relative to that of dry soil in accordance with coefficient of permeability [8]. In the analysis, the soil is idealized as a two phase material composed of soil skeleton and pore water. This figure clearly indicates that dynamic stiffness differs ac-

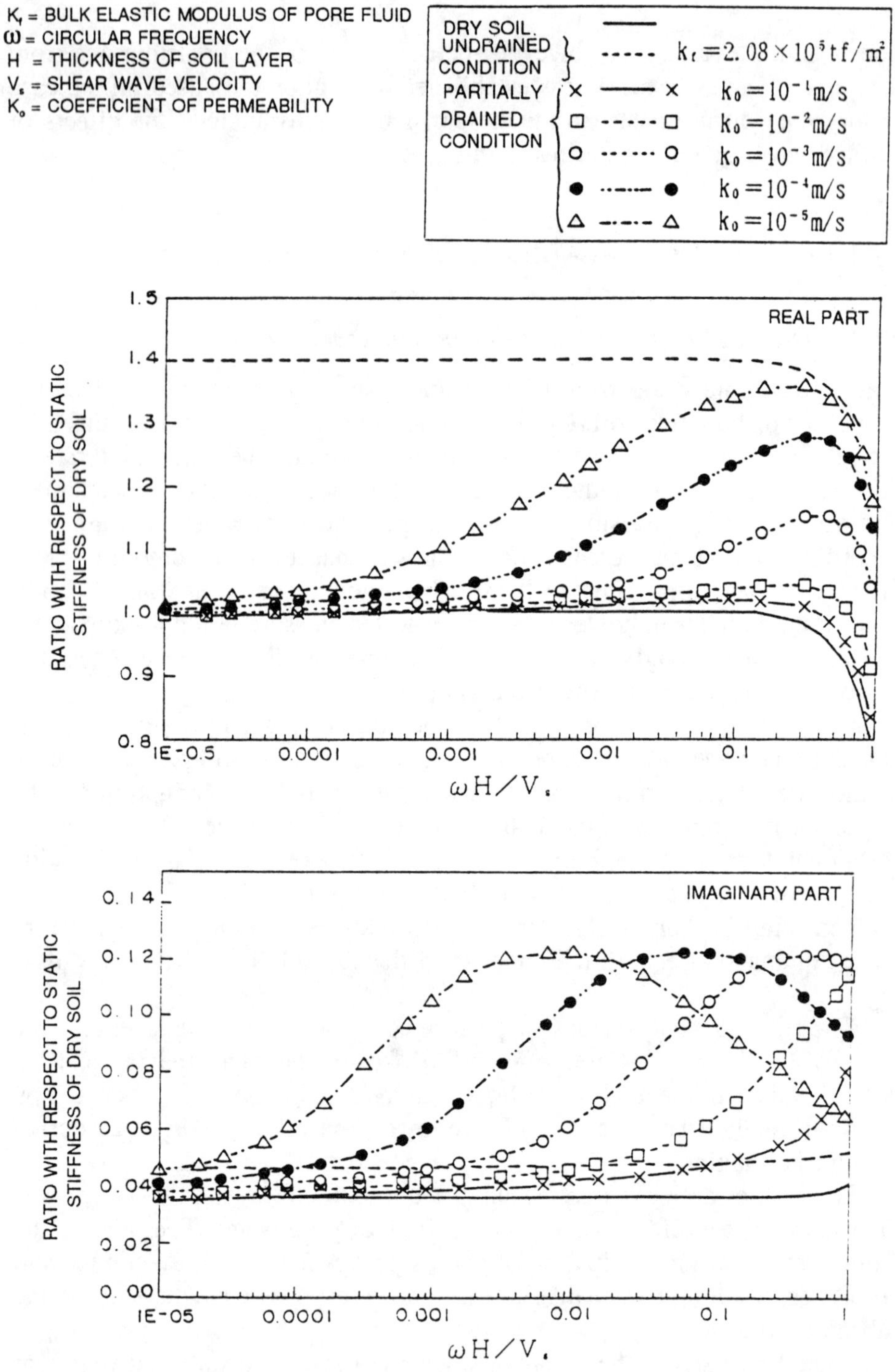

Figure T1.14. Effect of permeability on dynamic soil stiffness for a saturated soil.

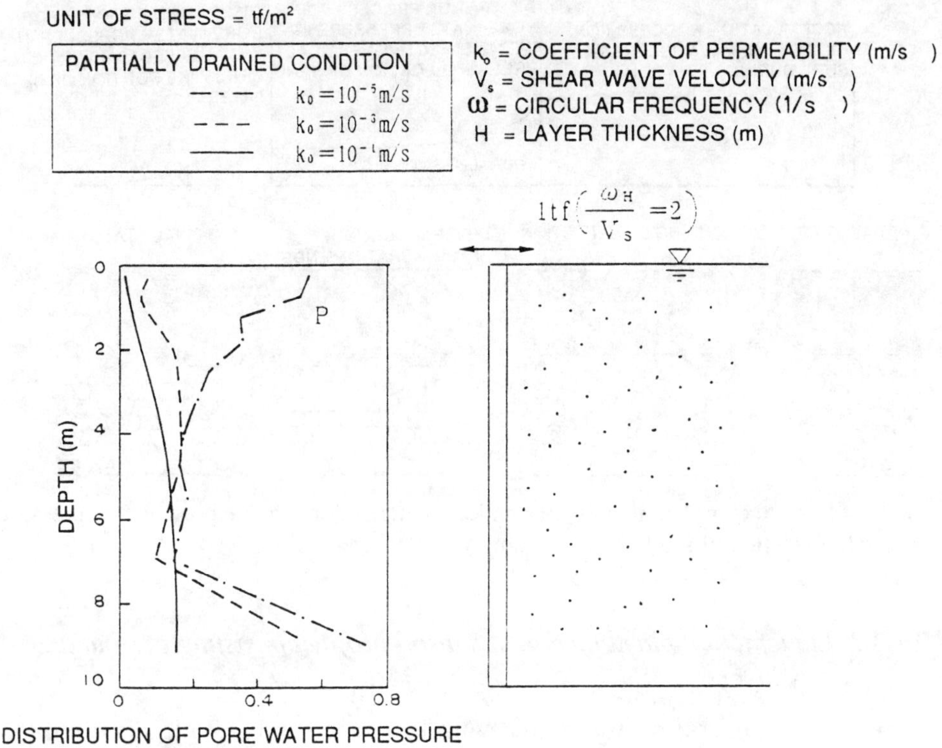

Figure T1.15. Distribution of pore water pressure.

cording to the coefficient of permeability; soil with a smaller coefficient of permeability becomes similar to the soil under an undrained state over a wider frequency range. This figure also indicates that the imaginary part of the complex dynamic stiffness, corresponding to the damping, becomes about 3 times, at a maximum, that of the undrained state. These differences may have significant effects on the earthquake response of the subsoil profile.

Figure T1.15 shows the distribution of pore water pressure around a pile when the head of the pile is excited horizontally by a unit load (1 tf/m) in a saturated soil [8]. The pore water pressure is computed at the mid point of each layer in the total subsoil profile divided into 10 layers in the analysis. When the coefficient of permeability is small, it is apparent that a large excess pore water pressure is generated even near the ground surface. When the coefficient of permeability is large, it approaches the hydrostatic pressure distribution.

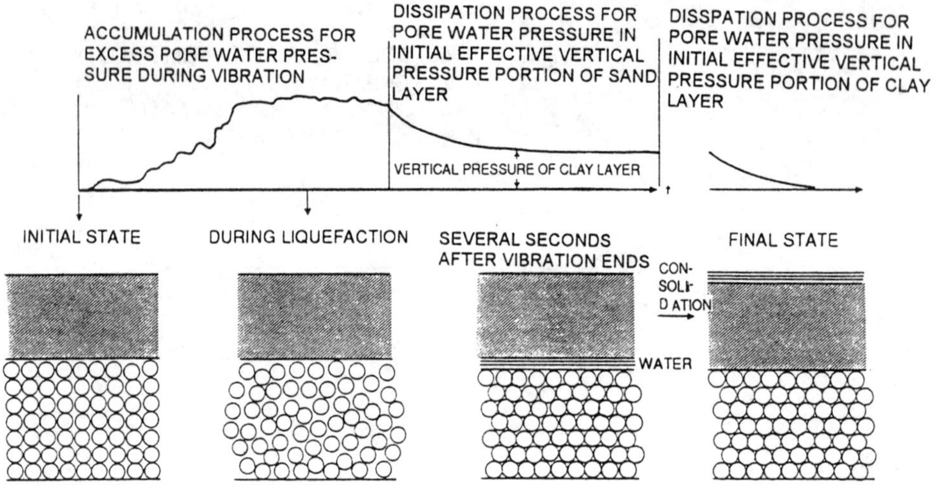

Figure T1.16. Accumulation and dissipation patterns for excess pore water pressure during liquefaction of sandy soil overlain by a clay layer.

T1.4.2 *Liquefaction phenomenon of subsoil profile consisting of sand and clay layers*

In practice, a subsoil profile mixed with layers of sand and clay is often encountered; it may originate from the fluctuation of the sea surface level during the glacial age. When sandy soil interposed between clay layers undergoes liquefaction, excess pore water pressure dissipation is prevented by these clay layers, leading to a possibility of serious damage long after the shaking stops. There are many unknown factors in the earthquake response characteristics of this type of soil; in particular, when the soft clay layer is thick, there is a tendency to amplify a predominant sinusoidal response of the subsoil.

Figure T1.16 shows schematically the process of the liquefaction phenomenon when there is a clay layer on top of the sand. When the sandy layer below the clay layer undergoes liquefaction, excess pore water pressure of the sand may dissipate at a rate corresponding to the coefficient of permeability of the sand. Excess pore water pressure in the clay layer, however, does not dissipate quickly because of its small coefficient of permeability. After a few tens of seconds, a thin layer of water may be created between the sandy subsoil and the clay layer. Because this thin layer of water does not have shear resistance, it will trigger flow failure or a large deformation of the subsoil induced by the presence of structures resting on the ground surface or with a slight slope of the ground surface. This type of phenomenon was simulated in centrifuge model tests [9, 10].

TECHNICAL NOTE 2

Recent advances and future trends in liquefaction remediation

Among various structures constructed in reclaimed land, structures such as quay walls and embankments may still continue to function although they suffer a certain deformation induced by liquefaction. In these cases, economical liquefaction remediation may be possible by incorporating an allowable deformation in the design procedure. Economical liquefaction remediation may also be possible for lifeline facilities such as buried pipelines if some degree of deformation is allowed in the improved soil by designing the embedded pipelines to be flexible at the pipe connections. Some examples of designs which incorporate an allowable deformation concept for liquefaction remediation are presented in the next chapter of this handbook.

T2.1 EXAMPLE OF DEFORMATION BASED DESIGN OF AN SOIL IMPROVEMENT AREA FOR A SHEET PILE QUAY WALL

The design of a soil improvement area for a sheet pile quay wall considers conditions for the seabed soil in front of the wall and the backfill soil behind it. The example [1] shown below is for a case where the improvement area of subsoil was in front of the sheet pile wall. The design was performed based on shaking table tests and effective stress analysis. The cross section of the sheet pile wall is shown in Figure T2.1. The displacement and bending moment of the sheet pile induced after shaking were related to the width of the area improved by compaction. The width of the improved area was designated as L as shown in the figure. In this figure, the portion designated as ‘loose sand’ corresponds to the sandy layer which undergoes liquefaction during earthquakes. For more details, see reference [1]. The measured and computed bending moments after excitation are shown in Figure T2.2. These results were further converted into a plot of the maximum bending moment related to the width of the improved area *L* as shown in Figure T2.3.

Similarly the measured and computed displacements in the sheet pile are

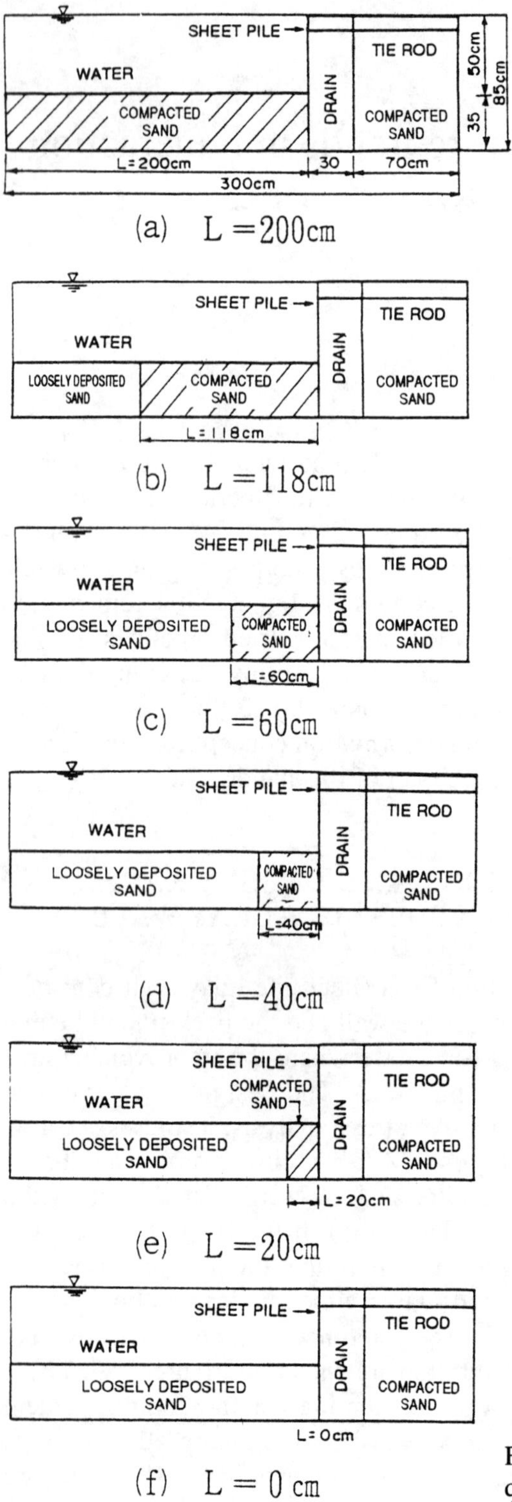

Figure T2.1. Cross sections of sheet pile quay walls.

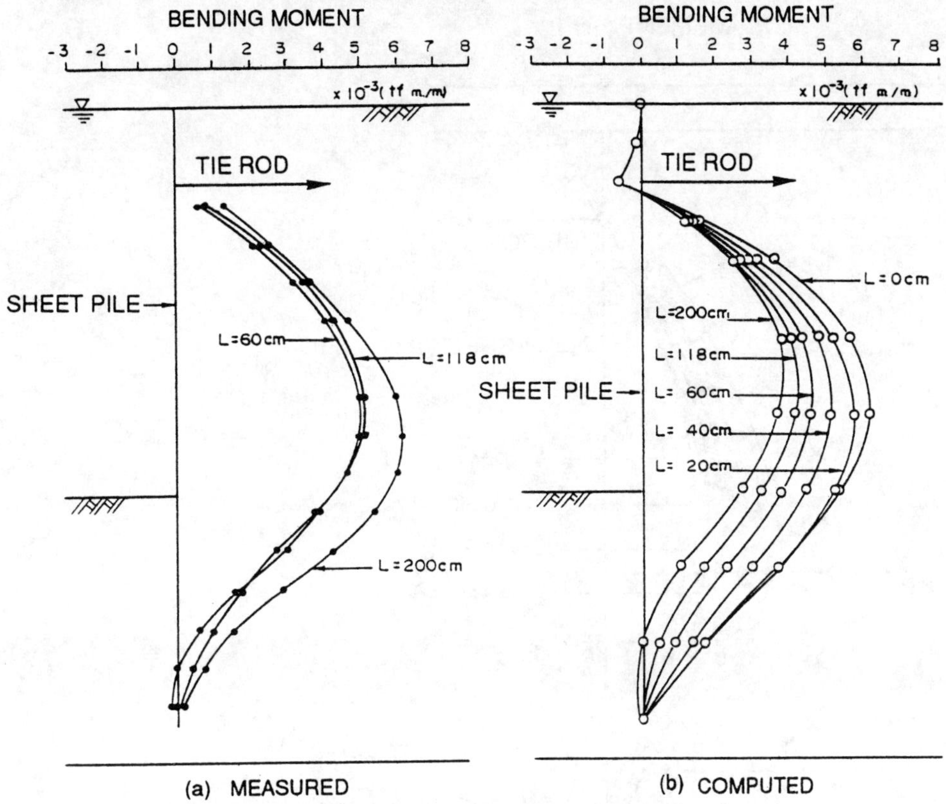

Figure T2.2. Bending moment of the sheet pile after excitation.

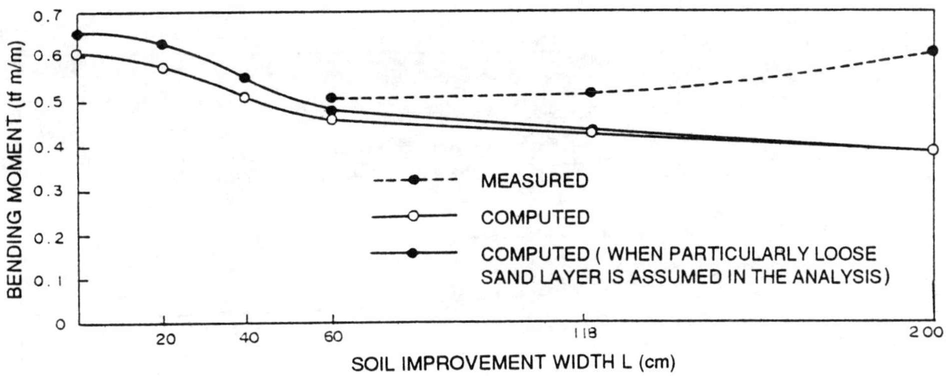

Figure T2.3. Relationship between soil improvement area and the maximum bending moment.

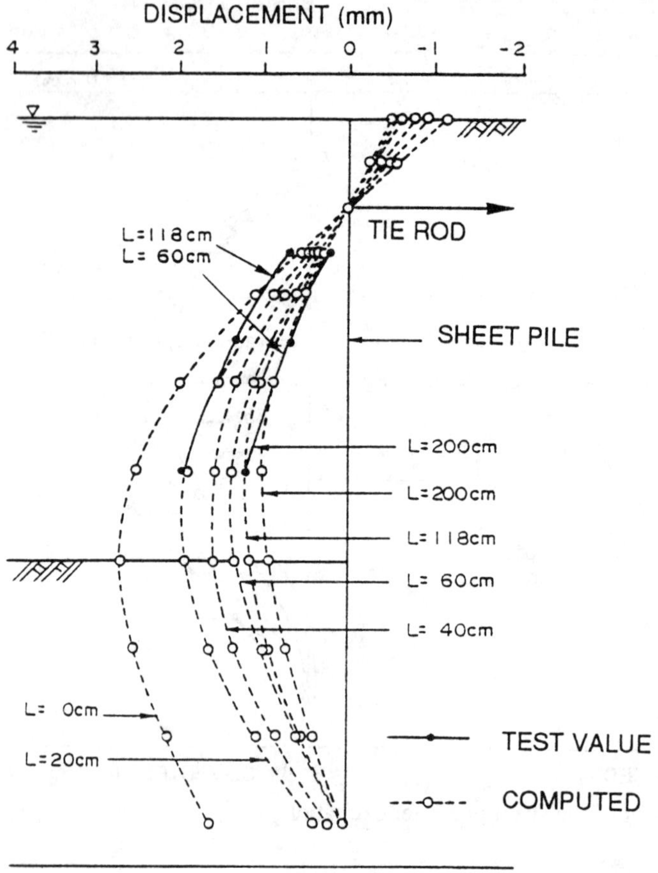

Figure T2.4. Displacement after excitation.

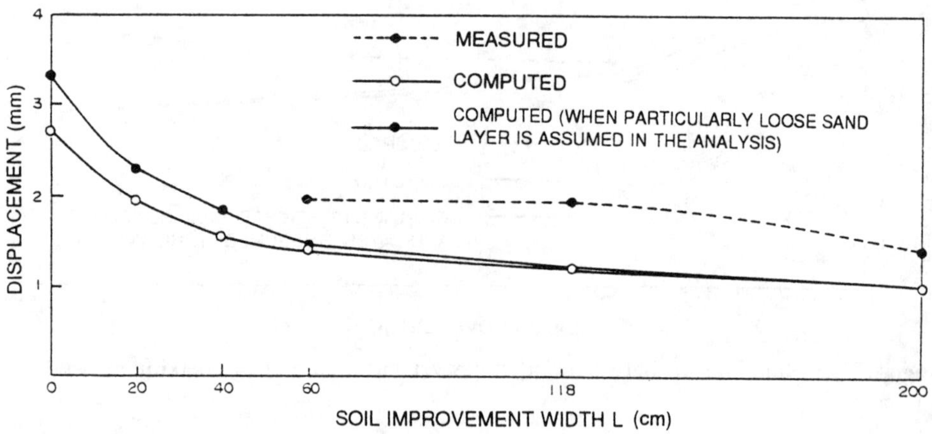

Figure T2.5. Relationship between the maximum displacement and the soil improvement area.

shown in Figure T2.4; the relationship between the maximum displacements and the width of the improved area are shown in Figure T2.5.

This type of study makes it possible to obtain the basic relationship necessary for determining the improved area from the allowable displacement and bending moment. It is anticipated that this approach will lead to a practical design of economical liquefaction remediation in accordance with allowable deformation of the structures.

T2.2 EXAMPLE OF DEFORMATION-BASED-DESIGN OF BURIED STRUCTURES

Heave of a buried structure due to liquefaction is governed by the deformation of the soil directly below the underground structure. An example [2]

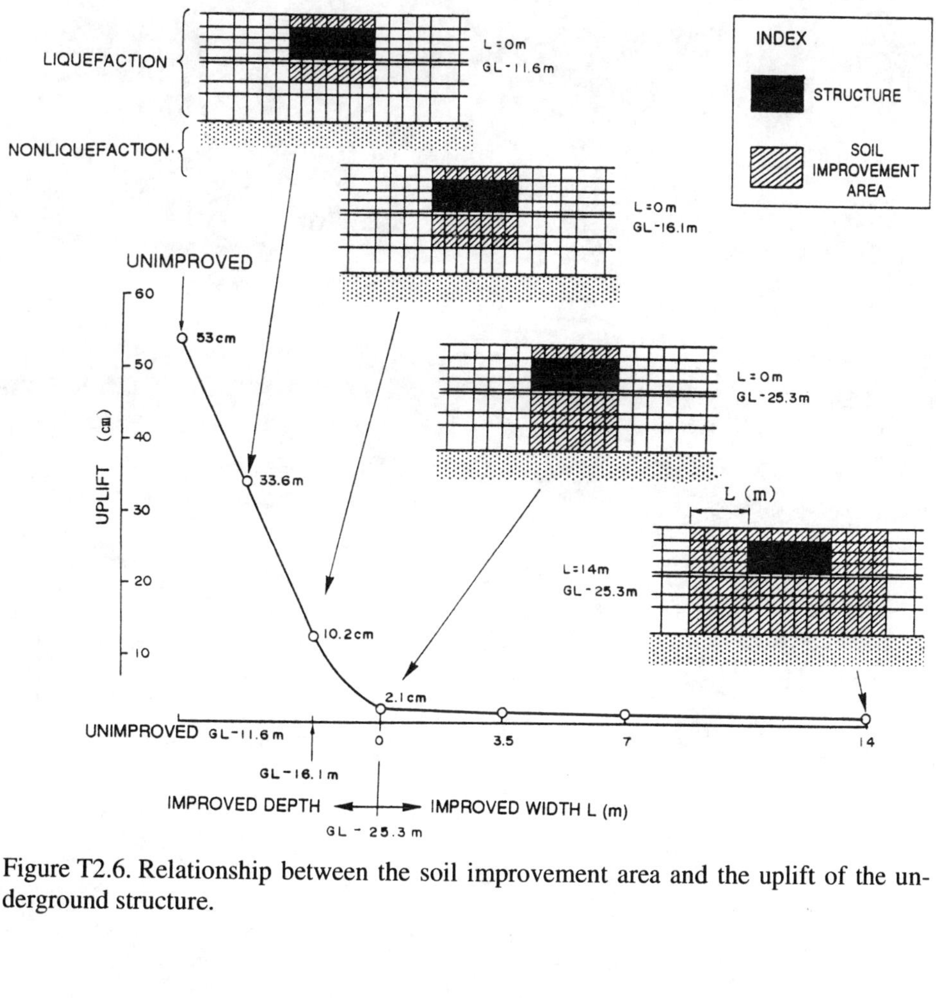

Figure T2.6. Relationship between the soil improvement area and the uplift of the underground structure.

shown in Figure T2.6 is a variation on the study noted in Technical note 1. With respect to liquefaction induced heave, the design based on a balance between the buoyancy and gravity forces is the only simplified procedure currently used in practice. The effective stress analysis provides a means of estimating the degree of heave. As shown in Figure T2.6, reduction in the heave is clearly shown with increasing improvement area below the buried structure.

APPENDIX 1

Summary of seismic design guidelines and standards in Japan

In this chapter, liquefaction prediction/determination methods specified in seismic design guidelines and standards are briefly described. As shown in Table A1.1, some of the design guidelines and standards have been updated several times. This section of the handbook covers not only the current state of design guidelines and standards in Japan but also their chronology.

A1.1 DESIGN STANDARDS FOR PORT AND HARBOUR STRUCTURES

A1.1.1 *Supplement for design standards for port and harbour structures (1971)*

This supplement indicates the procedure for liquefaction prediction as follows:

1. The possibility of liquefaction is determined based on gradation by using the zones shown Figure A1.1;
2. The SPT *N*-value of the soil near the ground water level is investigated at the site;
3. The maximum acceleration of the ground surface caused by a design earthquake is estimated;
4. The SPT *N*-value of the soil and critical *N*-value indicated in Figure A1.2 are compared using maximum acceleration of the ground surface in 3; if the SPT *N*-value of the soil is less than the critical *N*-value, the subsoil is considered to liquefy.

The broken line in Figure A1.2 is the SPT *N*-value corresponding to a relative density of 80%. This line indicates that a suitable determination must be made regarding the soil with an SPT *N*-value between the solid line and the broken line because the possibility of liquefaction is considered very low when the relative density is more than 80%.

Table A1.1. Seismic design guidelines and standards for evaluation of liquefaction potential.

No.		Title	Organization	Date
1	a)	Supplements for design standards for port and harbour structures	Japan Port and Harbour Association	April 1971
	b)	Technical standards for port and harbour facilities and commentaries		March 1979
	c)	Technical standards for port and harbour facilities and commentaries		February 1989
2	a)	Seismic design specifications for highway bridges and commentaries	Japan Road Association	May 1972
	b)	Specifications for highway bridges and commentaries (Volume V, Earthquake resistant design)		May 1980
	c)	Specifications for highway bridges and commentaries (Volume V, Earthquake resistant design)		May 1990
3	a)	Design standards of building foundations and commentaries	Architectural Institute of Japan	November 1974
	b)	Recommendations for design of building foundations		January 1988
4	a)	Design standards for railway structures (foundations and retaining structures)	Japan National Railway	June 1974
	b)	Design standards for national railway structures (foundations and retaining structures)	Japan Society of Civil Engineers	March 1986
5	a)	Road construction, specifications for remediation of soft ground	Japan Road Association	January 1977
	b)	Road construction, specifications for remediation of soft ground		November 1986
6		Standards on regulation of hazardous materials	Fire Defense Agency	February 1978
7		Recommended practice for LNG underground storage tanks	Japan Gas Association Safety Committee for Liquefied Natural Gas Storage Tanks	March 1979
8		Specifications for construction of tailing dams and commentaries	Japan Mining Industry Association	March 1980
9		Seismic design specifications on submerged tunnels (tentative)	Japan Society of Civil Engineers	March 1975
10		Technical specifications on underground oil storage facilities (tentative)	Japan Society of Civil Engineers	May 1980
11		Technical specifications on seismic design of nuclear power plants	Japan Electricity Association Committee on Technical Standards on Electricity	August 1987
12		Specifications for seismic design of water facilities and commentaries	Japan Water Association	October 1979
13		Specifications for seismic design of sewage facilities and commentaries	Japanese Sewage Association	September 1981
14		Design specifications for common utility ducts	Japan Road Association	March 1986

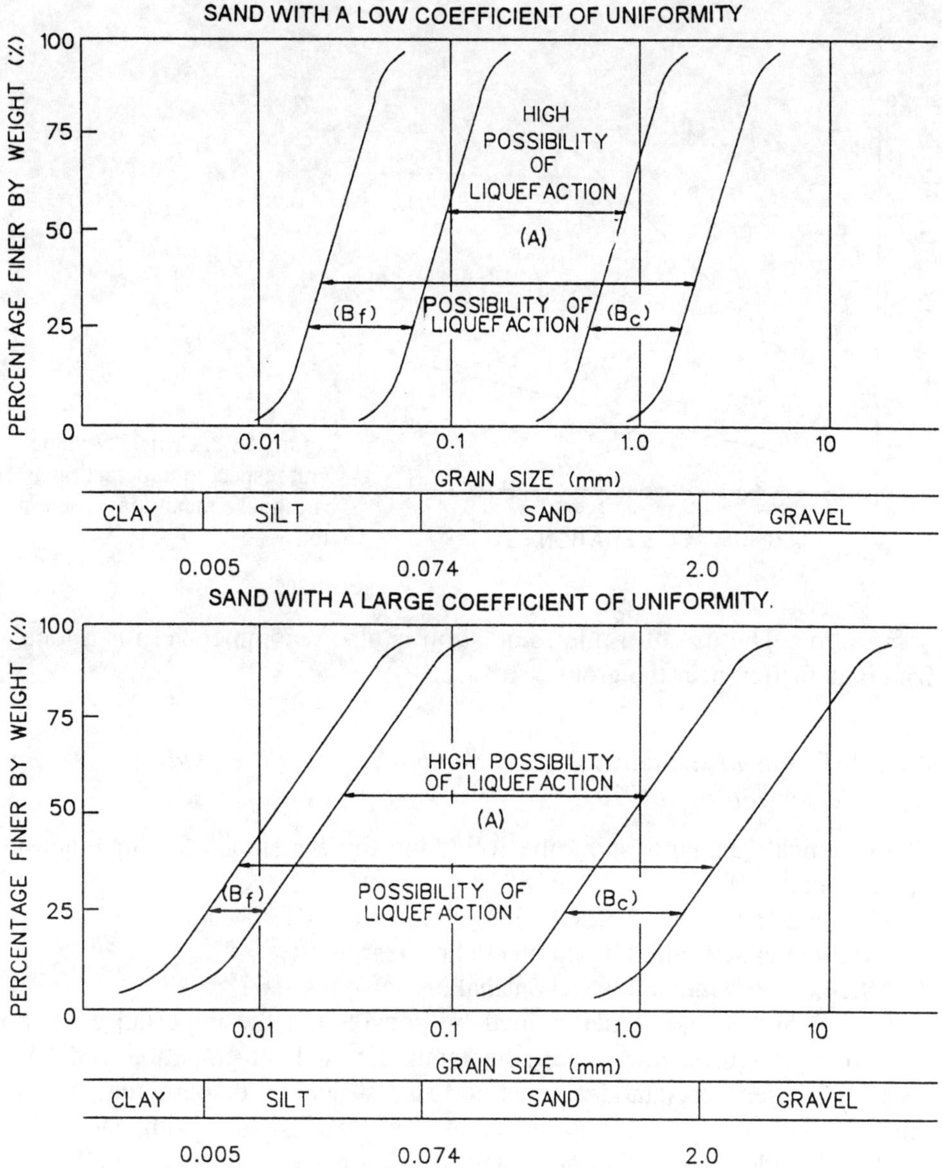

Figure A1.1. Gradation of soil having liquefaction possibility.

Figure A1.1 was obtained from the results of investigation for past case histories of liquefaction. The critical *N*-value in Figure A1.2 was obtained from an overall evaluation of the results obtained from the past experimental research and case histories of liquefaction [1].

When the subsoil profile is considered to liquefy, the subsoil should be improved so that the SPT *N*-value of soil exceeds the critical *N*-value by

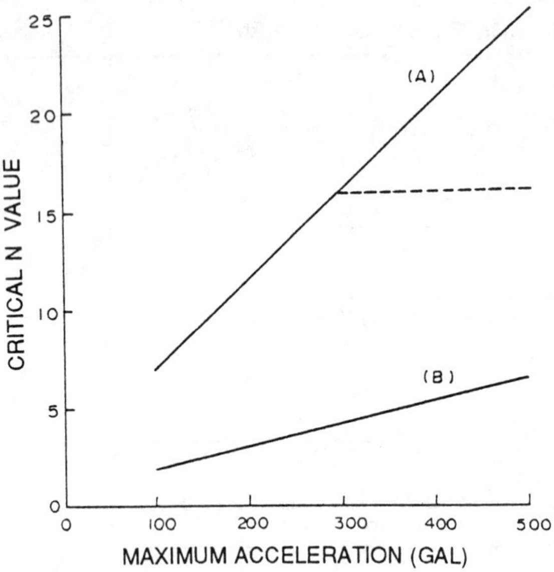

Figure A1.2. Critical *N*-value with respect to liquefaction and earthquake maximum acceleration .

compaction. The use of a pile foundation is also recommended for a localized liquefaction near the ground surface.

A1.1.2 *Technical standards for port and harbour facilities and commentaries (1979)*

The technical standards show the following three methods for liquefaction prediction:

- Method 1: A method based on gradation and SPT *N*-value;
- Method 2: A method based on cyclic shear tests;
- Method 3: A method based on shaking table tests.

The method to use is determined by considering the importance of the concerned structure; in general, the method based on gradation and SPT *N*-value is used for standard cases and, for structures of particular importance, it is preferable to use Method 2 in combination with Method 1. Method 1 is the same as that presented in Section A1.1.1.

In Method 2, shear stress in the subsoil during earthquakes is estimated, and whether or not liquefaction will occur is evaluated by the cyclic shear test. It is preferable that the shear stress in the subsoil profile during an earthquake is estimated from the earthquake response analysis of the subsoil profile. The liquefaction resistance of soils will be evaluated by cyclic triaxial tests, cyclic simple shear tests, or other types of cyclic shear tests.

Method 3 directly evaluates whether or not liquefaction of the sand layer occurs based on the performance of model saturated subsoil. Suitable correction is necessary for the difference in the stress state in the model and in-situ.

As remedial measures against liquefaction, two approaches are shown:

1. To improve the soil so that there is no possibility of liquefaction; and
2. To design the structure so that fatal damage does not occur to the structure even when liquefaction occurs.

In adopting the latter approach, shear resistance of the liquefied soil should be ignored in the design.

A1.1.3 *Technical standards for port and harbour facilities and commentaries (1989)*

The technical standards specify the following evaluation procedure for liquefaction prediction/determination.

1. Loosely deposited saturated sandy soil may liquefy during earthquakes, causing damage to structures, and therefore, whether or not liquefaction will occur with respect to the design earthquake should be predicted/determined in design.

– Whether or not liquefaction of the soil will occur with respect to the design earthquake is predicted/determined from the gradation and SPT *N*-value;

– When the liquefaction occurrence cannot be determined by the method based on the gradation and SPT *N*-value, the liquefaction occurrence is predicted/determined based on the results of cyclic triaxial tests of undisturbed sand.

2. Liquefaction prediction/determination based on the gradation and SPT *N*-value.

a) Initially, the soil is classified based on the gradation by using Figure A1.1. One of the two figures in Figure A1.1 is chosen based on the coefficient of uniformity of the soil. The coefficient of uniformity of $U_c = D_{60}/D_{10} = 3.5$ may be used as a criterion, where U_c = coefficient of uniformity, D_{60} = 60% diameter, and D_{10} = effective grain size (10% diameter). The soil gradation which falls in zones other than A, B_f, and B_c are considered nonliquefiable.

When difficult classification (e.g. when the gradation curve straddles the two zones of gradation) leads to a noticeable difference in the results of liquefaction prediction/determination according to the choice of the zones, suitable engineering judgement becomes necessary including the use of the prediction/determination method based on cyclic triaxial tests.

b) With regard to the soil layer with the gradation falling in zones A and B_c in Figure A1.1, the equivalent *N*-value is computed using the following equation.

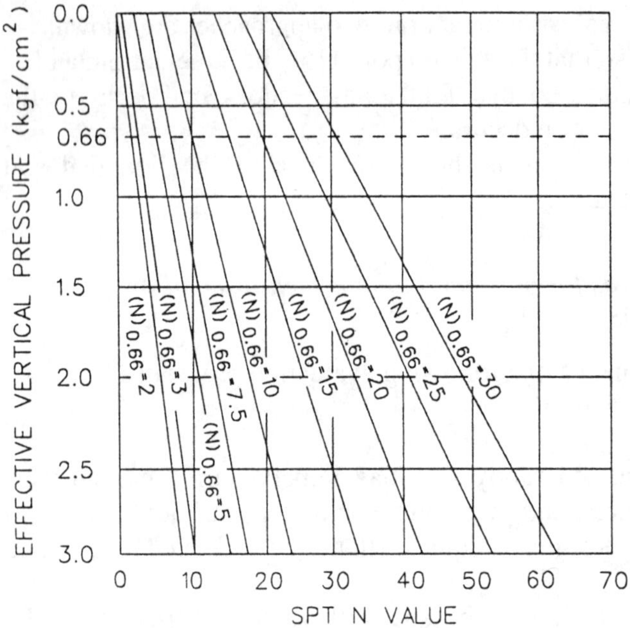

Figure A1.3. Chart for calculation of equivalent *N*-value.

$$(N)_{0.66} = \frac{N - 1.828\,(\sigma'_v - 0.66)}{0.399\,(\sigma'_v - 0.66) + 1} \tag{A1.1}$$

where: $(N)_{0.66}$ = equivalent *N*-value, N = SPT *N*-value of a soil layer, σ'_v = effective vertical pressure of a soil layer (kgf/cm^2) (the effective vertical pressure to be used here should be computed with respect to the ground surface elevation at the time of the standard penetration test).

Equation (A1.1) is applicable for the ranges $2 \le (N)_{0.66} \le 40$, $\sigma'_v \le 3$ kgf/cm^2. The relationship of Equation (A1.1) is shown in Figure A1.3.

The equivalent *N*-value refers to the SPT *N*-value corrected to the effective vertical pressure of 0.66 kgf/cm^2. This conversion reflects the practice that, in the former Technical standards for port and harbour facilities and commentaries (1979) before the 1989 revision, liquefaction prediction had been done with respect to the SPT *N*-value of the soil near the ground water table.

For soils with a gradation falling in zone B_f of Figure A1.1, the equivalent *N*-value is determined the same as the SPT *N*-value of the soil layer without correction.

c) Equivalent acceleration is computed for the soil layers with a gradation falling in zones A, B_f, and B_c of Figure A1.1. This corresponds to the maximum acceleration used in the prediction method noted in Section A1.1.2. The equivalent acceleration is computed using Equation (A1.2) based on the maximum shear stress obtained by earthquake response analysis of soil.

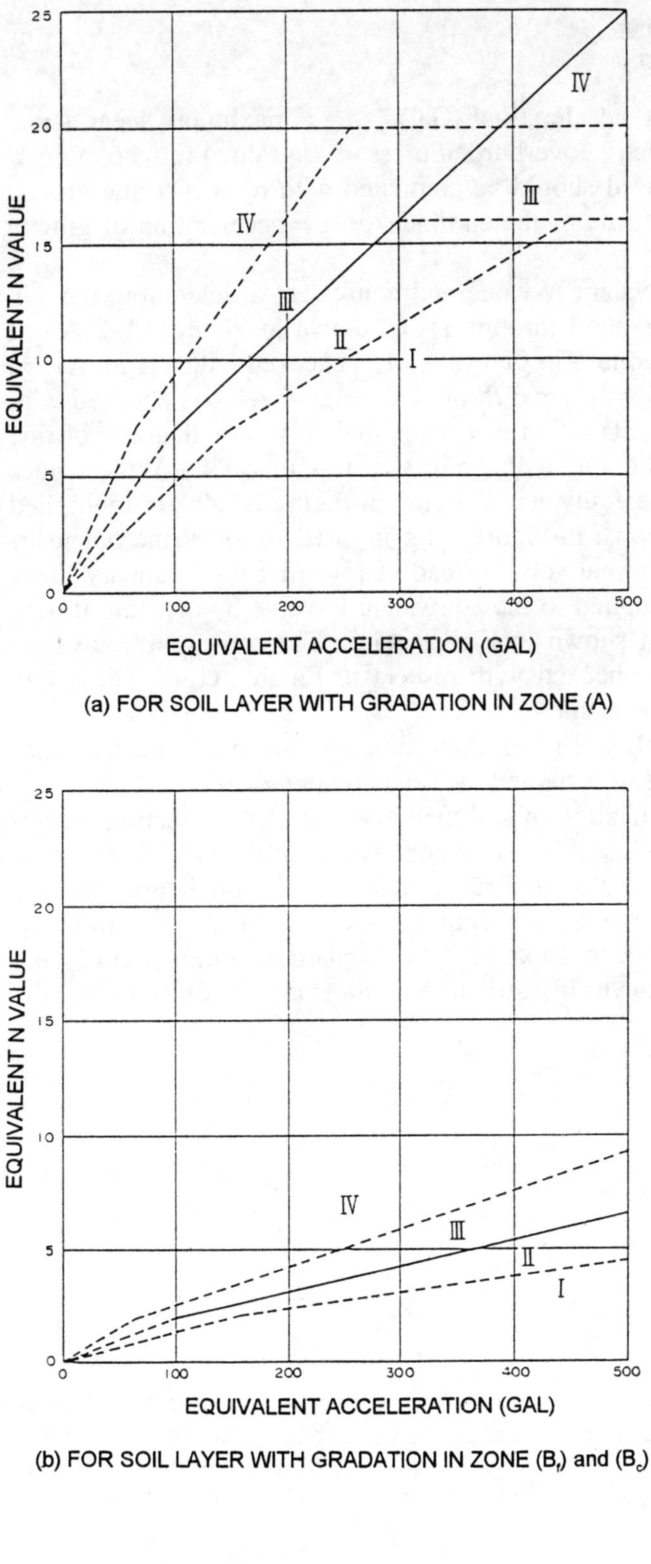

Figure A1.4. Classification of soil layer for liquefaction prediction based on equivalent *N*-value and equivalent acceleration.

$$\alpha_{eq} = 0.7 \times \frac{\tau_{max}}{\sigma'_v} \times g \tag{A1.2}$$

where: α_{eq} = equivalent acceleration (Gal), τ_{max} = maximum shear stress (kgf/cm^2), σ'_v = the effective overburden pressure (kgf/cm^2) (effective overburden pressure to be used should be computed with respect to the ground surface elevation at the time of the earthquake), g = acceleration of gravity (980 Gal).

d) Based on the equivalent *N*-value and equivalent acceleration, the soil layer is classified into zones I through IV as shown in Figure A1.4. A soil layer with gradation in zone *A* in Figure A1.1 is checked with Figure A1.4a; a soil layer with gradation in zone B_f or B_c in Figure A1.1 is checked with Figure A1.4b. If the soil classified in zone A includes more than 5% of fine grains (component of the soil with grain size less than 74 m), the values shown on the axis of the equivalent *N*-value in Figure A1.4a are multiplied by a reduction factor shown in Figure A1.5 in order to lower the boundary lines between zones. Alternatively, instead of lowering the boundary lines, the conversion can be applied to the equivalent *N*-value by dividing it with the reduction coefficient shown in Figure A1.5. The converted equivalent *N*-value can be directly checked with respect to Figure A1.4a. These two procedures yield the same result.

If a layer with zone B_c soil is overlain by a layer with low permeability such as clay or silt layers, it is treated the same as zone *A* soil.

e) Based on the classification of soil mentioned in d), liquefaction prediction/determination of each soil is performed using Table A1.2. Liquefaction determination is not simply related to the result of the prediction because other factors such as a required margin of safety factor for the structure should also be considered. In Table A1.2, a standard determination of each result of prediction is shown. If a structure requires a high safety factor, the

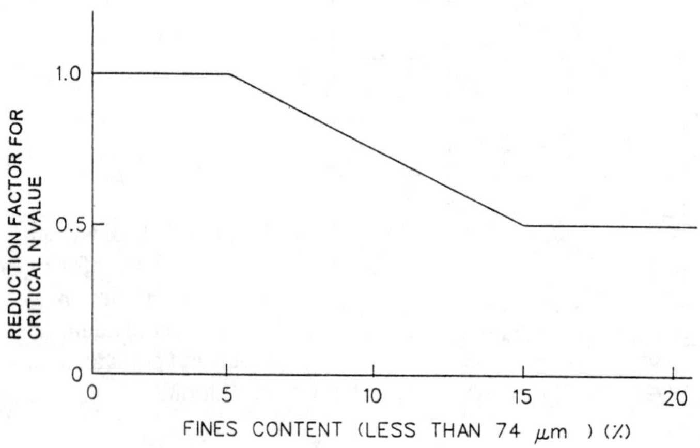

Figure A1.5. Reduction factor for critical *N*-value based on the fines content.

Table A1.2 Liquefaction prediction/determination for each soil layer based on the grain size distribution and SPT *N*-value.

Zones indicated in Figure A1.4	Liquefaction prediction based on gradation and SPT *N*-value	Liquefaction determination based on gradation and SPT *N*-value
I	Possibility of liquefaction occurrence is very high	Liquefaction will occur
II	Possibility of liquefaction occurrence is high	Decide either to determine that liquefaction will occur, or to conduct further evaluation based on cyclic triaxial tests
III	Possibility of liquefaction occurrence is low	Decide either to determine that liquefaction will NOT occur, or to conduct further evaluation based on cyclic triaxial tests When it is necessary to allow for a significant safety factor for a structure, decide either to determine that liquefaction will occur, or to perform further evaluation based on cyclic triaxial tests
IV	Possibility of liquefaction occurrence is very low	Liquefaction will not occur

assessment for zone III should be either to determine that liquefaction will occur or to evaluate based on the results of cyclic triaxial tests.

f) Based on the liquefaction determination for each soil layer mentioned in e), the liquefaction determination of the entire subsoil profile is assessed by considering the thickness and the depth of the liquefiable soil layers.

Liquefaction determination for the entire subsoil profile is made with an overall engineering evaluation of each site.

3. Prediction/determination based on cyclic triaxial tests.

a) When liquefaction occurrence cannot be determined based on gradation and SPT *N*-value, cyclic triaxial tests are performed on undisturbed samples of sand and earthquake response analysis of the subsoil profile is made. By comparing the liquefaction strength of the subsoil and shear stress in the subsoil profile during earthquakes, liquefaction of the subsoil is predicted/determined.

b) Suitable corrections must be made so that the results of the earthquake response analysis of the subsoil profile and cyclic triaxial tests reasonably represent in-situ conditions.

c) Overall liquefaction determination at a site consisting of soil layers is made by considering the thickness and depth of the liquefiable soil layers.

A1.2 DESIGN SPECIFICATIONS FOR HIGHWAY BRIDGES

A1.2.1 *Seismic design specifications for highway bridges and commentaries (1972)*

The specifications state that liquefaction occurs under the following conditions:

– The saturated sand soil layer is located between the ground surface and depth of 10 m;

– SPT *N*-value is less than 10;

– The coefficient of uniformity is less than 6;

– D_{20} (20% diameter) of the grain size distribution curve ranges from 0.04 to 0.5 mm.

A soil layer judged to liquefy should be treated as a soil layer in which bearing capacity in the earthquake design should be ignored.

A1.2.2 *Specifications for highway bridges and commentaries, volume V earthquake resistant design (1980)*

The specifications state that liquefaction determination should be accomplished during design for the following soil layers:

– Alluvium with ground water table less than 10 m from the ground surface;

– Saturated sand layers located less than 20 m depth below the ground surface with a mean grain size D_{50} between 0.02 mm and 2.0 mm.

With respect to this type of soil layer, the liquefaction resistance factor F_L defined below is obtained. In a soil layer in which this value is less than 1.0, liquefaction will occur.

$$F_L = \frac{R}{L} \tag{A1.3}$$

where: F_L = liquefaction resistance factor, L = shear stress ratio during earthquakes

$$L = r_d \times k_s \times \left(\frac{\sigma_v}{\sigma'_v}\right) \tag{A1.4}$$

R = dynamic shear strength ratio, determined as follows:

For 0.02 mm $\le D_{50} \le$ 0.05 mm

$$R = 0.0882\sqrt{\frac{N}{\sigma'_v + 0.7}} + 0.19 \tag{A1.5}$$

For 0.05 mm $\le D_{50} \le$ 0.6 mm

$$R = 0.0882\sqrt{\frac{N}{\sigma'_v + 0.7}} + 0.225\log_{10}\left(\frac{0.35}{D_{50}}\right) \qquad (A1.6)$$

For 0.6 mm $\leq D_{50} \leq$ 2.0 mm

$$R = 0.0882\sqrt{\frac{N}{\sigma'_v + 0.7}} - 0.05 \qquad (A1.7)$$

r_d = reduction factor for shear stress ratio during earthquakes

$$r_d = 1.0 - 0.015x \qquad (A1.8)$$

k_s = horizontal seismic coefficient at the ground surface used for liquefaction determination, σ_v = total overburden pressure (kg/cm^2), σ'_v = effective overburden pressure (kg/cm^2), x = depth from the ground surface (m).

If necessary, a response analysis of the subsoil profile, laboratory tests, detailed geological survey and soil investigation in-situ should be performed. Liquefaction determination based on existing data is considered advantageous.

The horizontal seismic coefficient for liquefaction potential evaluation is given by the following equation and used to obtain the shear stress ratio L during an earthquake.

$$k_s = \nu_1 \times \nu_2 \times \nu_3 \times k_{so} \qquad (A1.9)$$

where: k_s = design horizontal seismic coefficient used for liquefaction determination (rounded to two digits after the decimal point), ν_1 = regional factor (0.7-1.0), ν_2 = site factor (0.9-1.2), ν_3 = importance factor (0.8-1.0), k_{so} = standard horizontal seismic coefficient used for liquefaction determination (= 0.15).

According to the design specifications, for a sandy soil layer subject to liquefaction, it is necessary to reduce the soil constant for earthquake design according to the value of the resistance factor F_L. Reduction in the soil constant is obtained by multiplying the soil constant obtained for a nonliquefac-

Table A1.3. Reduction factor for the soil constant.

Range of F_L	Depth from the ground surface x (m)	Factor to multiply the soil constant with D_E
$F_L \leq 0.6$	$0 \leq x \leq 10$	0
	$10 < x \leq 20$	1/3
$0.6 < F_L \leq 0.8$	$0 \leq x \leq 10$	1/3
	$10 < x \leq 20$	2/3
$0.8 < F_L \leq 1.0$	$0 \leq x \leq 10$	2/3
	$10 < x \leq 20$	1

tion condition by the factor D_E indicated in Table A1.3. The weight of the liquefiable soil layer should be included in the design procedure as an overburden weight with respect to the soil beneath it.

A1.2.3 *Specifications for highway bridges and commentaries, volume V earthquake resistant design (1990)*

The specifications (1990) show the same conditions for the soil layer requiring liquefaction evaluation as those shown in Section A1.2.2.

The liquefaction resistance factor F_L is calculated from Equation (A1.3). In order to consider the influence of the fines content FC on dynamic shear strength ratio R, however, the dynamic shear strength ratio R_3 expressed in Equation (A1.13) is added as follows:

$$R = R_1 + R_2 + R_3 \tag{A1.10}$$

where:

$$R_1 = 0.0882\sqrt{\frac{N}{\sigma'_v + 0.7}} \tag{A1.11}$$

$$R_2 = \begin{cases} 0.19 & (0.02\ \text{mm} \le D_{50} \le 0.05\ \text{mm}) \\ 0.25\log_{10}(0.35 / D_{50}) & (0.05\ \text{mm} < D_{50} \le 0.6\ \text{mm}) \\ -0.05 & (0.6\ \text{mm} < D_{50} \le 2.0\ \text{mm}) \end{cases} \tag{A1.12}$$

$$R_3 = \begin{cases} 0.0 & (0\% \le FC \le 40\%) \\ 0.004\,FC - 0.16 & (40\% < FC \le 100\%) \end{cases} \tag{A1.13}$$

Applicability of the above equation for evaluating R is limited for soils whose relative density is less than 60%.

The procedure to determine the reduction factor on a soil constant and design seismic coefficient is the same as discussed in Section A1.2.2.

A1.3 DESIGN STANDARDS OF BUILDING FOUNDATIONS

A1.3.1 *Design standards of building foundations and commentaries (1974)*

The standards state that a soil layer requires evaluation of liquefaction under the following conditions:

– A location less than 15 or 20 m deep below the ground surface;

– A medium uniformly graded sand layer (the liquefaction risk is moderately high when the coefficient of uniformity U_c is less than 10, and the risk

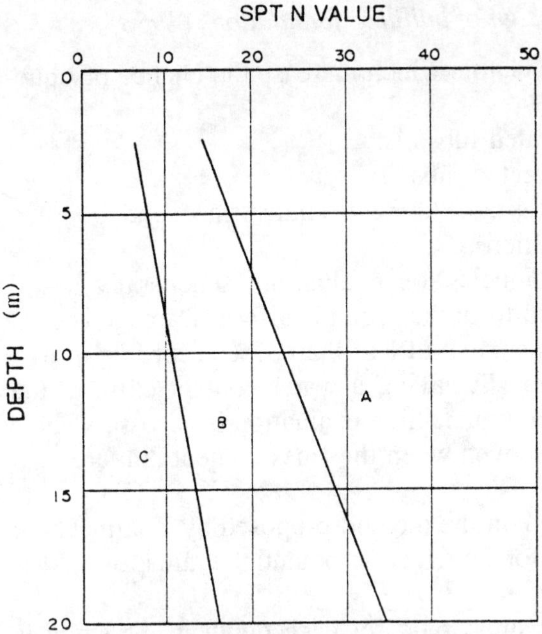

Figure A1.6. Liquefaction susceptibility and SPT *N*-value.

is high when U_c is less than 5. The liquefaction risk is also high if the silt/clay content is less than 10%, the mean grain size D_{50} ranges from 0.075 to 2.0 mm, and in particular, the soil is medium grain sand of approximately $D_{50} = 0.15$ to 1.0 mm);

– A water saturated location below the ground water table;

– A low degree of compaction and an SPT *N*-value is in zone *B* or *C* in Figure A1.6.

Accurate prediction of liquefaction is presently difficult, however, and further detailed investigation must be made regarding the individual site when necessary. The method proposed by Seed & Idriss (1971) (see Section 4.1.2) is introduced as a reference for further detailed study.

The following methods are suggested for remedial measures against liquefaction:

1. Increasing the relative density by compaction;
2. Replacing soil having a grain size which is susceptible to liquefaction; and
3. Lowering of ground water level.

With regard to 1, soil compaction has been proved to be effective in the 1964 Niigata earthquake and the 1968 Tokachioki earthquake. When piles were driven deeply through a liquefied layer to the nonliquefiable foundation layer, damage to structures was minimal in these earthquakes. There is also a report showing no damage caused by liquefaction when combining driven piles with compaction.

A1.3.2 *Recommendations for design of building foundations (1988)*

The recommendations state four factors which increase the chance of liquefaction as follows:

1. Lower fines content in saturated subsoils;
2. Smaller SPT *N*-value of saturated subsoil;
3. Proximity of the ground water table to the ground surface; and
4. Greater input earthquake motion.

A soil layer which requires liquefaction evaluation is generally a soil layer located less than about 20 m from the ground surface. Damage is concentrated in soil layers with fines content of less than 35% according to past case histories of liquefaction. For silt having a water content close to the liquid limit, or a low plasticity silt, liquefaction evaluation is necessary since there are examples of liquefaction even when the fines content is more than 35%.

The following procedure, based on the method proposed by Tokimatsu & Yoshimi (1983) [2], is specified for liquefaction potential evaluation with a slight modification.

a) The equivalent cyclic shear stress ratio for each depth in the study is calculated using the following equation.

$$\frac{\tau_d}{\sigma'_z} = r_n \frac{\alpha_{max}}{g} \frac{\sigma_z}{\sigma'_z} r_d \qquad \text{(A1.14)}$$

where: τ_d = equivalent cyclic shear stress amplitude applied on a horizontal plane with constant amplitude (tf/m^2), σ'_z = effective overburden pressure at each depth (effective vertical pressure) (tf/m^2), r_n = correction factor related to equivalent number of earthquake loading cycles which is formulated by $r_n = 0.1\,(M-1)$, where M is the magnitude of the design earthquake, α_{max} = maximum horizontal acceleration at the ground surface (Gal), g = acceleration of gravity (980 Gal), σ_z = total overburden pressure at each depth (total vertical pressure) (tf/m^2), r_d = reduction factor (= 1-0.015z) where z is the depth in meters from the ground surface.

b) The corrected SPT *N*-value (N_a) at each depth is calculated using the following equations and by Figure A1.7.

$$N_a = N_1 + N_f \qquad \text{(A1.15)}$$

$$N_1 = C_N \times N \qquad \text{(A1.16)}$$

$$C_N = \sqrt{\frac{10}{\sigma'_z}} \qquad \text{(A1.17)}$$

where: N_a = corrected SPT *N*-value, N_1 = equivalent SPT *N*-value, N_f = increment of equivalent SPT *N*-value depending on fines content as shown in

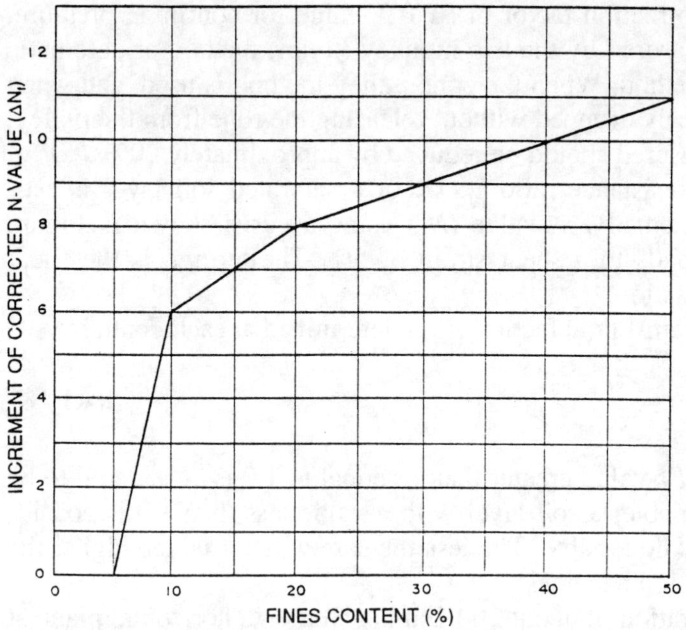

Figure A1.7. Relationship between fines content and increment of corrected N-value ΔN.

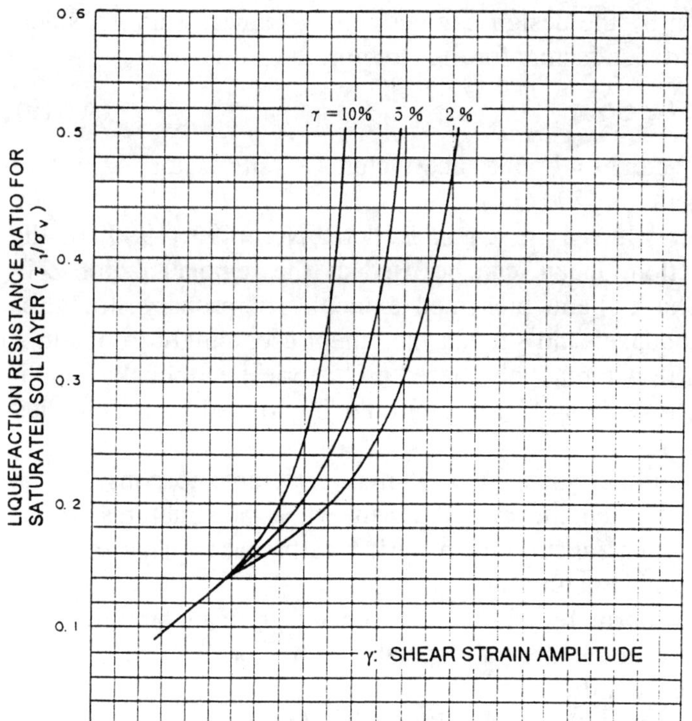

Figure A1.8. Relationship between corrected N-value (N_a) and liquefaction resistance ratio for a saturated soil layer.

Figure A1.7, C_N = correction factor of SPT *N*-value for confining pressure, N = SPT *N*-value measured by the trip monkey (tonbi) method or automatic free falling device method. When the cone pulley method is used, and when the hammer is not freely dropped without releasing the rope from the pulley, the SPT *N*-value measured should be reduced by approximately 20%.

c) A liquefaction resistance ratio τ_l / σ'_z of a saturated soil layer is estimated from a corrected SPT *N*-value (N_a) using a curve shown in Figure A1.8 which corresponds to a shear strain of 5%. The term τ_l is the shear stress to cause liquefaction.

d) Safety factor against liquefaction F_L is determined at each depth by

$$F_L = \frac{\tau_l / \sigma'_z}{\tau_d / \sigma'_z} = \frac{\tau_l}{\tau_d} \tag{A1.18}$$

A soil layer with an F_L-value greater than or equal to 1.0 is considered to be not potentially liquefiable; a soil layer with a value less than 1.0 is considered to be potentially liquefiable. The less the safety factor is, the higher the potential for liquefaction becomes.

The design acceleration of about 200 Gal is given on a horizontal plane at the ground surface for liquefaction determination. Alternatively, the value calculated by the following equation as a criterion is considered favorable because determination by the design engineers may be necessary considering the area and importance factors for the structure.

$$\alpha_{max} = Z \times I \times G \times \alpha_0 \tag{A1.19}$$

where: Z = regional factor, I = importance factor, G = site factor, α_0 = standard horizontal acceleration (200 Gal).

The regional factor which is given in Bulletin No. 1493 (1980) of the Ministry of Construction, Japan, can be used as the regional factor Z in Equation (A1.19). This bulletin presented examples of primary structures damaged during earthquakes and summarized them accordingly to obtain Z ranging from 0.7 to 1.0. Alternatively, based on historical earthquake data, it is possible to obtain the regional factor Z by referring to the results compiled by statistical methods.

With regard to the importance factor I, it is preferable to consider the factor based on the design engineers' judgement when the structures are required to maintain normal function even during earthquakes. This factor is fixed at 1.0 when normal structures are considered.

The site factor G for normal ground is 1.0. When significant amplification is anticipated such as in thick alluvial deposits, the factor G is evaluated in accordance with actual conditions.

For a pile foundation with the possibility of liquefaction, it is necessary to consider reduction of the coefficient of horizontal subgrade reaction for evaluation of horizontal resistance as shown in Table A1.4. When the lique-

Table A1.4. Reduction factor for coefficient of horizontal subgrade reaction.

Safety factor for liquefaction (Range of F_L)	Depth from ground surface z (m)	Reduction factor for coefficient of horizontal subgrade reaction r_k			
		$N_a \leq 8$	$8 < N_a \leq 14$	$14 < N_a \leq 20$	$20 < N_a$
$F_L \leq 0.5$	$0 \leq z \leq 10$	0	0	0.05	0.1
	$10 < z \leq 20$	0	0.05	0.1	0.2
$0.5 < F_L \leq 0.75$	$0 \leq z \leq 10$	0	0.05	0.1	0.2
	$10 < z \leq 20$	0.05	0.1	0.2	0.5
$0.75 < F_L \leq 1.0$	$0 \leq z \leq 10$	0.05	0.1	0.2	0.5
	$10 < z \leq 20$	0.1	0.2	0.5	1.0

fiable layer is in the middle of the subsoil profile, consideration of the layer above is necessary. In particular, when the layer thickness of the top nonliquefiable layer is thin and the liquefiable layer below is thick, reducing the coefficient of horizontal subgrade reaction of the top nonliquefiable layer is considered appropriate.

With respect to friction piles, a considerable reduction in pile friction on the circumferential surface is anticipated; thus, in the soil layer over the liquefiable layer, it is considered appropriate to make the circumferential surface friction value zero.

A1.4 DESIGN STANDARDS FOR RAILWAY STRUCTURES

A1.4.1 *Design standards for railway structures (Foundations and Retaining Structures) (1974)*

The method for liquefaction determination shown in the design standards is basically the same as in Section A1.2.1.

A1.4.2 *Design standards for national railway structures (Foundations and Retaining Structures) (1986)*

The design standards state that an alluvial soil layer, located less than 20 m below the ground surface with saturated sandy soil, ground water level less than 10 m depth, and having mean grain size D_{50} between 0.02 and 2.0 mm, has a possibility to liquefy. Liquefaction determination, therefore, should be made using the following method, similar to Section A1.2.2. The liquefaction strength R_1 is given slightly differently from the method in the specifications for highway bridges.

$$F_L = R / L \tag{A1.20}$$

$$R = R_1 + R_2 \tag{A1.21}$$

$$L = r_d \, K_s \frac{\sigma_v}{\sigma'_v} \tag{A1.22}$$

$$r_d = 1.0 - 0.015\, z \tag{A1.23}$$

$$K_s = \Delta_1 \, \Delta_2 \, K_{10} \tag{A1.24}$$

$$R_1 = 0.0042\, D_r^* \quad (D_r^* \leq 60\%) \tag{A1.25}$$

$$R_1 = \left(\frac{D_r^*}{123} \right)^2 \quad (D_r^* > 60\%) \tag{A1.26}$$

$$D_r^* = 21 \sqrt{\frac{N}{\sigma'_v + 0.7}} \tag{A1.27}$$

where: F_L = liquefaction resistance factor, R = dynamic shear strength ratio, L = shear stress ratio during earthquake, r_d = reduction factor for shear stress ratio during earthquake in the vertical direction, K_s = horizontal seismic coefficient at the ground surface used for liquefaction determination, σ_v = total overburden pressure (kgf/cm^2), σ'_v = effective overburden pressure (kgf/cm^2), R_1 = the first term of dynamic shear strength ratio defined as a function of relative density, D_r^* = estimated relative density of soil, z = depth from the ground surface (m), Δ_1 = regional factor (= 0.75 to 1.0), Δ_2 = site factor (= 1.0), K_{10} = standard horizontal seismic coefficient used in liquefaction determination (= 0.15).

This method is not considered applicable to a soil layer with a relatively large fines content, and in this case it is preferable to perform a detailed geotechnical investigation.

The reduction factor for subgrade reaction is basically the same as in Section A1.2.2. For liquefiable soil layers, the reduction factor in Table A1.3 should be multiplied with the soil constants, including modulus of deformation E_o of the ground and coefficient of subgrade factor K, internal friction angle of soil ϕ, shear elastic wave velocity V_s, and circumferential surface bearing capacity of pile f.

When the subsoil profile liquefying during earthquakes is located near a culvert, there should be both:

1. An examination related to the heave and settlement when the liquefiable subsoil profile is located near the bottom of the culvert; and
2. An examination of the shear stress on the side wall when the liquefiable subsoil profile is near the side wall of the culvert.

A1.5 DESIGN SPECIFICATIONS FOR ROADS

A1.5.1 *Road construction, specifications for remediation of soft ground (1977)*

The specifications state that the following factors determine the liquefaction potential for a sandy soil profile:
- Magnitude and duration of the earthquake;
- Density of sand (void ratio, relative density, SPT *N*-value, etc.);
- Grain size distribution; and
- Vertical overburden pressure, ground water level, and subsoil stratification.

The critical SPT *N*-value and zone of grain size in which liquefaction actually occurred are shown in a figure in this specification.

Section A1.2.1 presents a specific method for liquefaction determination. The specifications state that a combination of various complex factors causes liquefaction of sandy subsoils; thus, a determination of liquefaction should be made based on past examples of damage and past case history data.

Remedial measures against liquefaction are considered when there is a possibility of liquefaction; the basic remedial measures are:
- Increasing the relative density of the sandy subsoil to a high level;
- Increasing the effective stress; and
- Increasing overall permeability of the subsoil in order to quickly dissipate pore water pressure.

A1.5.2 *Road construction, specifications for remediation of soft ground (1986)*

The specifications state the same factors that cause liquefaction as shown in Section A1.5.1 with additional explanations.

For liquefaction determination, a) Rough determination based on topography/geology and b) Rough determination based on SPT *N*-value and gradation (critical SPT *N*-value ranging from 10 to 15), are made at the preliminary stage of the survey when there is limited information about the subsoil conditions. Thereafter, more detailed investigations are considered.

Approximate evaluation of liquefaction potential is made based on the critical SPT *N*-value indicated in Figure A1.9. $k_h = 0.18$ is used for evaluation of liquefaction potential following Section A1.2.2.

If a detailed investigation is considered necessary, cyclic triaxial tests are performed using the soil sample to determine the liquefaction resistance and characteristics of pore water pressure under cyclic load. Dynamic response analysis can also be performed.

In determining the design horizontal seismic coefficient k_h for a detailed evaluation, the following equation is used.

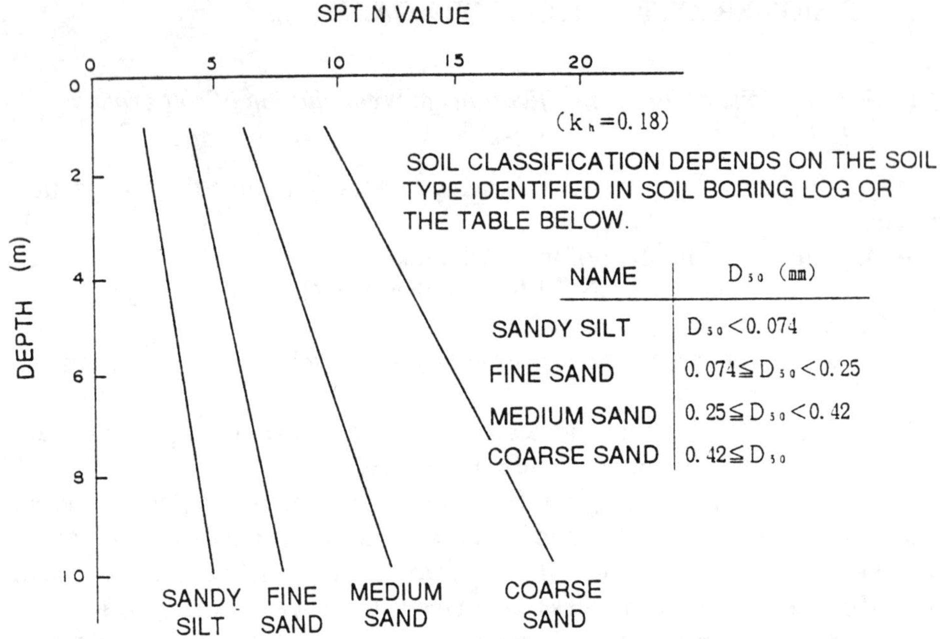

Figure A1.9. Critical *N*-value used for approximate determination.

$$k_h = \nu_1 \, \nu_2 \, k_0 \tag{A1.28}$$

where: k_0 = standard horizontal seismic coefficient used for liquefaction determination (= 0.15), ν_1 = regional factor (= 0.7 to 1.0), ν_2 = site factor (= 0.9 to 1.2).

There are two conditions which require evaluation related to liquefaction; stability analysis of an embankment resting on sandy subsoils during earthquakes and stability with respect to the heave of culverts under roads.

In evaluating the stability of the embankment, it is determined whether or not liquefaction will cause loss of stability of the embankment and then necessary remedial measures are considered. In evaluating the heave of a culvert, the safety factor F_U is obtained using the following equation.

$$F_U = \frac{W_S + W_B + Q_S}{U_S + U_D} \tag{A1.29}$$

where: W_S = dead weight of backfill soil above a buried structure (including the weight of water), W_B = dead weight of the road (including the dead weight of concrete slab and pavement), Q_S = frictional resistance along the side wall of a culvert or shear resistance of the soil put on an extending floor plate, U_S = buoyant force due to hydrostatic pressure at the bottom of a culvert, U_D = buoyant force due to excess pore water pressure.

Q_S should be zero when F_L is less than 1.0. The safety factor given by

Equation (A1.29) should be higher than 1.0 during earthquakes. For estimating excess pore water pressure of liquefied subsoils, refer to Section A1.14.

Attention should be given to the fact that the soil pressure on the side wall of a culvert increases during liquefaction.

A1.6 STANDARDS ON REGULATION OF HAZARDOUS MATERIALS (1978)

The standards state that the subsoil conditions should *not* be as follows:

– At a depth more than 3 m from the top surface of foundation and less than 15 m from the ground surface;

Table A1.5. Critical SPT *N*-value.

Fines content	SPT *N*-value	
	A	B
Less than 5%	12	15
From 5% to 10%	8	12
From 10% to 35%	6	7

SPT *N*-values for A and B are the values related to a sandy soil in the area shown in Figure A1.10.

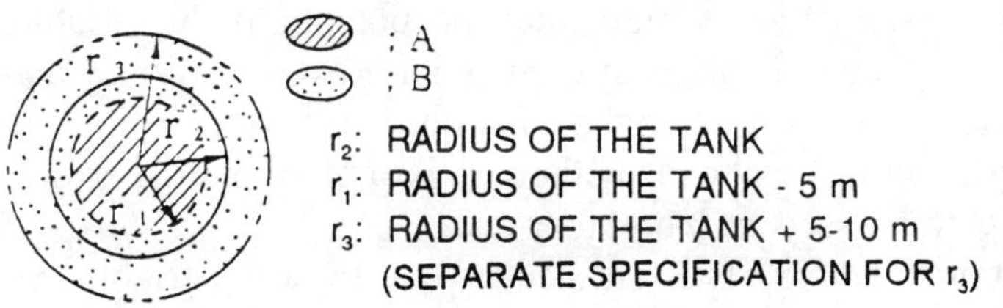

Figure A1.10. Installation area for remedial measure against liquefaction.

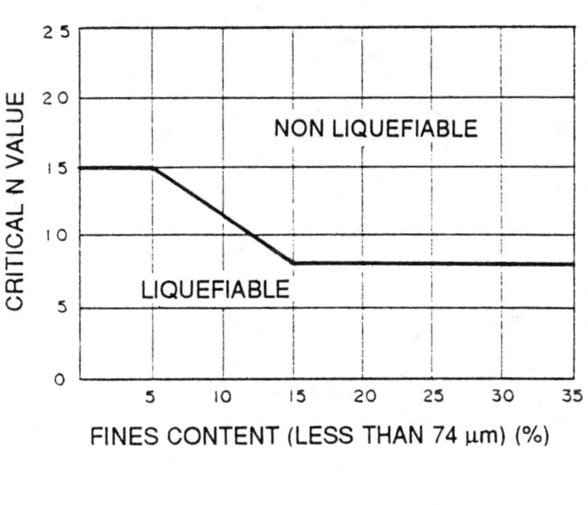

Figure A1.11. Fines content of liquefiable soil and critical *N*-value.

– Saturated by ground water;
– Mean grain size D_{50} is less than 2 mm;
– SPT *N*-value in the standard penetration test is less than the value indicated in Table A1.5 according to the fines content.

A1.7 RECOMMENDED PRACTICE FOR LNG UNDERGROUND STORAGE FACILITIES (1979)

This recommended practice states that a soil layer which satisfies all of the following conditions has a potential for liquefaction during an earthquake:
– The layer is saturated with ground water;
– The depth of the layer is less than 15 m from the ground surface;
– Mean grain size (D_{50}) is less than or equal to 2.0 mm;
– Sandy soil with fines content less than 35%;
– SPT *N*-value is less than the critical *N*-value shown in Figure A1.11. This critical *N*-value depends on the fines content. For natural soil deposits, the measured SPT *N*-value should be averaged. For artificially compacted subsoils using the sand compaction pile method, the value between sand compaction piles should be used.

Even if the soil layer is judged non-liquefiable based on the above criteria, attention should be paid to the fact that soils with a plasticity index less than 15 or soils which are classified to M_L (see Fig. A1.12) according to the Japanese unified soil classification system may be potentially liquefiable. The Japanese unified soil classification system is based on the liquid and plasticity limits of fine grained soils less than 74 μm. If the liquefaction determination is difficult using the above method, it is preferable to perform more detailed investigation based on laboratory tests.

The recommended practice for LNG tanks states that for a liquefiable soil layer, appropriate measures such as soil improvement against liquefaction

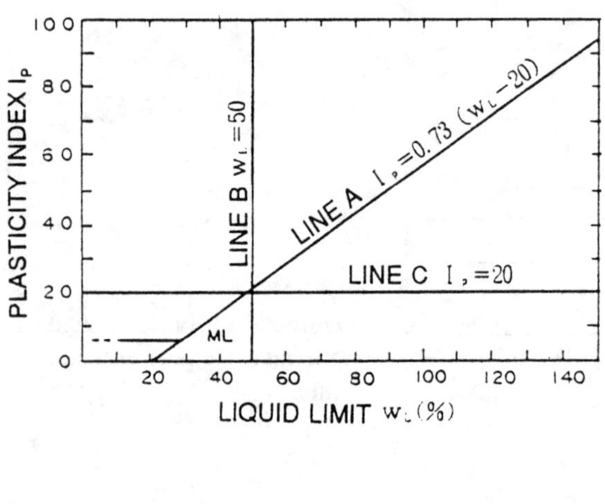

Figure A1.12. M_L in the plasticity chart of Japanese unified soil classification.

should be taken if necessary. The following measures can be considered as remedial measures against liquefaction:

1. Increasing the density of soil by vibration or impact;
2. The replacement with soil of non-liquefiable grain size; and
3. Lowering the ground water table.

In general, the compaction method is often used and in this case, the SPT *N*-value of the soil should be larger than the critical *N*-value given above.

Detailed statements are presented in a separate section on soil improvement methods.

A1.8 SPECIFICATIONS FOR TAILING DAMS AND COMMENTARIES (1980)

This specifications use F_L for liquefaction determination. Slope failure of tailing dams is the main concern.

The liquefaction resistance R_l is evaluated by the following equation for tailings or quarries except for the waste stones from black ore mines.

$$R_l = 0.088\sqrt{\frac{N}{0.1\sigma'_v + 0.7}} + 0.085\log_{10}\frac{0.50}{D_{50}} \qquad \text{(A1.30)}$$

where: R_l = cyclic stress ratio, N = *N*-value by standard penetration test, σ'_v = effective overburden pressure (tf/m^2), D_{50} = mean grain size of soil (mm); and the minimum value of R_l should be 0.15 in Equation (A1.30).

For sediment deposit, the liquefaction resistance is evaluated by the following equation.

$$R_l = 0.088\sqrt{\frac{N}{0.1\sigma'_v + 0.7}} + 0.10 \qquad \text{(A1.31)}$$

where: $R_l = 0$, if $N < 1$.

The shear stress ratio induced by earthquakes L is given as follows

$$L = \left(\frac{4}{3}\right)K_h\left(\frac{\sigma_v}{\sigma'_v}\right)(1 - 0.025\,z) \qquad \text{(A1.32)}$$

where: L = stress ratio during earthquakes, K_h = design earthquake intensity, σ_v = total overburden pressure (tf/m^2), σ'_v = effective overburden pressure (tf/m^2), z = depth from the ground surface (m).

If liquefaction potential is high at a tailing dam site, it is necessary to estimate the safety factor with respect to the slope failure of the foundation soil and the embankment.

A1.9 SEISMIC DESIGN SPECIFICATIONS ON SUBMERGED TUNNELS (TENTATIVE) (1975)

In the design specifications, Section A1.1.1 and Section A1.2.1, are shown as examples for liquefaction evaluation methods. The determination method based on the cyclic triaxial tests is also shown. The specifications state then that these determination methods are used with respect to structures other than tunnels and when considering a possibility of damage by liquefaction which may lead to the loss of lives, the liquefaction prediction of tunnels should be performed more carefully.

For seismic design for ventilation towers, shear resistance cannot be expected in liquefied sandy soil. Liquefaction must not occur in subsoils surrounding a submerged tunnel; therefore, if such a possibility exists, measures such as soil improvement must be considered.

A1.10 TECHNICAL SPECIFICATIONS ON UNDERGROUND OIL STORAGE FACILITIES (TENTATIVE) (1980)

The specifications state that 'the possibility of liquefaction in subsoil profile should be determined based on the results of the geotechnical investigation'. There is, however, no mention of specific methods for liquefaction determination.

The area around oil storage facilities below ground surface must be stable against liquefaction during earthquakes, and when there is a possibility of liquefaction, either soil improvement against liquefaction must be performed or a structure design should assure the safety of the oil storage facility even when liquefaction occurs.

A1.11 TECHNICAL SPECIFICATIONS ON SEISMIC DESIGN OF NUCLEAR POWER PLANTS (1987)

The specifications state that liquefaction determination of dense sand soil is preferably evaluated according to the flow chart shown in Figure A1.13 which takes into account the difference of dense sand from loose sand. The method shown in Figure A1.14 is used for evaluation of liquefaction. The equation which represents the figure is given by

$$R = a\left[N_L^{0.5} + (bN_L)^n + c - 14.8\, f\,(D_{50})\right] \qquad \text{(A1.33a)}$$

$$f\left(D_{50}\right) = 0.255 \log\left(\frac{D_{50}}{0.35}\right) \qquad : 0.04 \le D_{50} \le 0.6 \text{ mm} \qquad \text{(A1.33b)}$$

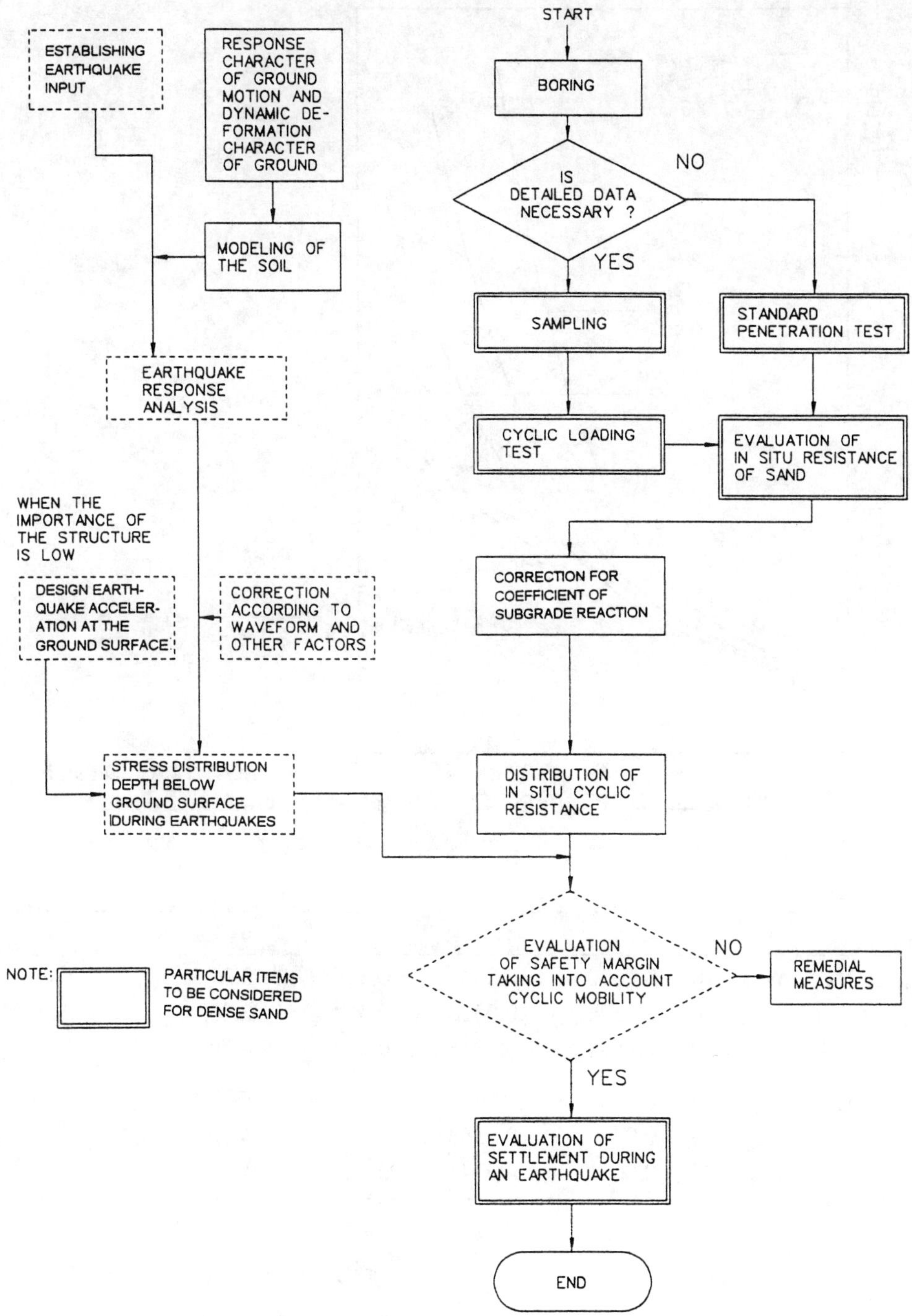

Figure A1.13. Flow chart for liquefaction determination for a dense sandy soil.

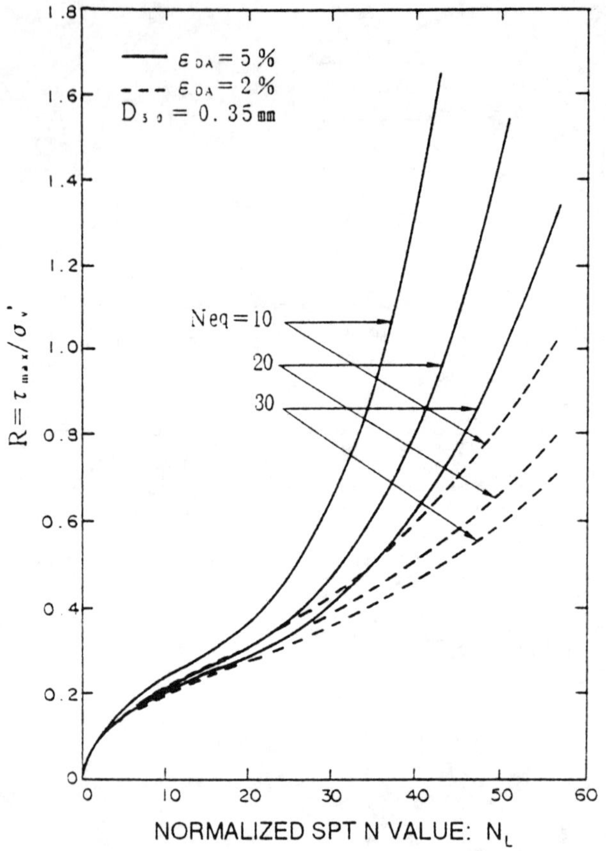

Figure A1.14. Liquefaction determination curve.

Table A1.6. Table of constants.

		a	b	c	n
$\varepsilon_{DA} = 2\%$	$N_{eq} = 10$	0.0630	0.0346	0.312	3.21
	$N_{eq} = 20$	0.0584	0.0336	0.312	2.80
	$N_{eq} = 30$	0.0554	0.0318	0.312	2.78
$\varepsilon_{DA} = 5\%$	$N_{eq} = 10$	0.0760	0.0420	0	4.77
	$N_{eq} = 20$	0.0676	0.0368	0	4.52
	$N_{eq} = 30$	0.0631	0.0345	0	4.00

$$f\left(D_{50}\right) = 0.05 \qquad : 0.6 \le D_{50} \le 1.5 \text{ mm} \qquad \text{(A1.33c)}$$

where: R = liquefaction stress ratio, N_L = normalized SPT N-value, a, b, c, n = constants (see Table A1.6).

When the SPT N-value is considerably large, c should be given as $c = 0$.

The earthquake force is not indicated in the liquefaction determination method. The following equation is used, however, for seismic resistant design of important structures.

$$K_h = n_1 \times n_2 \times n_3 \times k_0 \qquad \text{(A1.34)}$$

where: n_1 = regional factor (generally 1.0), n_2 = site factor depending on soil conditions (1.0 or 1.5 according to soil conditions), n_3 = correction factor depending on factors other than above (generally 1.0), k_0 = seismic coefficient (generally 0.2).

A1.12 SPECIFICATIONS FOR SEISMIC DESIGN OF WATER FACILITIES AND COMMENTARIES (1979)

The specifications state that the subsoils not described by 1-4 below and saturated sandy soil layers below the ground water table have some possibility of liquefaction during an earthquake, and therefore a liquefaction determination must be made.

1. Subsoils deeper than 20 m below the ground surface;
2. Sites having the ground water level deeper than 10 m below the ground surface;
3. Subsoils having mean grain size D_{50} below 0.02 mm or over 2 mm;
4. Subsoils deposited before Holocene era.

The liquefaction determination method is almost the same as in Section A1.2.2. The method for estimating the horizontal seismic coefficient is somewhat different as shown below.

The earthquake force is given as the horizontal seismic coefficient K_S for obtaining shear stress ratio L during an earthquake.

$$K_S = \Delta_1 \Delta_2 K_{S0} \qquad \text{(A1.35)}$$

where: K_{S0} = standard horizontal seismic coefficient (= 0.15) used for liquefaction determination, Δ_1 = regional factor based on past case histories of earthquakes, Δ_2 = site factor estimated from the subsoil conditions.

When liquefaction is predicted to occur, remedial measures such as ground improvement must be implemented if necessary.

A1.13 SPECIFICATIONS FOR SEISMIC DESIGN OF SEWAGE FACILITIES AND COMMENTARIES (1981)

In the specifications, the liquefaction determination method is indicated as a 'reference'. The method is similar to the F_L method specified in Section A1.2.2.

The earthquake force is given by a design horizontal seismic coefficient K_S for obtaining shear stress ratio L during an earthquake.

$$K_S = \nu_1 \, \nu_2 \, \nu_3 \, K_{S0} \tag{A1.36}$$

Each correction factor in Equation (A1.36) is the same as in Section A1.2.2.

When liquefaction is predicted to occur, remedial measures against liquefaction such as soil improvement must be implemented if necessary.

A1.14 DESIGN SPECIFICATIONS FOR COMMON UTILITY DUCTS (1986)

The specifications show basically the same procedures for liquefaction evaluation as those specified in Section A1.2.2. When the bottoms of common utility ducts are located in a soil layer requiring liquefaction evaluation or when the depth of embedment depth of a common utility duct into a clay layer below this liquefiable soil layer is not sufficient, liquefaction evaluation related to heave of the common utility duct should be made.

For liquefaction determination, the F_L method is used and the earthquake

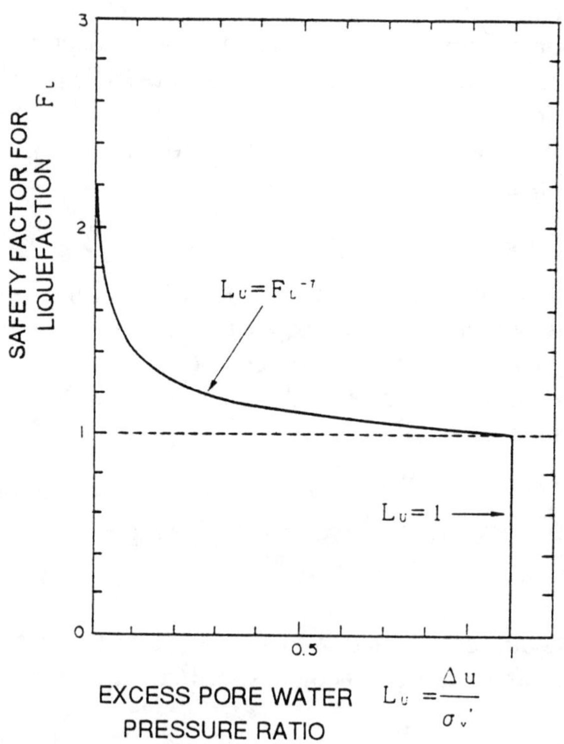

Figure A1.15. Relationship of F_L and L_u.

force is given by a seismic coefficient K_S to obtain shear stress ratio L during earthquake.

A common utility duct is a light structure, and therefore an evaluation of heave must be made when the soil has some potential for liquefaction. F_L is obtained using the same method as in Section A1.2.2, and then the safety factor F_U with respect to heave is obtained using the following equation.

$$F_U = \frac{W_S + W_B + Q_S + Q_B}{U_S + U_D} \qquad \text{(A1.37)}$$

where: W_S = dead weight of backfill soil above common utility duct (including the weight of water), W_B = dead weight of common utility duct, Q_S = shear resistance of backfill soil, Q_B = shear resistance along the side walls of common utility duct, U_S = buoyant force by hydrostatic water pressure at the bottom of common utility duct, U_D = buoyant force due to excess pore water pressure.

Q_S and Q_B should be zero when F_L is less than 1.0. Minimum safety factor to maintain stability of common utility duct is 1.1.

Buoyant force due to excess pore water pressure at the bottom of a common utility duct is calculated using the following equation.

$$U_D = \Delta_u \times B = L_U \times \sigma'_v \times B \qquad \text{(A1.38)}$$

where: Δ_u = excess pore water pressure, B = width of common utility duct, σ'_v = effective overburden pressure, L_U = excess pore water pressure ratio.

L_U is obtained from F_L using Figure A1.15.

APPENDIX 2

Soil improvement area for various structures

A2.1 SOIL IMPROVEMENT AREA FOR LIGHT, SMALL-SCALE STRUCTURES SUCH AS WOODEN HOUSES

When dealing with small scale structures that require only a relatively small bearing capacity, such as wooden houses, surveys of past case histories indicate that compacting only the upper portion of subsoil is often effective to mitigate damage and it is not necessary to compact the lower portion of the liquefiable subsoil [1]. This condition is shown in Figure A2.1.

The principal mechanism for establishing this condition is similar to that of a raft foundation on soft clay. The top nonliquefiable layer is considered as a raft if liquefaction occurs. The conditions shown in Figure A2.1 are important for practical use. With regard to applicability, however, the method still requires much more research and verification [1].

A2.2 GUIDELINES FOR OIL TANKS

For oil tanks, the soil improvement area L should be two thirds of the improvement depth (provided $5\ \mathrm{m} < L < 10\ \mathrm{m}$) as shown in Figure A2.2, as specified in the Standards of regulation for hazardous materials [2].

These guidelines are considered to be basically the same as the standard design approach discussed in Section 5.7.

A2.3 MEASURES FOR UNDERGROUND STRUCTURES

The basic design approach for a soil improvement area around underground structures has many points in common with those indicated in Section 5.7. For small structures, however, the concepts related to the passive resistance necessary for the stability of underground structures are not always consid-

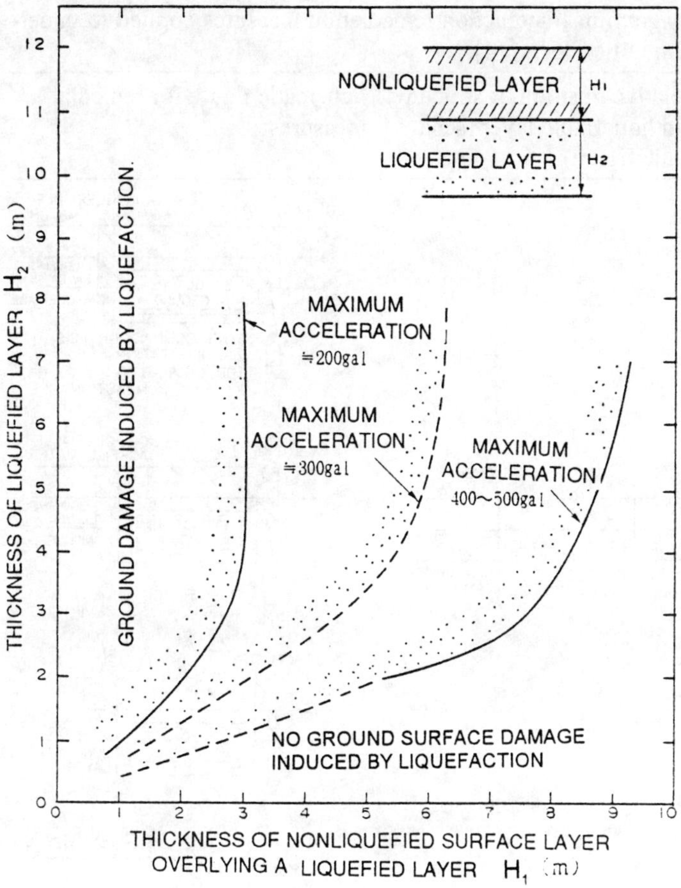

Figure A2.1. Conditions under which ground surface damage will be caused by liquefaction.

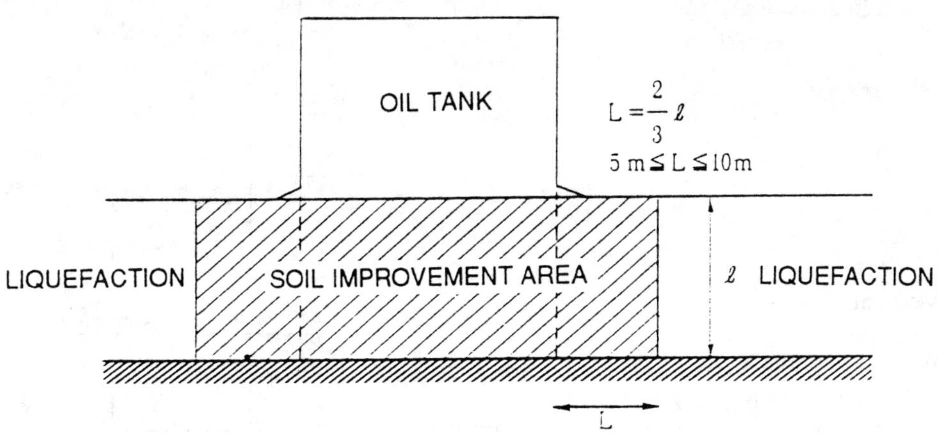

Figure A2.2. Design for soil improvement area of tank foundation.

Table A2.1. Schematic diagram of liquefaction remediation measures applied to underground structures (common utility ducts) [4].

Depth/thickness of liquefiable layer	Relative position of structure and liquefiable layer (schematic figure)	Schematic figure for remedial measures
Shallow When the bottom end of a liquefiable layer is located shallower than the bottom of a structure	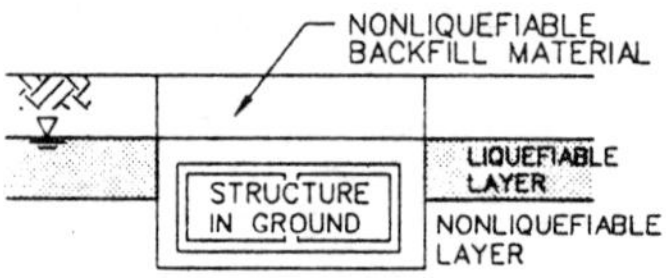	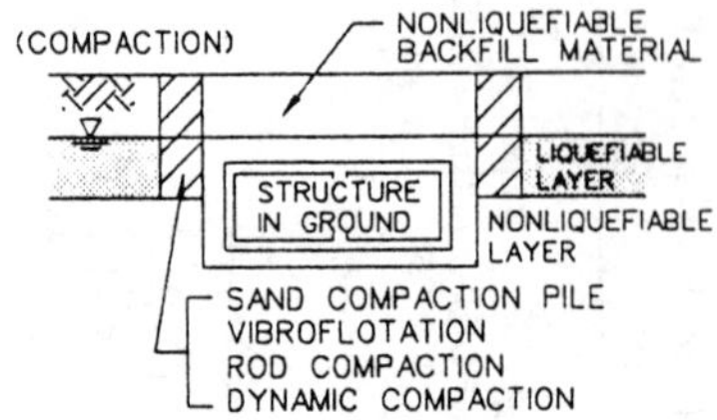 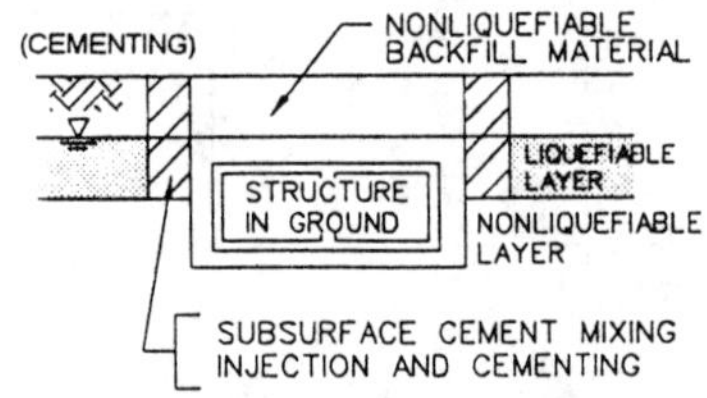 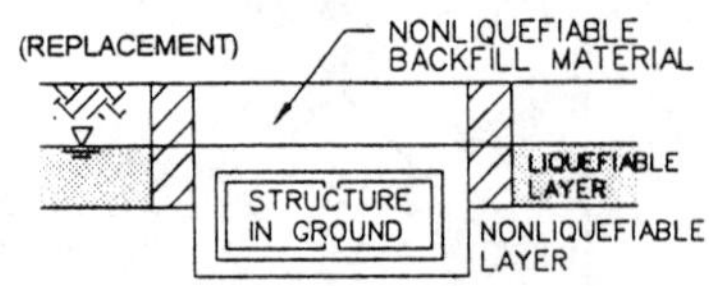 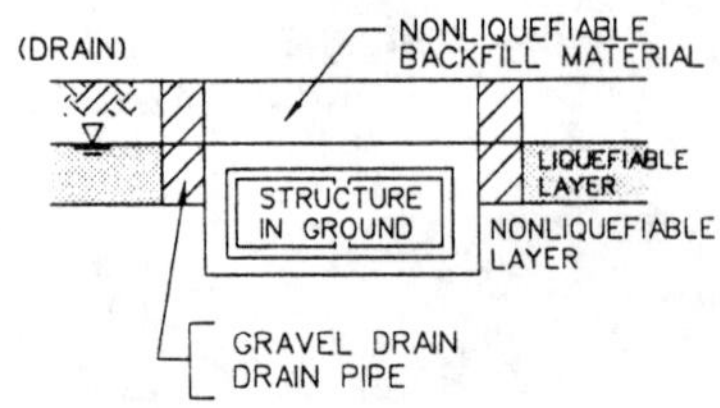 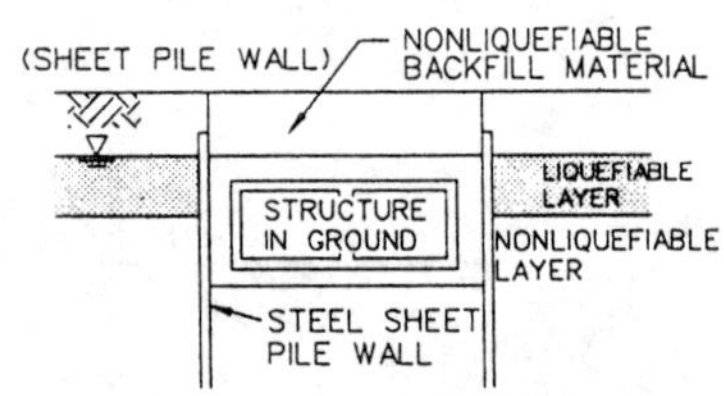
Medium When the bottom end of a liquefiable layer is located slightly deeper than the bottom of a structure	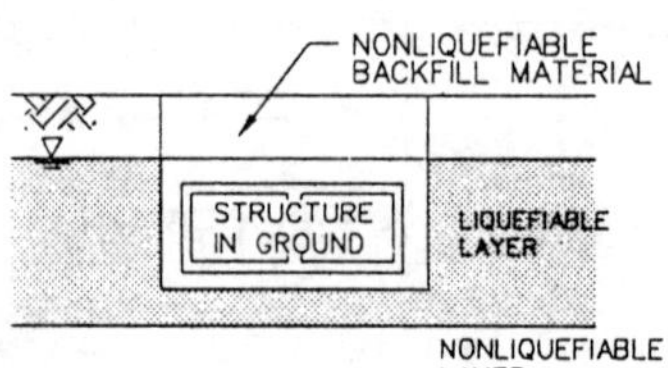	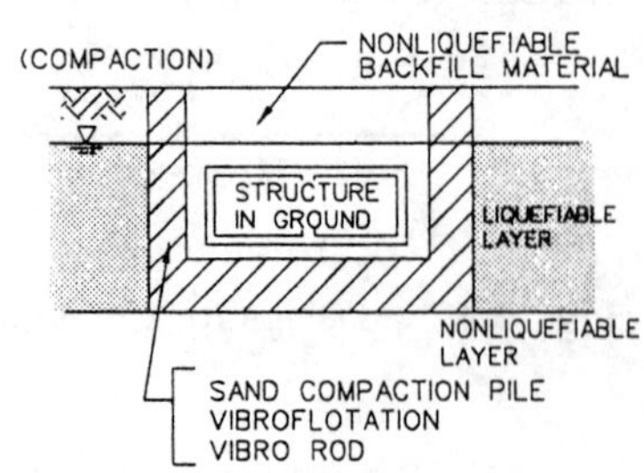

Table A2.1. Continued.

<table>
<tr><th>Depth/thickness of liquefiable layer</th><th>Relative position of structure and liquefiable layer (schematic figure)</th><th>Schematic figure for remedial measures</th></tr>
<tr><td>Medium

When the bottom end of a liquefiable layer is located slightly deeper than the bottom of a structure</td><td>

*It is preferable to improve the liquefiable layer below the bottom of the structure</td><td>
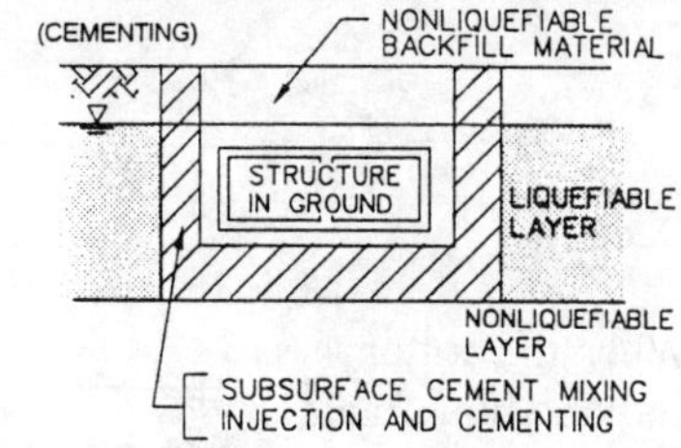

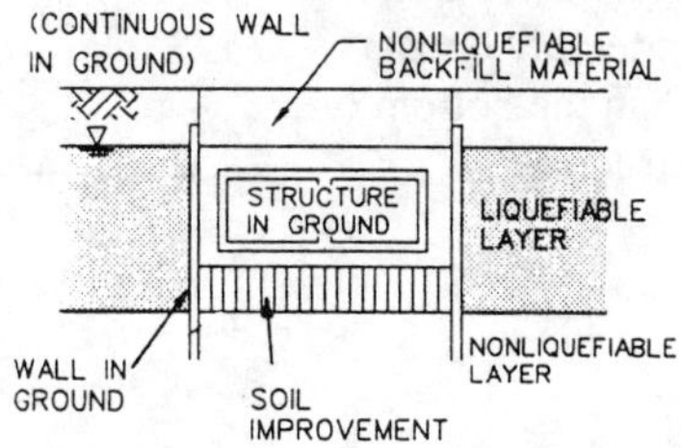

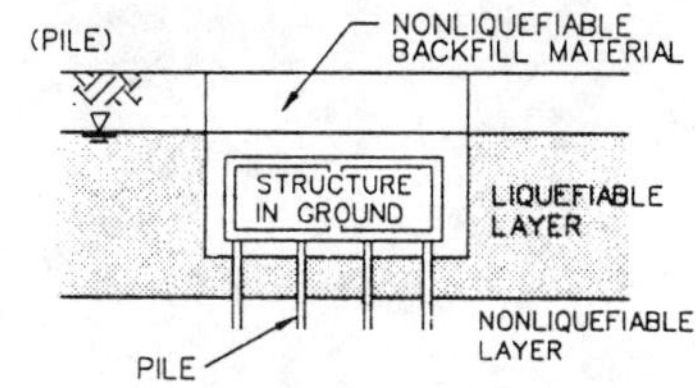

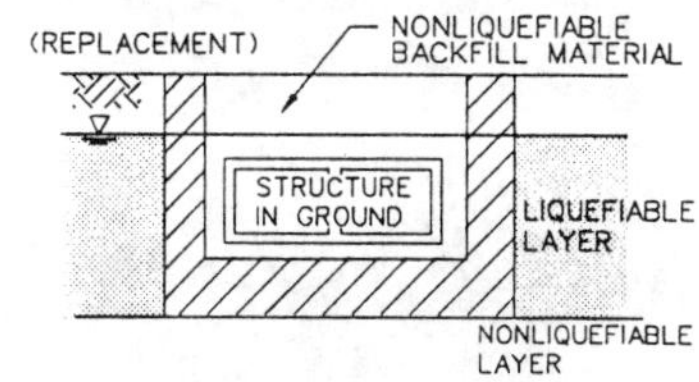

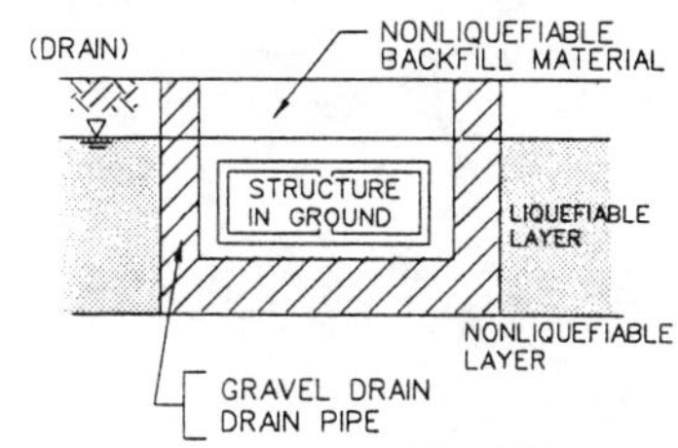

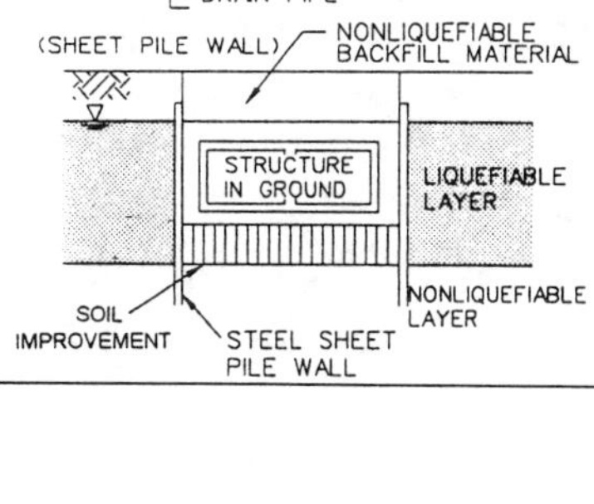

</td></tr>
</table>

Table A2.1. Continued.

Depth/thickness of liquefiable layer	Relative position of structure and liquefiable layer (schematic figure)	Schematic figure for remedial measures
Deep When the bottom end of a liquefiable layer is located considerably deeper than the bottom of a structure	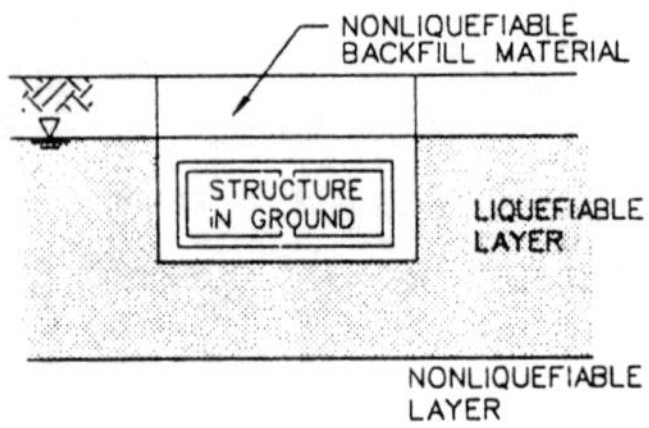	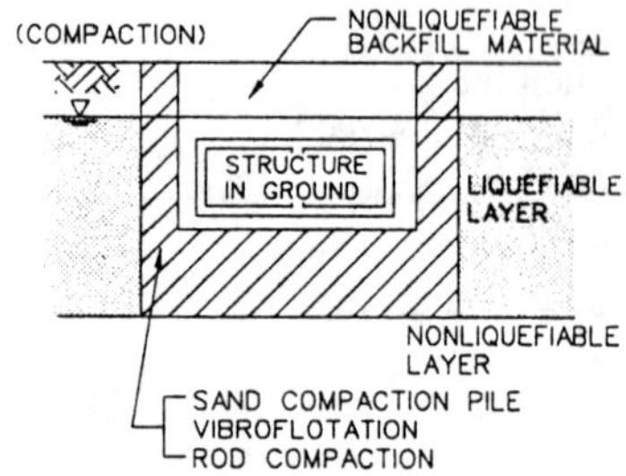
		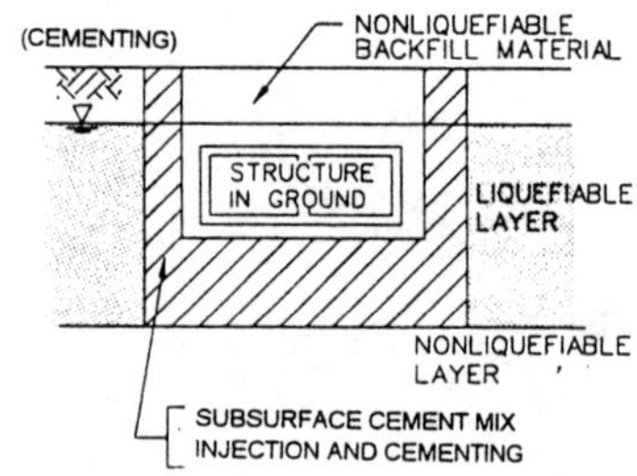
		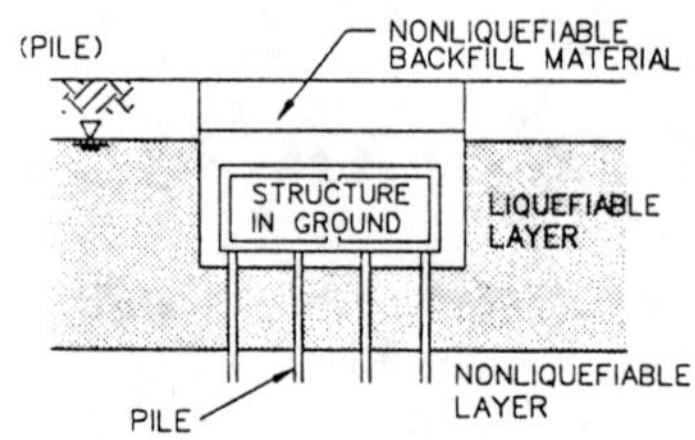
		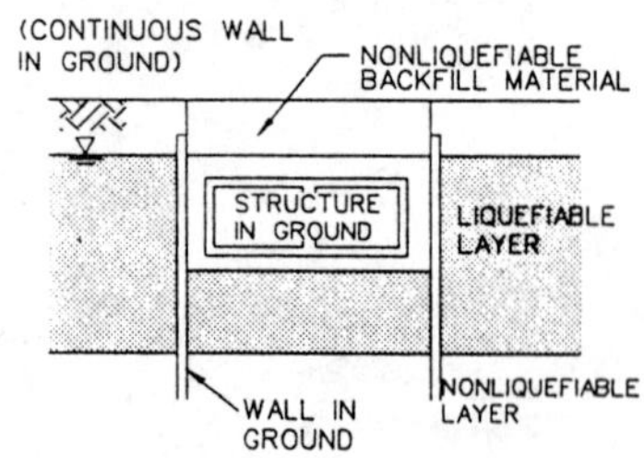

ered, and the improvement against liquefaction is considered only from the view point of safety measures regarding heave [3].

There is a manual compiled by the Port and Harbor Bureau, Tokyo Metropolitan Government [4] that indicates the measures against liquefaction for underground structures such as common utility ducts. Representative measures against liquefaction are shown in Table A2.1.

APPENDIX 3

Examples of remedial measures against liquefaction

The examples shown in this appendix are some of the case histories of remediation accomplished in the port and harbor areas in Japan. The following tables show three examples of the actual implementation of the remediation measures.

Table A3.1. Examples of remedial measures against liquefaction.

Example	Facility	Type of structure	Remedial measure
1	Wharf –13 m	Caisson	Vibro rod compaction method Gravel drain method
2	Wharf –10 m	Steel sheet pile wall with vertical anchor	Vibro rod compaction method Gavel drain method
3	Wharf –10 m	Steel sheet pile wall with vertical anchor	Vibro rod compaction method
4	Wharf –10 m	Steel sheet pile wall with vertical anchor	Vibro rod compaction method Gravel drain method
5	Wharf –12 m	Steel pipe pile wall with batter pile anchor	Vibro rod compaction method
6	Wharf –10 m	Steel pipe pile wall	Gravel drain method Plastic pipe drain method
7	Wharf –12 m	Steel pipe pile wall with batter pile anchor	Vibro rod compaction method
8	Wharf –10 m	Steel pipe pile wall	Vibro rod compaction method Sand compaction pile method Gravel drain method Plastic pipe drain method
9	Wharf –12 m	Steel pipe pile wall	Vibro rod compaction method
10	Wharf –10 m	Steel pipe pile wall with sheet pile anchor	Gravel drain method
11	Wharf –10 m	Steel pipe pile wall with sheet pile anchor	Sand compaction pile method Gravel drain method
12	Wharf –10 m	Caisson	Sand compaction pile method
13	Wharf –9 m	Trestle pier	Vibro rod compaction method Gravel drain method
14	Wharf –10 m	Steel pipe pile wall with sheet pile anchor	Sand compaction pile method
15	Wharf –12 m	Steel pipe pile wall with sheet wall anchor	Vibroflotation method
16	Wharf –12 m	Steel pipe pile wall with vertical pile anchor	Gravel compaction method Vibro rod compaction method

Table A3.2. Example of remedial measures against liquefaction (1).

Structure: Wharf –10 m steel sheet pile wall vertical anchor (example 2)
Remedial measure: Vibro rod compaction and gravel drain methods

Soil condition

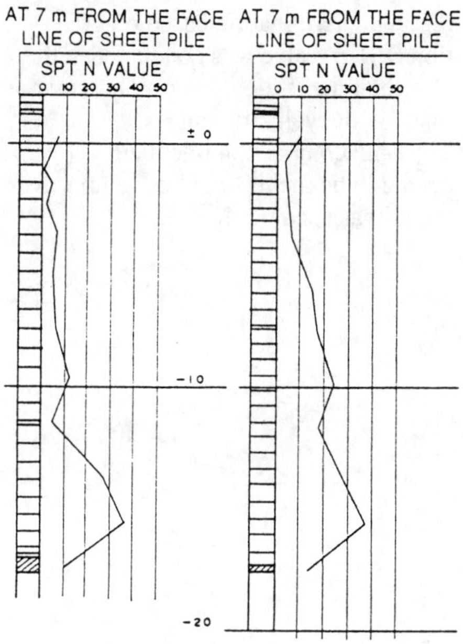

Consideration on remedial measures against liquefaction:

- As a result of a soil investigation after an earthquake, most SPT *N*-values for the back of the wharf were less than the liquefaction critical *N*-value. As a remedial measure against liquefaction, soil improvement was conducted for damage restoration work. For the soil improvement, the vibro rod compaction method was selected, which has a relatively small influence on the existing structures, is economical with favorable applicability. However, the construction area for the vibro rod compaction method was established 4 m away from the existing sheet pile because consideration was given to the vibration induced stress and deformation of the steel sheet pile. In areas less than 4 m from the existing sheet pile, the gravel drain method was used to dissipate excess pore water pressure in saturated sandy soil during an earthquake
- Liquefaction prediction/determination was conducted based on grain size distribution/SPT *N*-value. For this soil improvement, a test installation was implemented in order to determine improvement specifications

Schematic figure for the remedial measure against liquefaction

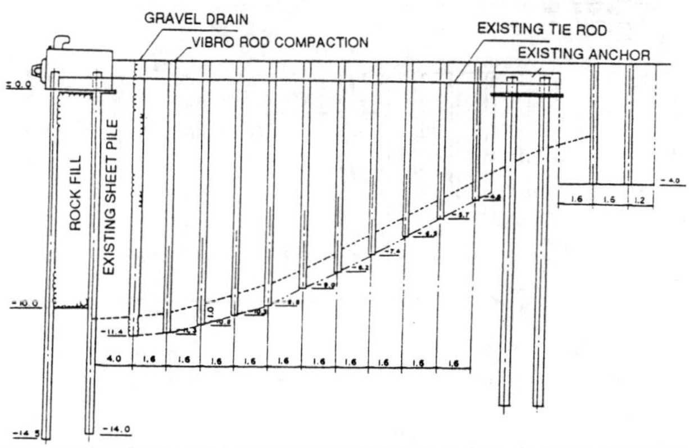

Table A3.3. Example of remedial measures against liquefaction (2).

Structure: Wharf –10 m caisson type (example 12)
Remedial measure: Sand compaction pile method

Soil condition

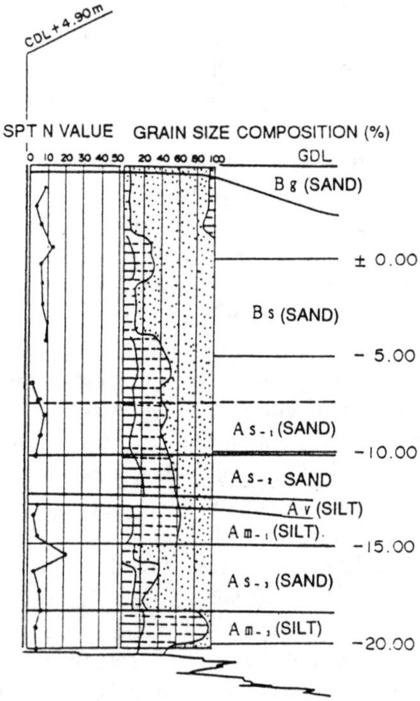

Consideration on remedial measures against liquefaction:

- In the liquefaction determination, prediction/determination based on gradation/SPT *N*-value was conducted and also the liquefaction determination was made with cyclic triaxial test
- The sand compaction pile method (expected to be the most effective improvement) was used

Schematic figure for the remedial measure against liquefaction

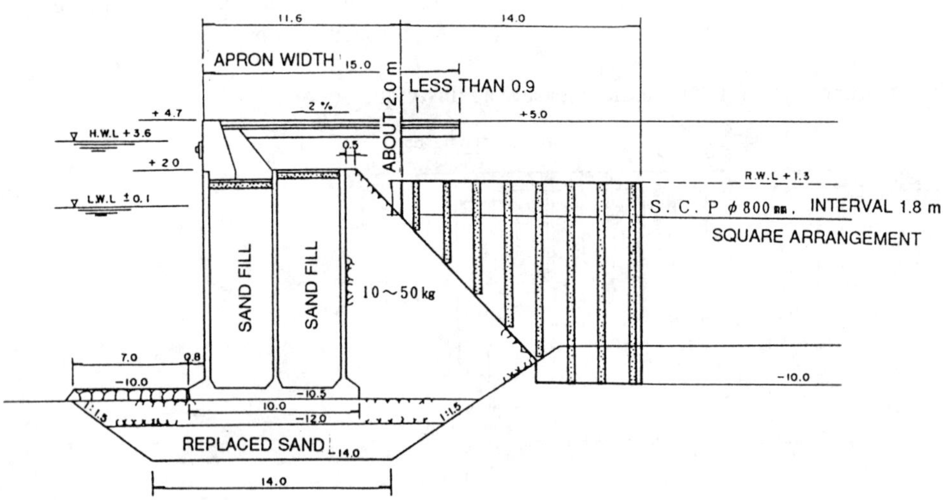

Table A3.4. Example of remedial measures against liquefaction (3).

Structure: Wharf –9 m trestle pier type (example 13)
Remedial measure: Vibro rod compaction method and gravel drain method

Soil condition

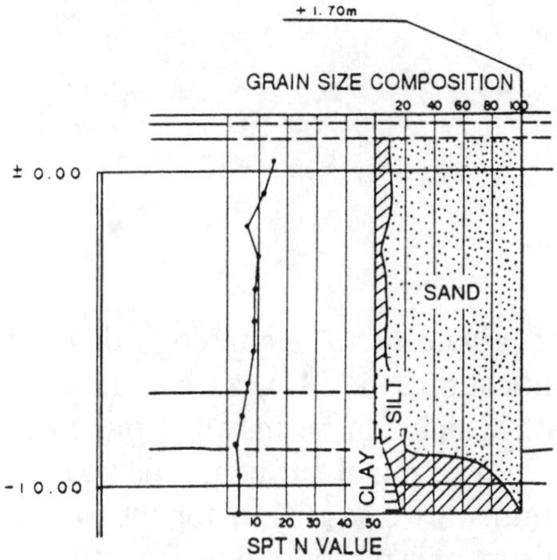

Consideration on remedial measures against liquefaction:

- Liquefaction prediction/determination was based on the gradation/SPT *N*-value
- In the selection of the remedial measure, five methods, i.e. vibro rod compaction method, sand compaction pile method, replacement method, chemical grouting method and gravel drain method were compared as to their applicability, economic factors, case histories and influence on existing structures. As a result, the vibro rod method, and the gravel drain method were selected

Schematic figure for the remedial measure against liquefaction

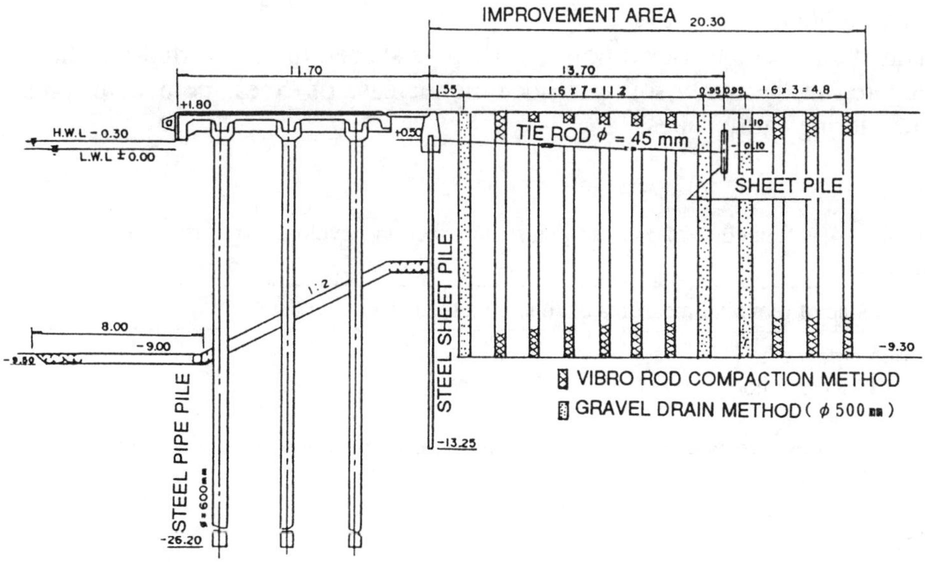

APPENDIX 4

Liquefaction remediation methods developed by private sector technology

Based on Article 5 in the promotion of private sector technology development (1989 Ministry of Transport Notification No. 341), an evaluation of 'remedial measures against liquefaction', which can be installed more efficiently than the existing measures or with minimal influence on existing structures during installation, was performed as one project for the year of 1990 (dated September 11, 1992, Ministry of Transportation Bulletin No. 468). The techniques which were awarded by the evaluation project are shown in Table A4.1.

A4.1 DRAIN PIPE METHOD

Outline of the technique

This drain method uses a polystyrene pipe-shaped drain inserted into holes augured in the sandy soil to reduce the increase of excess pore water pressure during earthquakes.

Table A4.1. Remedial measures against liquefaction developed by private sector technology.

The name of remedial measures against liquefaction
Drain pipe method
KS-HARD method
Deep vibro method
Remedial measure against liquefaction combining compaction and drain effects
Grid drain method
Spiral drain method
Mini-composer method
Gravel drain method with a double casing
Noise-free and vibration-free gravel drain method

Schematic diagram of the installation

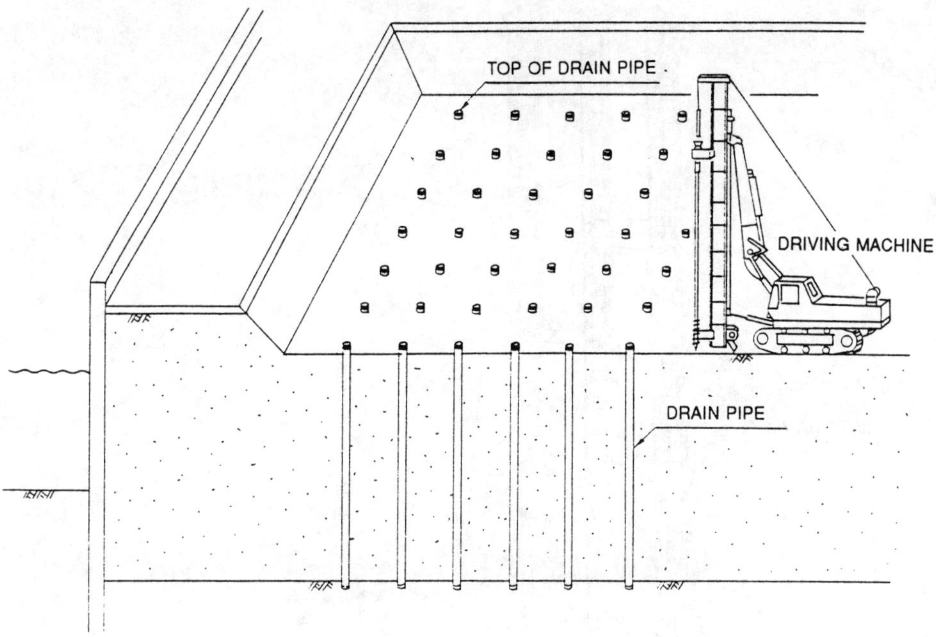

Figure A4.1. Schematic diagram of drain pipe method.

Evaluation of the technique

a) Construction close to existing facilities is possible.

b) The design method for the plastic drain is applicable.

c) The material is light, easy to handle, and has sufficient pressure resistance, durability, and drainage.

d) The drain can be formed accurately.

e) The same drainage effect is expected as the gravel drain method.

f) The installation equipment is compact and is operated efficiently compared to the gravel drain method.

g) Influence of noise and vibration on the surrounding area is minimal.

A4.2 KS-HARD METHOD

Outline of the technique

This sand compaction pile method achieves improvement in construction control by employing a vibro compaction hammer which can vary the diameter of the sand pile, functions to monitor the natural subsoil strength index during installation, and can achieve automatic sand feeding and pile formation.

Schematic diagram of the construction

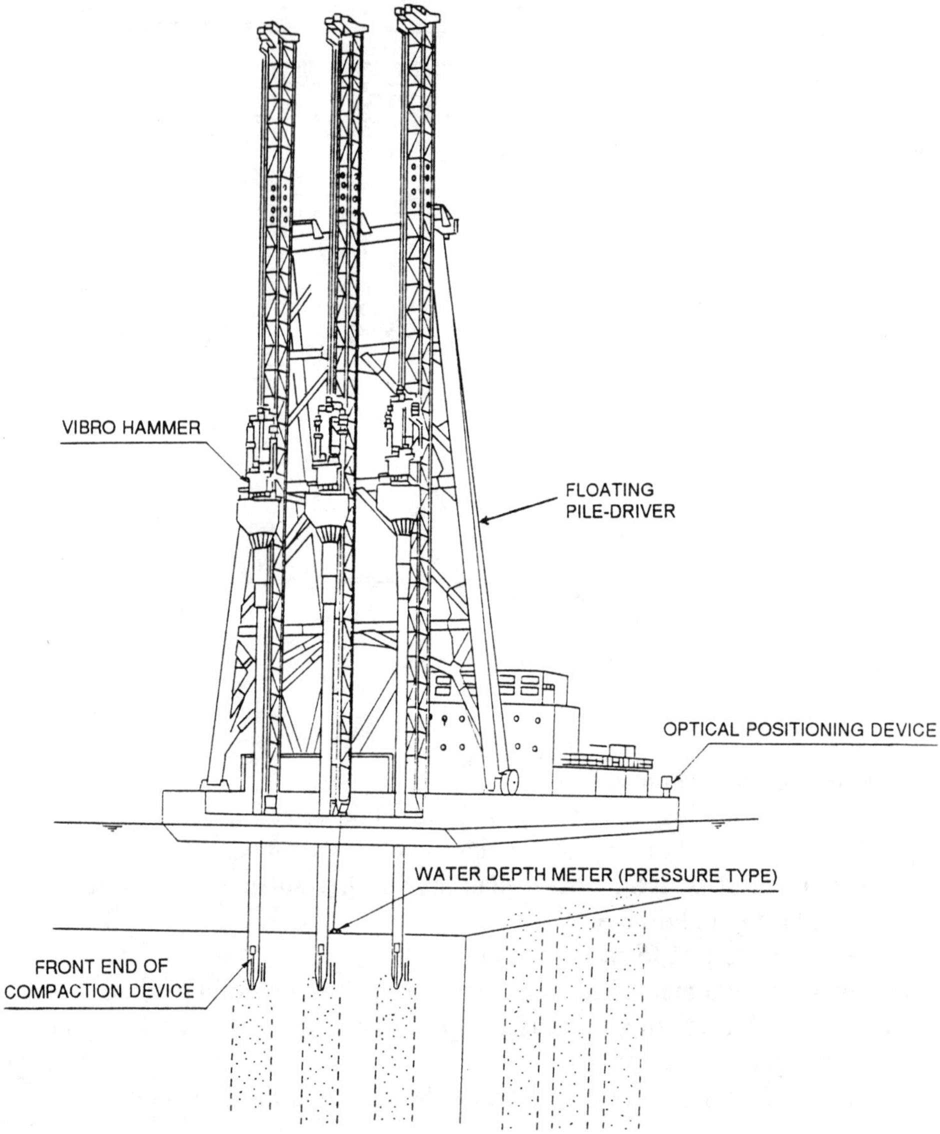

Figure A4.2. Schematic diagram of KS-HARD method.

Evaluation of the technique

a) The existing design method is applicable.

b) Reliable sand pile formation control can be performed with improvement.

c) The same compaction capacity is expected as in the existing sand compaction pile methods.

d) Efficient operation is expected due to enhanced automatic execution.

A4.3 DEEP VIBRO METHOD

Outline of the technique

This vibroflotation method reduces vibration and noise during construction and increases the depth of application by using a large capacity vibroflot.

Schematic diagram of the construction

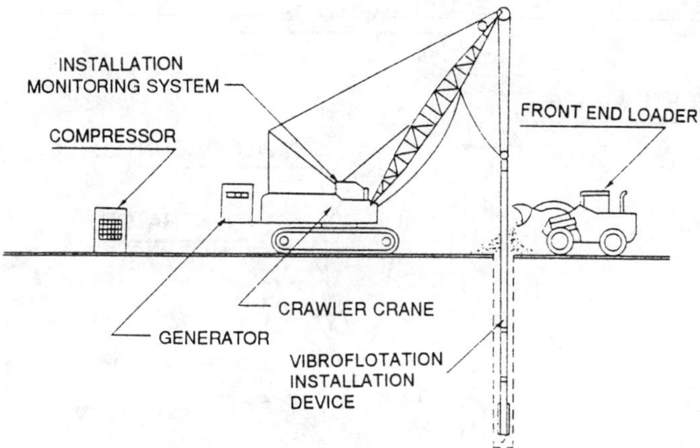

Figure A4.3. Schematic diagram of deep vibro method.

Evaluation of the technique

a) Construction closer to existing facilities is possible than with the existing sand compaction pile method.

b) The existing design method is applicable.

c) Sand piles can be formed accurately.

d) The same compaction capacity is expected as in the existing sand compaction pile methods.

e) The installation equipment is compact and can be operated efficiently compared with the existing sand compaction pile methods.

f) Influence of noise and vibration on the surrounding area is less than the existing sand compaction pile methods.

A4.4 REMEDIAL MEASURE AGAINST LIQUEFACTION COMBINING COMPACTION AND DRAIN EFFECT

Outline of the technique

This gravel drain method combines the effect of subsoil densification during installation by discharging gravel into sandy subsoil using a tamping rod built into a casing.

Schematic diagram of the construction

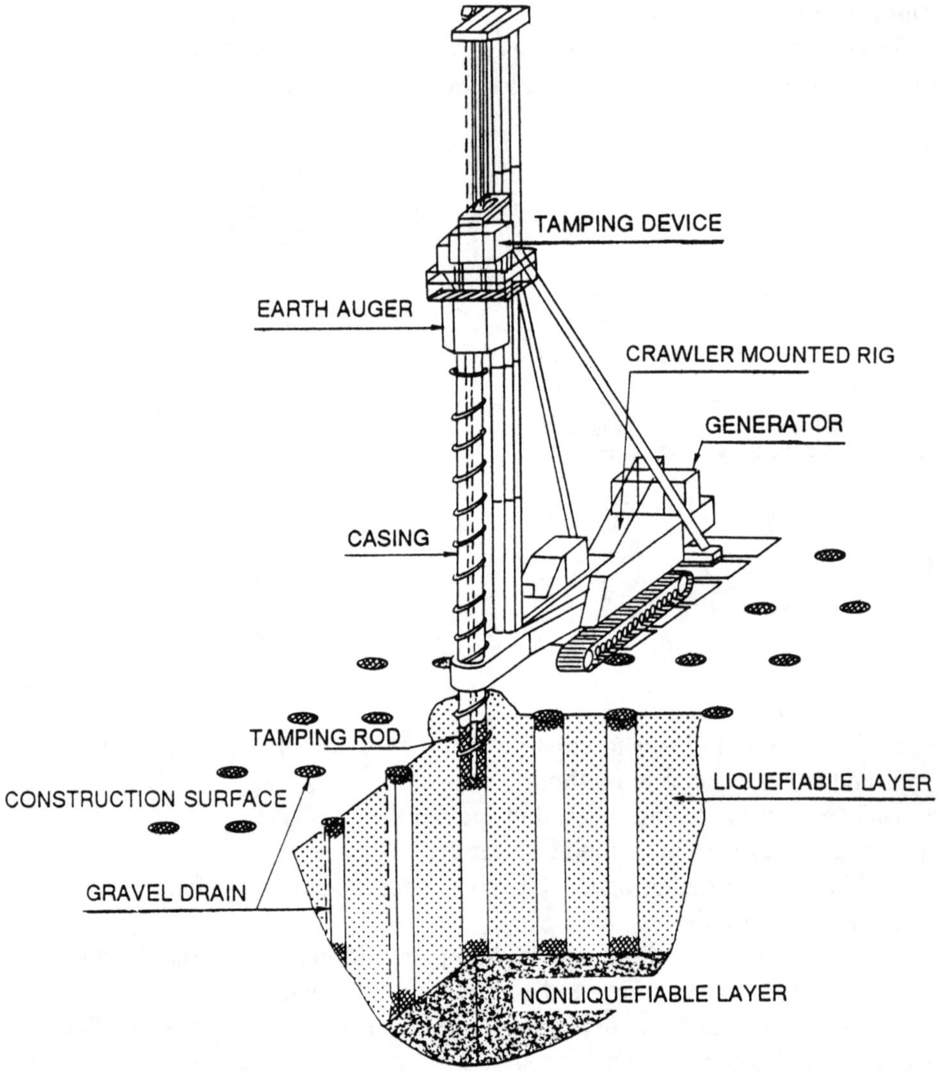

Figure A4.4. Schematic diagram of remedial measure against liquefaction combining compaction and drain effect.

Evaluation of the technique

a) Construction close to existing facilities is possible.

b) The existing design method for drains is applicable with consideration of the compaction effect on subsoils.

c) The drain can be formed accurately.

d) The same drainage effect is expected as with the existing gravel drain methods.

e) Influence of vibration and noise on the surrounding area is minimized.

A4.5 GRID DRAIN METHOD

Summary of the technique

This drain method uses a vinyl chloride band shaped drain with a rectangular cross section into sandy subsoils to reduce the excess pore water pressure increase during earthquakes.

Schematic diagram of the construction

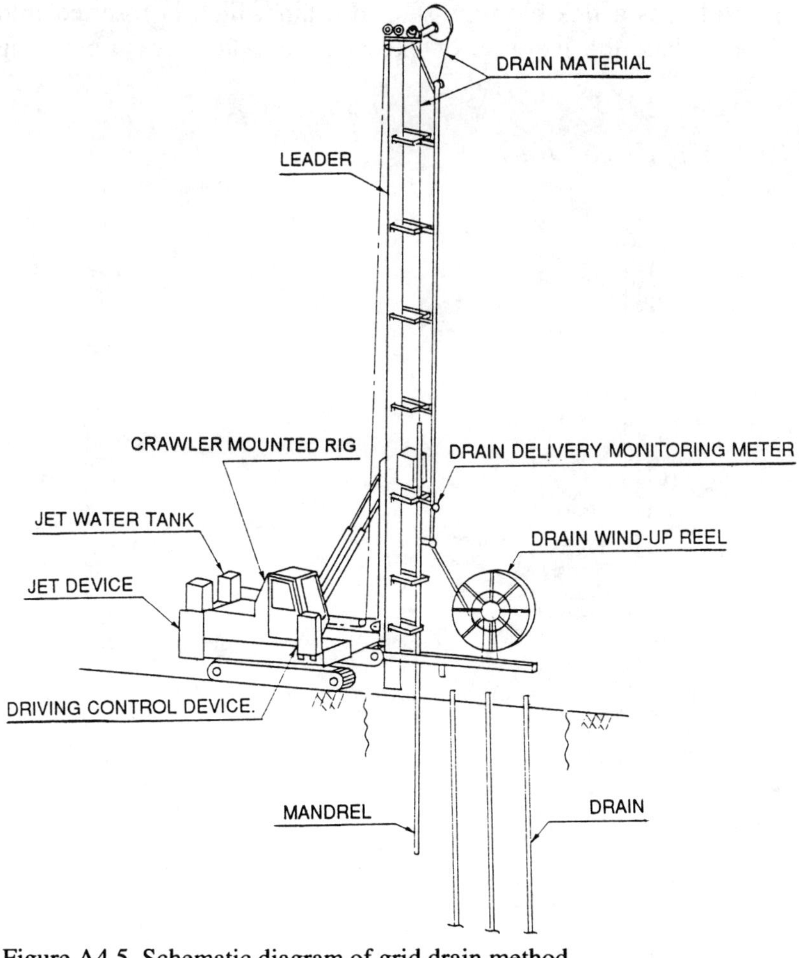

Figure A4.5. Schematic diagram of grid drain method.

Evaluation of the technique

a) Construction close to existing facilities is possible.

b) The design method for the plastic drain method is applicable.

c) The material is light and easy to handle, and has sufficient pressure resistance, durability, and drainage.

d) The drain can be formed accurately.

e) The same drainage capacity is expected as in the gravel drain method.

f) The installation equipment is compact and is operated efficiently compared to the gravel drain method.

g) Influence of noise and vibration on the surrounding area is minimized.

A4.6 SPIRAL DRAIN METHOD

Outline of the technique

This drain method uses a flexible pipe-shaped drain which is inserted into sandy subsoil to reduce the increase of excess pore water pressure during earthquakes.

Schematic diagram of the construction

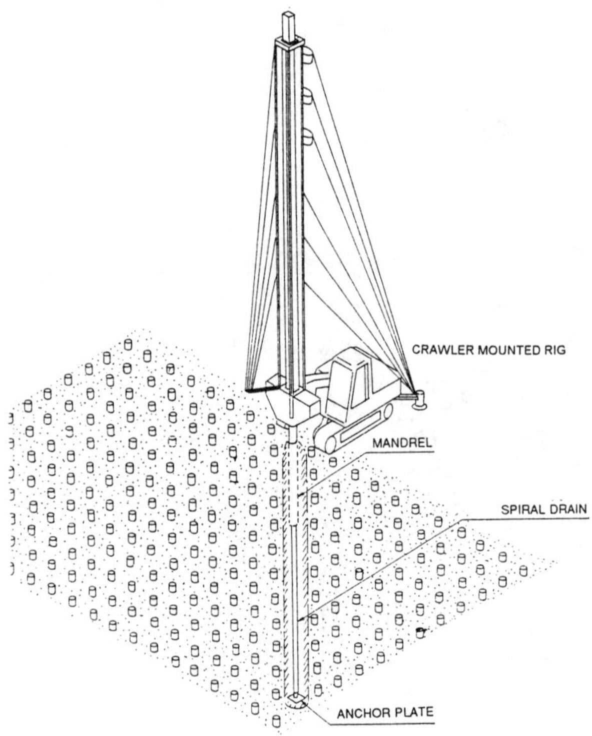

Figure A4.6. Schematic diagram of spiral drain method.

Evaluation of the technique

a) Construction near existing facilities is possible.

b) The design method for plastic drain is applicable.

c) The material is light, easy to handle, and has sufficient pressure resistance, durability, and drainage.

d) The drain can be formed accurately.

e) The same drainage effect is expected as in the gravel drain method.

f) The installation equipment is compact and can be operated efficiently compared to the gravel drain method.

g) Influence of noise and vibration on the surrounding area is minimized.

A4.7 MINI-COMPOSER METHOD

Summary of the technique

This sand compaction pile method achieves reduction in noise, vibration, and ground displacement by operating the vibro hammer at high frequencies and by making the diameter of the casing small compared to the existing methods.

Schematic diagram of the construction

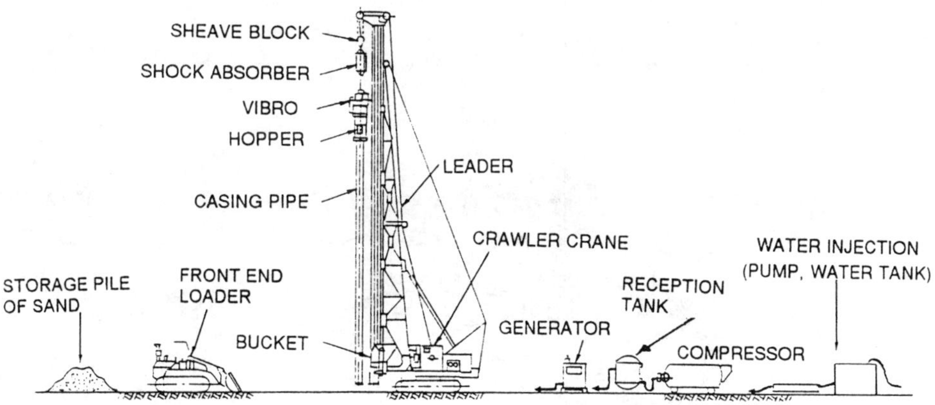

Figure A4.7. Schematic diagram of mini-composer method.

Evaluation of the technique

a) Construction closer to existing facilities than with the existing sand compaction pile method is possible.

b) The existing design method is applicable.

c) The sand pile can be formed accurately.

d) The same compaction effect is expected as in the existing sand compaction pile methods.

e) The installation equipment is compact and can operate efficiently compared to the existing sand compaction pile method.

f) Influence of noise and vibration on the surrounding area is minimal compared to the existing sand compaction pile method.

A4.8 GRAVEL DRAIN METHOD WITH DOUBLE CASING

Outline of the technology

This gravel drain method uses a double casing; the outer casing is used for auguring whereas the inner casing is for placement of gravel, thus providing an efficient installation.

Schematic diagram of the construction

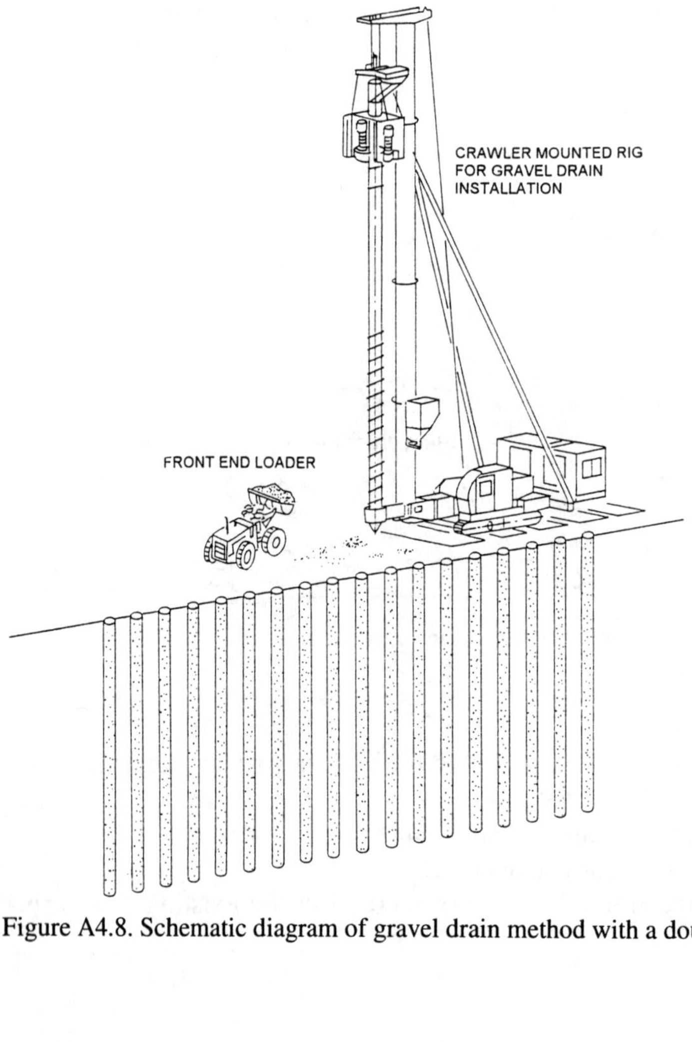

Figure A4.8. Schematic diagram of gravel drain method with a double casing.

Evaluation of the technique

a) Construction close to existing facilities is possible.

b) The existing design method is applicable.

c) The drain can be formed accurately.

d) The same drainage capacity is expected as in the existing gravel drain method.

e) Influence of noise and vibration on the surrounding area is minimized.

A4.9 NOISE-FREE AND VIBRATION-FREE GRAVEL DRAIN METHOD

Outline of the technique

This gravel drain method forms gravel columns by applying pressure to and placing gravel into sandy subsoil layers with a screw built in the casing.

Schematic diagram of the construction

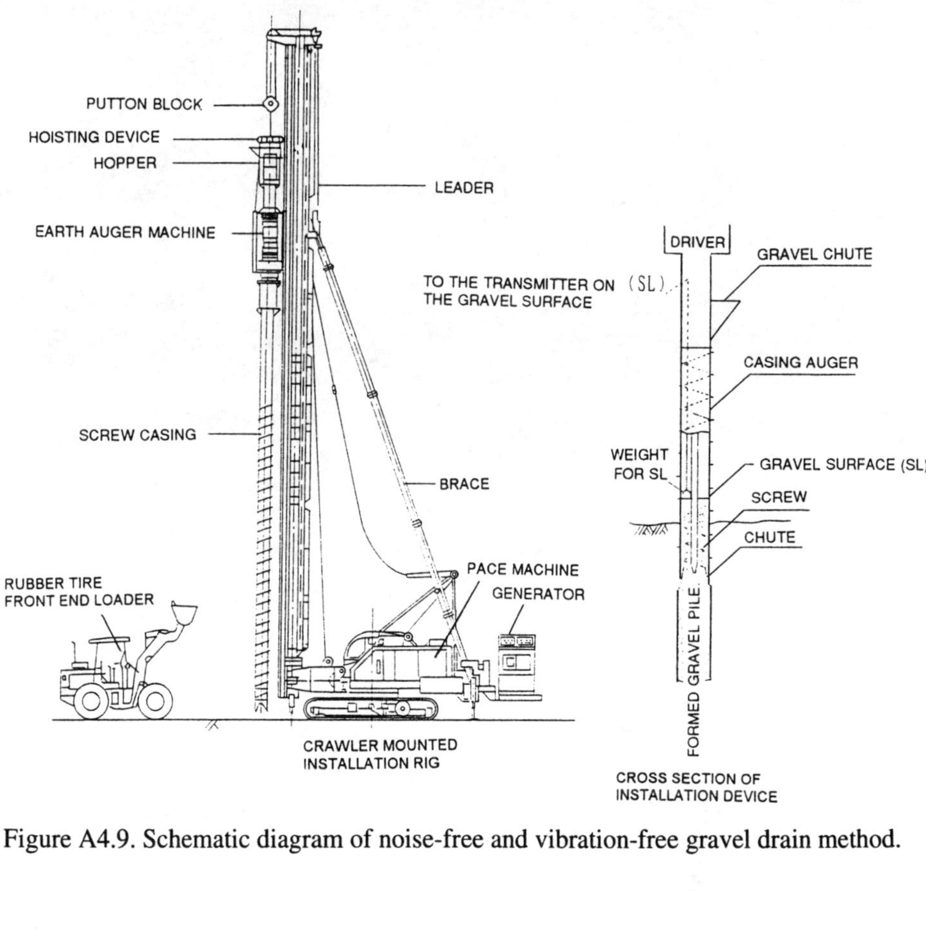

Figure A4.9. Schematic diagram of noise-free and vibration-free gravel drain method.

Evaluation of the technique

a) Construction close to existing facilities is possible.

b) The existing design method is applicable.

c) The drain can be formed accurately.

d) The same draining capacity is expected as in the existing gravel drain method.

e) Influence of noise and vibration on the surrounding area is minimized.

APPENDIX 5

Design calculation examples

A5.1 LIQUEFACTION PREDICTION/DETERMINATION EXAMPLES BASED ON GRADATION AND SPT *N*-VALUE

The results of a subsoil survey for a waterfront area are shown in Figures A5.1 and A5.2. Based on the results of earthquake response analysis performed for the subsoil profile, the maximum shear stress τ_{max} of each layer was estimated as shown in Figure A5.1. Under these conditions, an example of a calculation for a soil layer 6 m deep is given below.

1. At a depth of 6 m, an SPT *N*-value of 12 is noted from Figure A5.1, coefficient of uniformity is $D_{60}/D_{10} = 0.17/0.1 = 1.7$ from Figure A5.2, and the fines content (component with grain size less than 0.074 mm) is zero.

2. When the grain size distribution in Figure A5.2 is compared with the zones shown in the figure for a small coefficient of uniformity in Figure 4.39 shown in Section 4.5, it is apparent that the sand is included in zone *A* with a high possibility of liquefaction.

3. Effective overburden pressure at this depth is

$$\sigma'_v = 1.8 \times 2.0 + 1.0 \times (6.0 - 2.0) = 7.6 \text{ tf/m}^2 = 0.76 \text{ kgf/m}^2 \quad \text{(A5.1)}$$

and therefore the equivalent *N*-value is obtained from Equation (4.23) as

$$(N)_{0.66} = \frac{\{12 - 1.828 \times (0.76 - 0.66)\}}{\{0.399 \times (0.76 - 0.66) + 1\}} = 11.4 \quad \text{(A5.2)}$$

4. Usually, earthquake response analysis is performed with respect to the concerned site in order to obtain the maximum shear stress τ_{max} for each soil layer. In this example calculation, however, the result is shown in Figure A5.1 as noted at the beginning, so that the equivalent acceleration is obtained by using Equation (4.24). The maximum shear stress τ_{max} of the soil layer with a depth of 6 m is 0.304 kgf/cm^2. From the above, the equivalent acceleration is obtained as follows

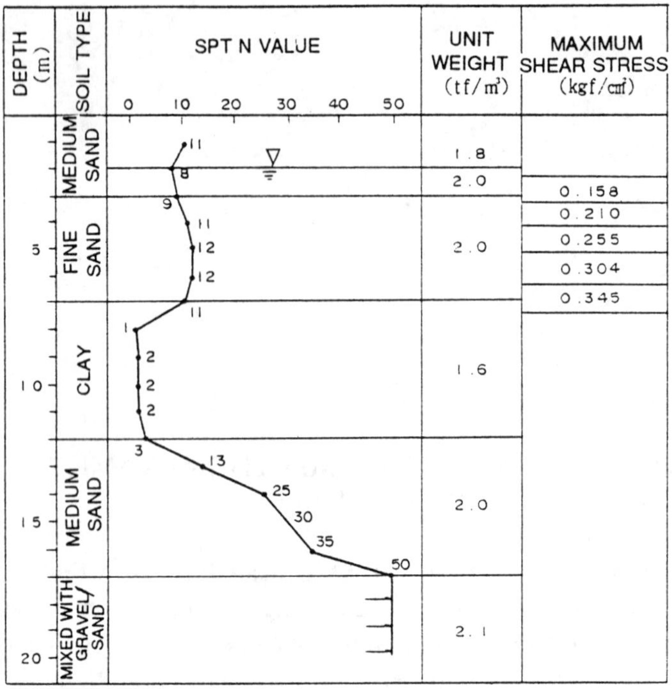

Figure A5.1. Soil boring log.

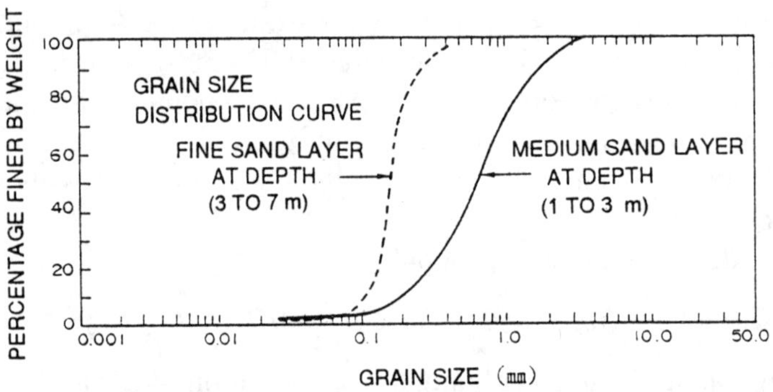

Figure A5.2. Gradation of the soil in each layer.

$$\alpha_{eq} = 0.7 \times \frac{0.304}{0.76} \times 980 = 274 \text{ Gal} \tag{A5.3}$$

(If the ground height differs from the height where the SPT *N*-value was measured, the effective overburden pressure is recalculated and an equivalent acceleration is calculated using this value.)

5. Based on the results of 3 and 4, i.e. equivalent *N*-value of 11.4 and equivalent acceleration of 274 Gal, the soil layer is categorized in zone II in Figure 4.41. Since the fines content is zero, it is apparent from Figure 4.42 that it is not necessary to consider the reduction in critical *N*-value in this evaluation.

6. As a result of liquefaction prediction from Table 4.17, the possibility of liquefaction is high and the decision is either to determine that liquefaction will likely occur or to perform cyclic triaxial tests.

A5.2 AN EXAMPLE OF PREDICTION/DETERMINATION BASED ON CYCLIC TRIAXIAL TEST RESULTS

A5.2.1 *The cyclic triaxial test results*

From the cyclic triaxial test results on undisturbed samples obtained from sandy subsoil, the relationship between cyclic shear stress ratio τ_l / σ'_c and the number of load cycles to liquefaction N_l is presented as shown in Figure A5.3. The cyclic shear stress ratio $(\tau_l / \sigma'_c)_{Nl=20}$ (when double amplitude $2\varepsilon_a$ of axial strain is 5%) with respect to the number of load cycles of 20 is determined. Table A5.1 shows the result at each depth from the results shown in Figure A5.3.

Table A5.1. Cyclic shear stress ratio for number of load cycles to liquefaction $N_l = 20$.

Sample name KU Harbour	Depth (m)	SPT *N*-value	Dr (%)	τ_l / σ'_c $2\varepsilon_a = 5\%$
No. 1 site				
C – 2	2.76	6	52.1	0.216
C – 3	3.76	3	41.0	0.193
C – 4	5.46	4	34.4	0.208
C – 5	6.26	7	65.7	0.196
C – 6	7.06	7	51.3	0.231
C – 8	8.76	10	30.7	0.211

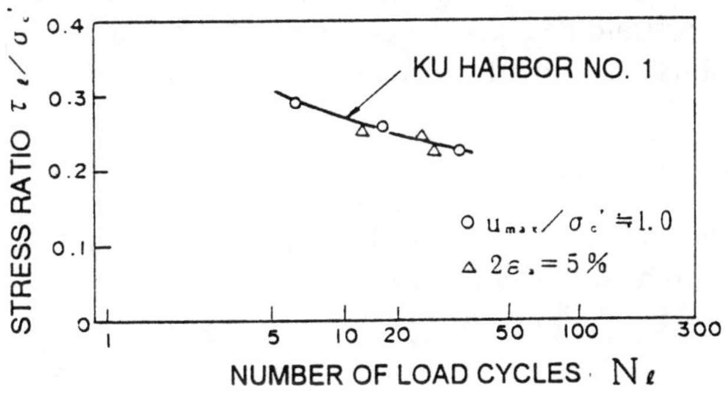

Figure A5.3. Cyclic shear stress ratio and number of load cycles to liquefaction.

Table A5.2. Soil profile modeled for earthquake response analysis.

Layer number	Soil type	Layer thickness (m)	Depth (m)	Effective overburden pressure σ'_v (tf/m^2)	Shear modulus (tf/m^2)	Unit weight (tf/m^3)	Shear wave velocity (m/s)
1	Sand	1.65	0.83	1.48	2222	1.8	110
2	Sand	0.67	1.99	3.31	2469	2.0	110
3	Sand	0.67	2.66	3.98	2469	2.0	110
4	Sand	0.67	3.34	4.66	2469	2.0	170
5	Sand	0.67	4.01	5.33	2469	2.0	170
6	Sand	0.61	4.66	5.97	5898	2.0	170
7	Sand	0.61	5.27	6.58	5898	2.0	170
8	Sand	0.61	5.88	7.19	5898	2.0	170
9	Sand	0.61	6.49	7.80	5898	2.0	170
10	Sand	0.61	7.10	8.41	5898	2.0	170
11	Sand	0.61	7.71	9.02	5898	2.0	170
12	Sand	0.61	8.32	9.63	5898	2.0	170
13	Sand	0.61	8.92	10.24	5898	2.0	170
14	Sand	0.61	9.53	10.85	5898	2.0	170
15	Sand	0.61	10.14	11.46	5898	2.0	170
16	Sand	3.80	12.35	13.67	17163	2.0	290
17	Sand	11.85	20.17	21.49	31041	2.0	390
18	Bedrock				34145	2.2	390

Number of layers 17; layer number below water table 2; depth to bedrock 26.10 m; depth to water table 1.65 m.

A5.2.2 *Modeling of the subsoil profile for earthquake response analysis*

Earthquake response analysis is performed using the computer program SHAKE and the cyclic shear stress generated by an earthquake is computed. In performing the earthquake response analysis, the specified subsoil profile is modeled by referring to Section 4.4.4. Table A5.2 shows the results of a soil deposit model idealized for response analysis consisting of 18 layers. The depth for each layer is defined at the center of each layer. The depth to the bedrock is 26.1 m. A strain dependence curve $G/G_0 \sim \gamma$ for the normalized shear modulus and a strain dependence curve $h \sim \gamma$ of the damping factor of each layer are established with reference to Section 4.3.1.

A5.2.3 *Earthquake motion at bedrock*

The maximum acceleration and waveform of the earthquake motion at bedrock are established according to Section 4.4.2.

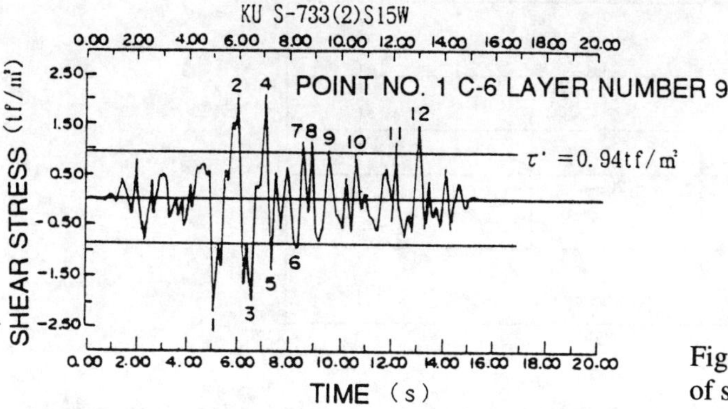

Figure A5.4. Time history of shear stress.

A5.2.4 *Calculation of maximum shear stress ratio L_{max} and liquefaction strength ratio R_{max}*

Figure A5.4 is the computed time history of shear stress in layer number 9 of Table A5.2 using the computer program SHAKE. The maximum shear stress $\tau_{d\max} = 0.229$ kgf/cm^2 is obtained, which is given number 1 in Figure A5.4. When this value is divided by the effective vertical pressure $\sigma'_v = 0.78$ kgf/cm^2 as indicated in Equation (4.26), the maximum shear stress ratio is obtained as $L_{\max} = 0.294$.

The time history shown in Figure A5.4 exhibits the waveform pattern of the 'impact type' (to be defined when less than 2 peaks with an amplitude of over 60% of $\tau_{\max}$ exist before $\tau_{d\max}$ occurs), so that the conversion coefficient of $C_k = 0.55$ is used with respect to an irregular waveform. When the conditions $(\tau_l / \sigma'_c)_{Nl=20} = 0.231$ in C-6 of Table A5.1, $C_k = 0.55$, and $K_o = 0.5$ are substituted into Equation (4.27), the liquefaction strength is obtained as $R_{\max} = 0.252$.

A5.2.5 *Liquefaction prediction/determination*

When $L_{\max}$ and $R_{\max}$ are substituted into Equation (4.25), $F_L = 0.252/0.294 = 0.86$ and for this layer, liquefaction is predicted to occur.

A5.3 AN EXAMPLE OF PREDICTION/DETERMINATION BY THE METHOD BASED ON CUMULATIVE DAMAGE THEORY

The method based on the cumulative damage theory is applied for the example shown in Section A5.2 (see Table A5.3.)

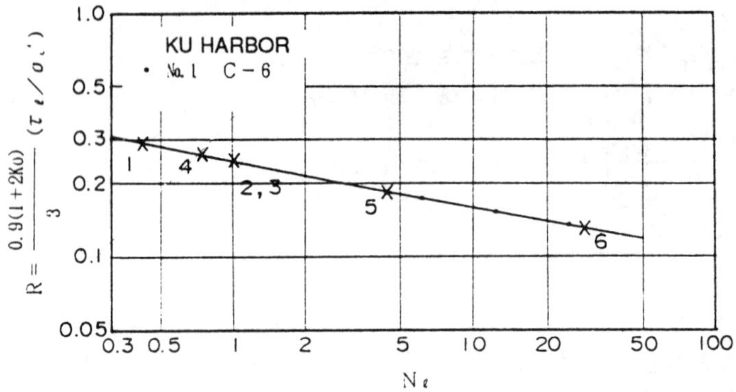

Figure A5.5. Relationship between the number of load cycles for liquefaction and the liquefaction strength ratio plotted at logarithmic scale.

A5.3.1 *The cyclic triaxial test results*

(τ_l / σ'_c) in Figure A5.3 is corrected using Equation (4.34) and the result is plotted on a log-log coordinate as shown in Figure A5.5. Test results cannot be obtained when the number of load cycles to liquefaction is less than 6 cycles and therefore a value is obtained by linear extrapolation in the log-log coordinate.

A5.3.2 *Calculation of equivalent shear stress*

The threshold cyclic shear stress τ' which triggers excess pore water pressure increase is obtained from Figure A5.5, where $\tau' = 0.094$ kgf/cm^2 is obtained from liquefaction strength ratio $R = 0.121$ at $N_l = 50$ cycles. This value is shown in Figure A5.4, the unit being in tf/m^2. The load cycles below τ' do not influence liquefaction and are omitted from the evaluation procedure.

The equivalent shear stress τ_e, which becomes the reference when obtaining the equivalent number of load cycles, is calculated as 0.65 times the maximum shear stress $\tau_{dmax} = 0.229$ kgf/cm^2, resulting in $\tau_e = 0.149$.

A5.3.3 *Calculation of shear stress ratio L*

The shear stress ratio L is calculated using Equation (4.33), resulting in $L = \tau_e / \sigma'_v = 0.191$. The equivalent number of load cycles N_{ef} corresponding to L is read off from Figure A5.5, resulting in $N_{ef} = 4$.

A5.3.4 *Calculation of equivalent number of load cycles N_{eq}*

τ_i / σ'_v is obtained from shear stress τ_i corresponding to each half-wave

Table A5.3. Calculation using the method based on cumulative damage theory.

Boring location / KU Port (site No. 1)	Sample / C – 6
Depth (m) / 6.48	Earthquake waveform / KU S-733(2)S15W
Maximum acceleration of the bedrock (Gal)	188.4
Effective overburden pressure σ'_v (kgf/cm^2)	0.780
Threshold cyclic stress ratio τ' / σ'_v	0.121
Threshold cyclic shear stress τ' (kgf/cm^2)	0.094
Maximum shear stress τ_{max} (kgf/cm^2)	0.229
Equivalent shear stress $\tau_e = 0.65\ \tau_{max}$	0.149
Cyclic shear stress ratio $L = \tau_e / \sigma'_v$	0.191
N_{ef} (N_l with respect to L)	4.0
Number of half-wave	12

Half-wave No.	τ_i	τ_i / σ'_{vo}	N_{if}	$1/N_{if}$	Half-wave No.	τ_i	τ_i / σ'_{vo}	N_{if}	$1/N_{if}$
1	0.229	0.294	0.4	2.5	9	0.103	0.132	34	0.03
2	0.194	0.249	1.0	1.0	10	0.098	0.126	36	0.03
3	0.196	0.251	1.0	1.0	11	0.119	0.153	12	0.08
4	0.205	0.263	0.72	1.39	12	0.142	0.182	5	0.20
5	0.145	0.186	4.2	0.24					
6	0.102	0.131	29	0.03	$\Sigma (1/N_{if}) = 6.65$				
7	0.120	0.154	12	0.08					
8	0.117	0.150	14	0.07					

Equivalent conversion wave number $N_{eq} = N_{ef} \times 0.5 \times \Sigma (1/N_{if})$	13.3
$R (N_{eq})$ (R with respect to N_{eq})	0.151
Liquefaction resistance factor $F_L = R (N_{eq}) / L$	0.79
Liquefaction potential evaluation	Liquefaction will occur

(numbers 1-12) in Figure A5.4. N_{if} is read off from Figure A5.5. The results determined are shown in Table A5.3. Equivalent number of load cycles N_{eq} is calculated as

$$N_{\mathrm{eq}} = 0.5\, N_{ef} \sum \left(\frac{1}{N_{if}} \right) \tag{A5.4}$$

and therefore $N_{\mathrm{eq}} = 0.5 \times 4.0 \times 6.65 = 13.3$.

A5.3.5 *Calculation of $R(N_{eq})$*

The calculated N_{eq} is used to obtain $R\,(N_{\mathrm{eq}})$ from Figure A5.5 as $R\,(N_{\mathrm{eq}}) =$ 0.151.

A5.3.6 *Liquefaction prediction/determination*

F_L is obtained by substituting L and $R\,(N_{\mathrm{eq}})$ in Equation (4.36), resulting in $F_L = 0.151/0.191 = 0.79$. Thus, liquefaction is predicted to occur.

A5.3.7 *Consideration in applying this method*

The F_L values obtained in the previous and present subsections are not the same. The F_L value in this subsection is smaller. Difficulties are often encountered because the number of soil samples is limited for obtaining the cyclic shear stress ratio with a small number of load cycles to liquefaction. Cyclic triaxial test results are, therefore, often extrapolated for one to five cycles of load as shown in Figure A5.5. This extrapolation usually results in a smaller liquefaction strength than the actual one. The F_L obtained in this section, therefore, is conservative. When using the cumulative damage approach shown in this section, it is necessary to obtain test results with reasonable accuracy at a small number of load cycles to liquefaction by using as many samples as possible.

References

REFERENCES TO CHAPTER 2

[1] Kawamura, S., Nishizawa, T. & Wada, T. 1985. Damage to pile foundations due to liquefaction found during excavation 20 Years after the 1964 Niigata earthquake, *Nikkei Architecture, 7.* pp. 130-134 (in Japanese).

[2] Hamada, M., Yasuda, S., Isoyama, R. & Emoto, K. 1986. Observation of permanent ground displacements induced by soil liquefaction, *Proc. of Japan Society of Civil Engineers, No. 376/III-6.* pp. 211-220 (in Japanese).

[3] Tada, Y. (ed.) 1964. Report on Niigata earthquake, Report of Railway Technology Research Institute, No. 448 (in Japanese).

REFERENCES TO CHAPTER 3

[1] Wakamatsu, K. 1991. Liquefaction history of reclaimed land, *Tsuchi-to-Kiso, Japanese Society of Soil Mechanics and Foundation Engineering, Vol. 39, No. 1.* pp. 78-84 (in Japanese).

[2] Shimizu, K. 1973. Characteristics of the reclamation ground in Tokyo Harbour, *Tsuchi-to-Kiso, Japanese Society of Soil Mechanics and Foundation Engineering, Vol. 21, No. 6.* pp. 45-52 (in Japanese).

REFERENCES TO CHAPTER 4

[1] Kuribayashi, E. & Tatsuoka, F. 1975. Brief review of soil liquefaction during earthquakes in Japan, *Soils and Foundations, JSSMFE, Vol. 15, No. 4.* pp. 81-92.

[2] Wakamatsu, K. 1991. Maps for historic liquefaction sites in Japan, *Tokai University Press* (in Japanese with English abstract and captions).

[3] Kotoda, K., Wakamatsu, K. & Midorikawa, S. 1988. Seismic microzoning on soil liquefaction potential based on geomorphological land classification, *Soils and Foundations, JSSMFE, Vol. 28, No. 2.* pp. 127-143.

[4] Fire Defense Agency 1978. Rules for regulation of hazardous materials (in Japanese).

[5] Japan Road Association 1990. Specifications for highway bridges, Part V Earthquake Resistant Design (in Japanese).
[6] Architectural Institute of Japan 1988. Recommendations for design of building foundations (in Japanese).
[7] Japan Ports and Harbours Association 1989. Technical standards for port and harbour facilities and commentaries, Revised edition (in Japanese).
[8] Seed, H.B. & Idriss, I.M. 1967. Analysis of soil liquefaction: Niigata earthquake, *Journal of Soil Mechanics and Foundations Division, ASCE, Vol. 93, No. SM3.* pp. 83-108.
[9] Ishihara, K. 1976. Fundamentals of soil dynamics, Kajima Publication (in Japanese).
[10] Finn, W.D.L., Lee, K.W. & Martin, G.R 1977. An effective stress model for liquefaction, *Journal of Geotechnical Engineering Division, ASCE, Vol. 103, No. GT6.* pp. 517-533.
[11] Iai, S., Matsunaga, Y. & Kameoka, T. 1992. Strain space plasticity model for cyclic mobility, *Soils and Foundations, JSSMFE, Vol. 32, No. 2.* pp. 1-15.
[12] Iai, S., Koizumi, K. & Tsuchida, H. 1986. A new criterion for assessing liquefaction potential using grain size accumulation curve and *N*-value, *Report of the Port and Harbour Research Institute, Vol. 25, No. 3.* pp. 125-234 (in Japanese).
[13] Iai, S., Tsuchida, H. & Koizumi, K. 1989. A liquefaction criterion based on field performances around seismograph stations, *Soils and Foundations, JSSMFE, Vol. 29, No. 2.* pp. 52-68.
[14] Ports and Harbours Bureau, Ministry of Transport 1971. Port and harbour survey manual, Japan Association of Ports and Harbours (in Japanese).
[15] Ports and Harbours Bureau, Ministry of Transport 1989. Technical standards for port and harbour facilities and commentaries, Japan Association of Ports and Harbours (in Japanese).
[16] Florin, V.A. & Ivanov, P.L. 1961. Liquefaction of saturated sandy soils, *Proc. of the 5th International Conference on Soil Mechanics and Foundation Engineering.* pp. 107-111.
[17] Zen, K., Yamazaki, H. & Umehara, Y. 1987. Experimental study on shear modulus and damping ratio of natural deposits for seismic response analysis, *Report of the Port and Harbour Research Institute, Vol. 26, No. 1.* pp. 41-113 (in Japanese).
[18] Japanese Society of Soil Mechanics and Foundation Engineering 1982. Soil survey method, Chapter 7: Sampling. pp. 275-342 (in Japanese).
[19] Zen, K. & Yamazaki, H. 1988. Comparison of liquefaction strength of frozen and non-frozen undisturbed samples, *Proceedings of the Symposium on Cyclic Undrained Test, Japanese Society of Soil Mechanics and Foundation Engineering.* pp. 143-148 (in Japanese).
[20] Sato, H., Yasuda, S. & Sakagami, S. 1979. A study on freezing and transportation of sand sample, *Proceedings of the 14th Japan National Conference on Soil Mechanics and Foundation Engineering.* pp. 41-44 (in Japanese).
[21] Tokimatsu, K. & Ohara, J. 1990. Lecture note: Thermal problem in the soil, 8.2 frozen sample, *Tsuchi-to-Kiso, Japanese Society of Soil Mechanics and Foundation Engineering, Vol. 38, No. 11.* pp. 61-68 (in Japanese).
[22] Zen, K., Matsumoto, K. & Ishii, I. 1982. Quality evaluation of soil within sampling tube by X-ray transmission test, *Technical Note of the Port and Harbour Research Institute, No. 438.* p. 93 (in Japanese).

[23] Yamazaki, H., Zen, K. & Kagaya, H. 1988. Quality evaluation of undisturbed sample by use of X-ray photography and liquefaction test result, *Proceedings of the Symposium on Cyclic Undrained Test, Japanese Society of Soil Mechanics and Foundation Engineering*. pp. 137-142 (in Japanese).

[24] Zen, K., Matsumoto, K., Umehara, Y. & Okumura, T. 1981. Sand sampling and an evaluation of sample quality, *Proceedings of the Sampling Symposium, Japanese Society of Soil Mechanics and Foundation Engineering*. pp. 55-62 (in Japanese).

[25] Japanese Society of Soil Mechanics and Foundation Engineering 1990. Soil testing method and explanation, shear test, Chapter 10: Cyclic undrained triaxial test. pp. 421-450 (in Japanese).

[26] Japanese Society of Soil Mechanics and Foundation Engineering 1990. Soil testing method and explanation, physical test, Chapter 9: Wet density test. pp. 116-124 (in Japanese).

[27] Imai, T. & Tonouchi, K. 1982. Correlation of SPT *N*-value with *S*-wave velocity and shear modulus, *Proc., 2nd ESOPT*.

[28] Hardin, B.O. & Richart, F.E. 1963. Elastic wave velocities in granular soils, *Proc., ASCE, Vol. 89, No. SM1*. pp. 33-65.

[29] Silver, M.L., Chan, C.K., Ladd, R.S., Lee, K.L., Tiedemann, D.A., Townsend, F.C., Valera, J.E. & Wilson, J.H. 1976. Cyclic triaxial strength of standard test sand, *Proc., ASCE, Vol. 102, No. GT5*. pp. 511-523.

[30] Annaki, M. & Lee, K.K. 1977. Equivalent uniform cyclic concept for soil dynamics, *Proc., ASCE, Vol. 103, No. GT6*. pp. 549-564.

[31] Yoshimi, Y. 1980. Liquefaction of sandy soil, Gihodo Publication (in Japanese).

[32] Seed, H.B., Pyke, R.M. & Martin, G.R. 1978. Effects of multidirectional shaking on pore pressure development in sands, *Proc., ASCE, Vol. 104, No. GT1*. pp. 27-44.

[33] Ishihara, K. 1977. Simple method of analysis for liquefaction of sand deposits during earthquakes, *Soils and Foundations, JSSMFE, Vol. 17, No. 3*. pp. 1-8.

[34] Tsuchida, H. 1970. Prediction and remedial measures against liquefaction of sandy soil, *Annual Seminar of Port and Harbour Research Institute*. pp. (3)1-33 (in Japanese).

[35] Umehara, Y., Zen, K. & Hamada, K. 1976. Liquefaction characteristics of saturated sand based on cyclic triaxial test, *Report of the Port and Harbour Research Institute, Vol. 15, No. 4*. pp. 49-74 (in Japanese).

[36] Kitazawa, S., Uwabe, T. & Higaki, N. 1984. Estimation of maximum acceleration expectancy of bedrock in coastal regions, *Technical Note of the Port and Harbour Research Institute, No. 486*. (in Japanese).

[37] Noda, S., Uwabe, T. & Chiba, T. 1975. Relation between ground acceleration and seismic coefficient for gravity wharf, *Report of the Port and Harbour Research Institute, Vol. 14, No. 4*. pp. 67-111 (in Japanese).

[38] Shnabel, P.B., Lysmer, J. & Seed, H.B. 1972. SHAKE, A computer program for earthquake response analysis of horizontally layered sites, Report No. EERC 72-12, University of California at Berkeley.

[39] Lysmer, J., Udaka, T., Tsai, C.F. & Seed, H.B. 1975. FLUSH, A computer program of approximate 3-D analysis of soil structure interaction problems, Report No. EERC 75-30, University of California at Berkeley.

[40] Inatomi, T., Yokoyama, I. & Kazama, M. 1986. One-dimensional nonlinear dynamic ground response analyses, *Technical Note of the Port and Harbour Research Institute, No. 561* (in Japanese).

[41] Towhata, I. 1992. Modeling of dynamic characteristics of soil material, *Tsuchi-to-Kiso, Vol. 40, No. 5, Japanese Society of Soil Mechanics and Foundation Engineering*. pp. 75-82 (in Japanese).

[42] Ishihara, K., Yoshida, N. & Tsujino, S. 1985. Modeling of stress-strain relations of soils in cyclic loading, *Proc. of 5th Conf. on Numerical Methods in Geomechanics, Nagoya, Vol. 1*. pp. 373-380.

[43] Finn, W.D.L., Martin, G.R. & Lee, M.K.W. 1978. Comparison of dynamic analysis for saturated sands, *Proc. ASCE, Earthquake Engineering and Soil Dynamics, Vol. 1*. pp. 472-491.

[44] Ishihara, K. 1981. Liquefaction of soil 4, response analysis of soil with consideration given to liquefaction, *Tsuchi-to-Kiso, Vol. 29, No. 11, Japanese Society of Soil Mechanics and Foundation Engineering*. pp. 85-91 (in Japanese).

[45] Japan Society of Civil Engineers 1989. Dynamic analysis and earthquake resistant design, Vol. 2 'Dynamic analysis', *Gihodo Publication*. p. 99 (in Japanese).

[46] Uwabe, T. & Noda, S. 1974. Ground acceleration during large earthquakes in the Tokyo coastal region, Part 2, *Technical Note of the Port and Harbour Research Institute, No. 194*. (in Japanese).

[47] Iai, S., Matsunaga, Y. & Kameoka, T. 1990. Strain space plasticity model for cyclic mobility, *Report of the Port and Harbour Research Institute, Vol. 29, No. 4*. pp. 27-56.

[48] Iai, S., Matsunaga, Y. & Kameoka, T. 1990. Parameter identification for a cyclic mobility model, *Report of the Port and Harbour Research Institute, Vol. 29, No. 4*. pp. 57-83.

[49] Iai, S. 1989. Analysis by FLIP, Japanese Society of Soil Mechanics and Foundation Engineering, *Symposium on Performance of Ground and Soil Structures during Earthquakes*. pp. 94-103 (in Japanese).

[50] Iai, S. & Matsunaga, Y. 1991. Analysis by FLIP, Japanese Society of Soil Mechanics and Foundation Engineering, *Symposium on Liquefaction Remediation of Soil*. pp. 135-144 (in Japanese).

[51] Iai, S. & Kameoka, T. 1993. Analysis of liquefaction induced damage to sheet pile quay walls, *Report of the Port and Harbour Research Institute, Vol. 31, No. 5*. pp. 221-250.

[52] Iai, S. & Matsunaga, Y. 1992. Analysis of liquefaction induced uplift of underground structures, *Proceedings from the Fourth Japan-US Workshop on Earthquake Resistant Design of Lifeline Facilities and Countermeasures for Soil Liquefaction, Technical Report NCEER-92-0019*. pp. 425-238.

[53] Uwabe, T., Tsuchida, H. & Kurata, E. 1983. Coupled hydrodynamic response analysis based on strong motion earthquake records of fill type breakwater in deep sea, *Report of the Port and Harbour Research Institute, Vol. 22, No. 2*. p. 312 (in Japanese).

[54] Kokusho, T., Sakurai, A. & Esashi, Y. 1979. Development of soil dynamic testing method from very small strain to large strain using a triaxial testing device and application in physical testing of sand, *14th Japan National Conference on Soil Mechanics and Foundation Engineering*. pp. 513-516 (in Japanese).

[55] Usui, T., Kazama, M. & Inatomi, T. 1989. The influence of confining pressure de-

pendence of shear modulus on the results of 1D equivalent linear seismic response analysis, *Japanese Society of Soil Mechanics and Foundation Engineering, Symposium on the Behavior of Soil and Soil Structure during Earthquakes,* pp. 219-224 (in Japanese).

[56] Uwabe, T. & Osada, M. 1990. Dynamic deformation characteristics of gravel material based on large shear ring soil tank vibration test, *Technical Note of the Port and Harbour Research Institute, No. 663* (in Japanese).

[57] Iai, S. & Matsunaga, Y. 1990. Simplified procedure for evaluating shear stress ratio for liquefaction prediction, *25th Japan National Conference on Soil Mechanics and Foundation Engineering.* pp. 773-776 (in Japanese).

[58] Japan Port and Harbour Association 1979. Technical standard for port and harbour facilities and commentaries.

[59] Matsunaga, Y., Iai, S. & Ito, H. 1991. Liquefaction of sand mound type breakwater, *21st National Conference of Japan Society of Civil Engineers.* pp. 237-240.

[60] Seed, H.B. & Idriss, I.M. 1971. Simplified procedure for evaluating soil liquefaction potential, *J. SMFE, ASCE, Vol. 97, No. SM9.* pp. 1249-1273.

[61] Japanese Society of Soil Mechanics and Foundation Engineering 1982. Soil engineering handbook, Chapter 10: Dynamic properties of soil. p. 362 (in Japanese).

[62] Finn, W.D.L., Pickering, D.J. & Bransby, P.L. 1971. Sand liquefaction in triaxial and simple shear tests, *J. SMFE, ASCE, Vol. 96, No. SM4.* pp. 639-659.

[63] Ishihara, K. & Li, S. 1972. Liquefaction of saturated sand in triaxial torsion shear test, *Soils and Foundations, JSSMFE, Vol. l2, No. 2.* pp. 19-39.

[64] Ishihara, K. & Perlea, V. 1984. Liquefaction-associated soil damage during the Vrancea earthquake of March 4, 1977, *Soils and Foundations, JSSMFE, Vol. 24, No. 1.* pp. 90-112.

[65] Zen, K., Umehara, Y. & Yamazaki, H. 1988. Case study of liquefaction nonliquefaction ground based on cyclic triaxial method, *Technical Note of the Port and Harbour Research Institute, No. 632* (in Japanese).

[66] Seed, H.B. 1979. Soil liquefaction and cyclic mobility evaluation for level ground during earthquakes, *J. GED, ASCE, Vol. 105, No. GT2.* pp. 201-255.

[67] Iwasaki, T., Tatsuoka, F., Tokida, K. & Yasuda, S. 1978. A practical method for assessing soil liquefaction potential based on case studies at various sites in Japan, *Proceedings of 5th Japan Earthquake Engineering Symposium.* pp. 641-648 (in Japanese).

REFERENCES TO CHAPTER 5

[1] Japanese Society of Soil Mechanics and Foundation Engineering 1988. Soft ground improvement methods – From investigation, design to construction. pp. 228-232 (in Japanese).

[2] Japanese Society of Soil Mechanics and Foundation Engineering 1991. Symposium on liquefaction remediation of ground. pp. 15-40 and pp. 215-290 (in Japanese).

[3] Mizuno, Y., Suematsu, N. & Okuyama, K. 1987. Design method of sand compaction pile for sandy soils containing fines, *Tsuchi-to-Kiso, Japanese Society of Soil mechanics and Foundation Engineering, Vol. 35, No. 5.* pp. 21-26 (in Japanese).

[4] Sato, S., Mori, H. & Okuyama, K. 1988. Study on the improvement of sandy soil by the vibro-rod method, *43rd National Conference of Japan Society of Civil Engineers, III.* pp. 106-107 (in Japanese).

[5] Shimizu, T. & Takano, Y. 1979. Design and construction of LNG tank foundation on loose sand – Indonesia, *Tsuchi-to-Kiso, Vol. 27, No. 1, Japanese Society of Soil Mechanics and Foundation Engineering.* pp. 21-26 (in Japanese).

[6] Tanaka, S. & Sasaki, C. 1989. Sandy ground improvement against liquefaction at Noshiro thermal power station, *Tsuchi-to-Kiso, Vol. 37, No. 3, Japanese Society of Soil Mechanics and Foundation Engineering.* pp. 86-90 (in Japanese).

[7] Iai, S. & Koizumi, K. 1986. Estimation of earthquake induced excess pore water pressure in gravel drains, *7th Japan Earthquake Engineering Symposium.* pp. 679-684.

[8] Yoshimi, Y. 1990. Ductility criterion for evaluating liquefaction remediation measures, *Tsuchi-to-Kiso, Japanese Society of Soil Mechanics and Foundation Engineering, Vol. 38, No. 6.* pp. 33-38 (in Japanese).

[9] Noda, S., Iida, T., Kida, H. & Saimura, Y. 1988. Shaking table tests on a countermeasure for liquefaction by drainage piles to quaywalls, *23rd Japan National Conference on Soil Mechanics and Foundation Engineering.* pp. 1015-1016 (in Japanese).

[10] Suzuki, Y., Tokito, K., Suzuki, Y. & Babasaki, J. 1989. Applied examples of anti-liquefaction foundation ground improvement method by solidification method, *Foundation Engineering, Vol. 17, No. 9* (in Japanese).

[11] Zen, K. 1990. Development of premixing method as a measure to construct a liquefaction-free reclaimed land, *Tsuchi-to-Kiso, Japanese Society of Soil Mechanics and Foundation Engineering, Vol. 38, No. 6.* pp. 27-32 (in Japanese).

[12] Japanese Society of Soil Mechanics and Foundation Engineering 1982. *Handbook of soil mechanics and foundation engineering.* pp. 1000-1002 (in Japanese).

[13] Omori, H. 1988. Design/construction example for soft soil, tank foundation (soil strengthening measure by pore water pressure dissipation method), *Foundation Engineering, Vol. 16, No. 12.* pp. 122-129 (in Japanese).

[14] Yamanouchi, K., Kumagai, T., Suzuki, Y., Miyasaki, Y. & Ishii, M. 1990. Liquefaction remedial measures using grid-shaped continuous underground wall, *25th Japan National Conference on Soil Mechanics and Foundation Engineering.* pp. 1037-1038 (in Japanese).

[15] Matsuoka, H. 1985. Liquefaction resistance of overconsolidated sand and simple liquefaction prevention method, *43rd National Conference of Japan Society of Civil Engineers, III.* pp. 15-16 (in Japanese).

[16] Nasu, M. 1984. Liquefaction remedial measures for Tokaido bullet train embankment, *Foundation Engineering, Vol. 12, No. 7.* pp. 64-68 (in Japanese).

[17] Yasuda, S., Hakuno, M., Kobayashi, S., Nagase, H. & Yoshida, T. 1991. Effects of countermeasure for liquefaction using top-shaped concrete blocks, *Symposium on Remediation of Liquefiable Ground, Japanese Society of Soil Mechanics and Foundation Engineering.* pp. 285-290 (in Japanese).

[18] Mitchell, J.K. 1981. State of the art on soil improvement, *Proc. 10th ICSMFE, Vol. 4.* pp. 510-520.

[19] Matsuo, S. (ed.) 1972. Handbook on soil stabilization methods, *Nikkan Kogyo News Press.* pp. 133-164 (in Japanese).

[20] Ogawa, M. & Ishido, M. 1965. On the use of vibro composer method to sandy

soil, *Tsuchi-to-Kiso, Japanese Society of Soil Mechanics and Foundation Engineering, Vol. 13, No. 2.* pp. 77-82 (in Japanese).

[21] Murayama, S. & Osaki, N. 1964. Foundation engineering handbook, *Asakura Publication.* pp. 750-760 and p. 756 (in Japanese).

[22] Inouye, T. (ed.) 1980. Design data book of installation works, *Morikita Publication.* p. 300 (in Japanese).

[23] Yashima, T. & Nakajima, T. 1964. New design of foundation installation method, *Kindai Publication.* p. 393 (in Japanese).

[24] Hirama, K. 1981. A few remarks on the use of relative density, *Symposium on Engineering Property and Relative Density of Sand, Japanese Society of Soil Mechanics and Foundation Engineering.* p. 53 (in Japanese).

[25] Meyerhoff, G.G. 1957. Discussion of session 1, *Proc. 4th ICSMFE, Vol. 4*, London. p. 110.

[26] Smoltezyk, J. 1983. Deep compaction - General report, *Proc. 8th ECSMFE, Vol. 3.* pp. 1105-1116.

[27] Japanese Society of Soil Mechanics and Foundation Engineering 1978. From insitu survey and design to installation of soil improvement. p. 171 (in Japanese).

[28] Brown, R.E. 1977. Vibroflotation compaction of cohesionless soils, *Proc. ASCE, GT12.* pp. 1437-1451 and p. 1441.

[29] Mogami, T. 1960. New construction methods for soils and foundations, *Japanese Society of Soil Mechanics and Foundation Engineering.* pp. 151-159 and p. 153 (in Japanese).

[30] Watanabe, T. 1962. Study on vibroflotation, Kajima Research Institute, Publication Department. p. 87 and p. 30 (in Japanese).

[31] Tokyo Soil Research 1981. Vibroflotation method (in Japanese).

[32] D'Appolonia, E., Miller, C.E., Jr. & Ware, T.M. 1955. Sand compaction by vibroflotation, *Trans. ASCE, Vol. 20.* pp. 154-168 and pp. 163-164.

[33] Fukuda, H. & Sagawa, Y. 1960. Vibroflotation installation record, *Kajima Research Institute, Annual Report, No. 9.* p. 93 (in Japanese).

[34] Ishii, Y. 1961. Construction methods for foundation, Sankaido. p. 165 (in Japanese).

[35] Webb, D.L. & Hall, R.I. 1969. Effects of vibroflotation on clayey sands, *Proc ASCE, No. SM6.* pp. 1365-1378.

[36] Basroe, C.E. & Boitano, J.D. 1969. Sand densification by piles and vibroflotation, *Proc. ASCE, No. SM6.* pp. 1303-1323.

[37] Seed, R.B. & Booker, J.R. 1977. Stabilization of potentially liquefiable sand deposits using gravel drains, *Journal of Geotechnical Engineering Division, ASCE, Vol. 103, No. GT7.* pp. 757-568.

[38] Sasaki, Y. & Taniguchi, E. 1982. Shaking table tests on gravel drains to prevent liquefaction of sand deposits, *Soils and Foundations, JSSMFE, Vol. 22, No. 4.* pp. 1-14.

[39] Noda, S., Kida, H. & Iida, T. 1990. Experimental study on attachment of drain function to steel members as liquefaction remediation measures, *8th Japan Earthquake Engineering Symposium.* pp. 885-890 (in Japanese).

[40] Kishida, T. 1983. On the soil improvement work on second plant in Nagoya port, Nisshin Seifun K.K. (Liquefaction remediation measure for existing wharf by gravel compaction method), *29th National Port and Harbour Construction Conference, Japanese Association of Ports and Harbours.* pp. 82-94 (in Japanese).

[41] Yoshimi, Y. & Tokimatsu, K. 1991. Ductility criterion for evaluating remedial measures to increase liquefaction resistance of sands, *Soils and Foundations, JSSMFE, Vol. 31, No. 1.* pp. 162-168.

[42] NKK corporation/Konoike Co. 1984. Permeability test result of gravel (in Japanese).

[43] Iai, S. 1988. Large-scale model tests and analysis of gravel drains, *Report of Port and Harbour Research Institute, Vol. 27, No. 3.* pp. 25-160.

[44] Zen, K., Oneda, H. & Irisawa, K. 1985. Experimental study on the evaluation of the clogging limits in gravel drain, *Technical Note of Port and Harbour Research Institute, No. 534.* p. 20. (in Japanese).

[45] Japan Large Dam Conference 1978. Design standard for dams (second edition), Chapter 5 (in Japanese).

[46] Ono, Y., Ito, K., Minagawa, Y. & Okita, K. 1984. On short-term clogging limit in gravel drains, *19th Japan National Conference on Soil Mechanics and Foundation Engineering, JSSMFE.* pp. 191-192 (in Japanese).

[47] Sherard, J., Dunnigan, L.P. & Talbot, J.R. 1984. Basic properties of sand and gravel filters, *Journal of Geotechnical Engineering Division, ASCE, Vol. 110, No. GT6.* pp. 684-700.

[48] Japanese Society of Soil Mechanics and Foundation Engineering 1982. *Handbook on soil mechanics and foundation engineering (1982 ed.).* pp. 69-70 (in Japanese).

[49] Terzaghi, K. & Peck, P.B. 1948. Soil mechanics in engineering practice, John Wiley & Sons. p. 44.

[50] Lee, K.L. & Albaisa, A. 1974. Earthquake induced settlements in saturated sand, *Journal of Geotechnical Engineering Division, ASCE, No. GT4.* pp. 387-406.

[51] Ishihara, K. & Okada, S. 1978. Effects of stress history on cyclic behavior of sand, *Soils and Foundations, JSSMFE, Vol. 18, No. 4.* pp. 31-45

[52] Ono, Y., Ito, K. & Okita, K. 1983. On coefficient of volume compressibility for excess pore water pressure dissipation analysis in the gravel drain method, *18th Japan National Conference on Soil Mechanics and Foundation Engineering, JSSMFE.* pp. 575-576 (in Japanese).

[53] Iai, S., Koizumi, K. & Noda, S. 1985. Large model tests and analyses of gravel drains, *18th National Conference on Earthquake Engineering, JSCE.* pp. 389-392 (in Japanese).

[54] Tanaka, Y., Kokusho, T., Esashi, Y. & Matsui, I. 1984. On liquefaction prevention of horizontal ground with gravel piles, *Proceedings of JSCE, III, No. 352.* pp. 88-98 (in Japanese).

[55] Iai, S. & Kajitani, T. 1989. Shaking table tests and analysis of plastic drain for liquefaction remediation, *Technical Note of the Port and Harbour Research Institute, No. 647.* (in Japanese).

[56] Ministry of Transport, First District Port and Harbour Construction Bureau 1984. Earthquake damage to Akita port and restoration – 1983 Nihonkai-Chubu earthquake (in Japanese).

[57] Sasaki, K. 1987. Test installation of improvement works on Chiba port, Funabashi West District Wharf (–10 m), *4th National Port and Harbour Technology Conference, Coastal Development Institute of Technology.* pp. 123-146 (in Japanese).

[58] Zen, K., Yamazaki, H. & Sato, Y. 1990. Strength and deformation characteristics of cement-treated sands used for premix method, *Report of the Port and Harbour Research Institute, Vol. 29, No. 2.* pp. 85-118 (in Japanese).

[59] Zen, K., Yamazaki, H. & Nagasawa, K. 1992. Model test on static earth pressure of cement-treated backfill, *47th National Conference of Japan Society of Civil Engineers, III.* pp. 974-975 (in Japanese).

[60] Zen, K., Yamazaki, H. & Nagasawa, K. 1991. Seismic response characteristics of backfill behind quaywall constructed with a cement-treated sand, *26th Japan National Conference on Soil Mechanics and Foundation Engineering.* pp. 1005-1006 (in Japanese).

[61] Zen, K., Yamazaki, H., Watanabe, A., Yoshizawa, H. & Tamai, A. 1987. Study on a reclamation method with cement-mixed sandy soils, *Technical Note of the Port and Harbour Research Institute, No. 579* (in Japanese).

[62] Zen, K., Yamazaki, H. & and Nagasawa, K. 1992. Dynamic strength and deformation characteristics of a cement-treated sand, *27th National Conference on Soil Mechanics and Foundation Engineering.* pp. 933-934 (in Japanese).

[63] Zen, K., Yamazaki, H., Kagaya, H. & Nagasawa, K. 1991. Shaking table test on liquefaction resistance of cement-treated sand, *46th National Conference of Japan Society of Civil Engineers, III.* pp. 216-217 (in Japanese).

[64] Coastal Development Institute of Technology 1989. Design procedure for premix method (Tentative) (in Japanese).

[65] Yoshizawa, H., Katano, H. & Kuroyama, H. 1988. Development of premix method, *43rd National Conference of Japan Society of Civil Engineers, VI.* pp. 8-9 (in Japanese).

[66] Katano, H., Kuroyama, H. & Wada, K. 1987. Research related to landfill method for improved sandy material mixing test with belt conveyor, *42nd National Conference of Japan Society of Civil Engineers, III.* pp. 826-827 (in Japanese).

[67] Mori, T., Kubo, H. & Toriihara, M. 1991. Development of premix method – Experiment on reclamation by a chute, 5th Report (Strength and deformation characteristics of cement-treated sand of young age), *46th National Conference of Japan Society of Civil Engineers, III.* pp. 1038-1039 (in Japanese).

[68] Kuroyama, H., Tamai, A. & Yoshizawa, H. 1987. Study on reclamation method with cement-treated material (4th Report) effect on water quality, *14th National Conference of Kanto Chapter of Japan Society of Soil Mechanics and Foundation Engineering.* pp. 164-165 (in Japanese).

[69] Yamazaki, H., Zen, K. & Kagaya, H. 1992. Liquefaction characteristics of overconsolidated sand, *27th Japan National Conference on Soil Mechanics and Foundation Engineering.* pp. 845-846 (in Japanese).

[70] Ishihara, K. & Takatsu, H. 1979. Effects of overconsolidation and K_O consolidation on the liquefaction characteristics of sand, Soils and Foundations, *JSSFME, Vol. 19, No. 4.* pp. 59-68.

[71] Ishihara, K. 1976. Fundamentals of soil dynamics, Kajima Publication. pp. 258-260 (in Japanese).

[72] Japanese Society of Soil Mechanics and Foundation Engineering 1988. Soft ground improvement methods – From investigation, design to construction pp. 59-83 (in Japanese).

[73] Iai, S., Koizumi, K. & Kurata, E. 1987. Basic consideration for designing the area of the ground compaction as a remedial measure against liquefaction, *Technical Note of Port and Harbour Research Institute, No. 590.* pp. 1-66 (in Japanese).

[74] Iai, S., Koizumi, K. & Kurata, E. 1991. Ground compaction area as a remedial measure against liquefaction, *Tsuchi-to-Kiso, JSSMFE, Vol. 39, No. 2.* pp. 35-40 (in Japanese).

measure against liquefaction, *Tsuchi-to-Kiso, JSSMFE, Vol. 39, No. 2.* pp. 35-40 (in Japanese).

[75] Ministry of Transport, Japan (ed.) 1989. Technical standards for port and harbour facilities and commentaries, Japanese Association of Ports and Harbours (in Japanese).

[76] Toyota, Y., Nakazato, S. & Nakajima, Y. 1988. Consideration on behavior of existing steel pipe pile wall wharf during installation of vibro rod method, *43th National Conference of JSCE, III.* pp. 765-757 (in Japanese).

[77] Kameoka, T. & Iai, S. 1993. Numerical analyses on the effects of initial stress conditions on seismic performance of sheet pile quay walls, *Technical Note of the Port and Harbour Research Institute, No. 751.* p.29 (in Japanese).

REFERENCES TO TECHNICAL NOTE 1

[1] Japanese Society of Soil Mechanics and Foundation Engineering 1989. Effective stress analysis of ground and soil structures, *Symposium on Seismic Performance of Ground and Soil Structures.* pp. 50-136 (in Japanese).

[2] Japanese Society of Soil Mechanics and Foundation Engineering 1991. Comparative study on effective stress analyses of liquefaction, *Symposium on Liquefaction Remediation Measures.* pp. 77-198 (in Japanese).

[3] Kobayashi, M. 1988. Stability analysis of geotechnical structures by adaptive finite element procedure, *Report of the Port and Harbour Research Institute, Vol. 27, No.* 2. pp. 3-22.

[4] Iai, S., Matsunaga, Y. & Kameoka, T. 1992. Strain space plasticity model of cyclic mobility, *Soils and Foundations, JSSMFE, Vol. 32, No. 2.* pp. 1-15.

[5] Iai, S., Matsunaga, Y. & Kameoka, T. 1990. Parameter identification for a cyclic mobility model, *Report of the Port and Harbour Research Institute, Vol. 29, No. 4.* pp. 57-83.

[6] Iai, S. & Kameoka, T. 1993. Finite element analysis of earthquake induced damage to anchored sheet pile quay walls, *Soils and Foundations, JSSMFE, Vol. 33, No. 1.* pp. 71-91.

[7] Iai, S. & Matsunaga, Y. 1992. Analysis of liquefaction induced uplift of underground structures, *Proceedings from the Fourth Japan-US Workshop on Earthquake Resistant Design of Lifeline Facilities and Countermeasures for Soil Liquefaction, Vol. II, National Center for Earthquake Engineering Research*, Technical Report NCEER-92-0019. pp. 425-438.

[8] Kazama, M. & Nogami, T. 1992. Dynamic response analysis of the saturated two-phase layered media by thin layered element method, *Proceedings of JSCE, No. 446/I-19.* pp. 187-196 (in Japanese).

[9] Kazama, M., Inatomi, T. & Murakami, Y. 1992. Centrifuge model tests on the effects of surface layer drainage condition on liquefaction, *27th Japan National Conference on Soil Mechanics and Foundation Engineering, JSSMFE.* pp. 1147-1150 (in Japanese).

[10] Arulanandan, K. Yogachandran, C., Muraleetharan, K.K., Kutter, B.L. & Chang G.S. 1988. Laboratory flow slide during earthquake simulation, *Proceedings of International Conference on Geotechnical Centrifuge Modeling*, Balkema. pp. 539-544.

REFERENCES TO TECHNICAL NOTE 2

[1] Iai, S. & Matsunaga, Y. 1991. Soil improvement area against liquefaction, Geo-Coast '91, *Proceedings of the International Conference on Geotechnical Engineering for Coastal Development,* Yokohama, Port and Harbour Research Institute and Coastal Development Institute of Technology. pp. 453-458.

[2] Iai, S. & Matsunaga, Y. 1992. Analysis of liquefaction induced uplift of underground structures, *Proceedings from the Fourth Japan-US Workshop on Earthquake Resistant Design of Lifeline Facilities and Countermeasures for Soil Liquefaction,* Vol. II, National Center for Earthquake Engineering Research, Technical Report NCEER-92-0019. pp. 425-438.

REFERENCES TO APPENDIX 1

[1] Tsuchida, H. 1970. Prediction and remedial measures against liquefaction of sandy soil, *Annual Seminar of Port and Harbour Research Institute.* pp. (3)1-33 (in Japanese).

[2] Tokimatsu, K. & Yoshimi, Y. 1983. Empirical correlation of soil liquefaction based on SPT *N*-value and fines content, *Soil and Foundations, JSSMFE, Vol. 23, No.4.* pp. 56-74.

REFERENCES TO APPENDIX 2

[1] Ishihara, K. 1985. Stability of natural deposits during earthquakes, *11th International Conference on Soil Mechanics and Foundation Engineering,* Vol. 1. pp. 321-376.

[2] Fire Defense Agency 1978. Standards on regulation of hazardous materials (in Japanese).

[3] Tanaka, Y., Komine, H., Tochigi, H., Tohma, J., Ohtomo, K. & Fukuda, S. 1991. Study on liquefaction remediation of underground pipe by compaction (3) – Examination on Improvement Area by Numerical Analysis, *21st National Conference Earthquake Engineering,* JSCE. pp. 257-260. (in Japanese).

[4] Port and Harbour Bureau, Tokyo Metropolitan Government 1990. Manual for remediation against liquefaction in reclaimed land of Tokyo harbour (in Japanese).

Subject index